SUPPLY-SIDE SUSTAINABILITY

Complexity in Ecological Systems Series

SUPPLY-SIDE
SUSTAINABILITY

T. F. H. Allen
Joseph A. Tainter
Thomas W. Hoekstra

Columbia University Press • New York

COLUMBIA UNIVERSITY PRESS
Publishers Since 1893
New York   Chichester, West Sussex

Library of Congress Cataloging-in-Publication Data

Allen, T. F. H.
Supply-side sustainability / Timothy F. H. Allen,
Joseph A. Tainter, Thomas W. Hoekstra.
p. cm.
Includes bibliographical references (p.).
ISBN 0–231–10586–X (cl. : alk. paper)
ISBN 0–231–10587–8 (paper)
1. Ecology. 2. Sustainable development.
I. Tainter, Joseph A. II. Hoekstra, T. W.
III. Title

QH541 .A457   2002
577—dc21   2002073814

∞

Columbia University Press books
are printed on permanent and durable
acid-free paper.

Printed in the United States of America

c 10 9 8 7 6 5 4 3 2 1
p 10 9 8 7 6 5 4 3 2 1

For Josephine, Gwynedd, Harrison, Eleanor, and Jamie
—T. F. H. Allen

For Malaika, Emmet, Stephanie, David, and Stephen
—J. A. Tainter

For Joyce VanDeWater
—T. W. Hoekstra

CONTENTS

Contents

Contents

University Press for seeing the book through production. This project was funded by the USDA Forest Service's Rocky Mountain Research Station and Inventory and Monitoring Institute and by National Science Foundation grants DEB 97039908, DEB 0083545, and DEB 9632853, all administered through the University of Wisconsin, Madison.

This book is of general interest to anyone with a role to play in promoting the sustainability of the modern world, including professionals
in science, government, and business. We outline a strategy for dealing
with the new challenges of sustaining natural resources and human
institutions. We have tried to be scientifically objective, but we also
argue that sustainability is a matter of human values. Because it affects
many people, discussions of sustainability inevitably enter the political
arena. We adamantly pursue no political orientation, but we find that
various parts of this book will appeal to either conservatives or liberals. Conservatives may agree with our finding that larger government
is not the solution to promoting sustainability. Indeed, we note that
growth in the complexity of government may combine with other factors to reduce sustainability, and reducing the scale of government may
promote it. Moreover, as we have discussed elsewhere (Allen, Tainter,
and Hoekstra 1999), we see commerce as having a central role in promoting sustainability. Liberals may be pleased by the conservative posture we adopt with regard to the ecological system and our environment. Furthermore, although the title of the book derives from the
conservative economic program of the early 1980s, we approach our
topic with explicit concern for many factors that conservative economists dismiss as externalities.

Although its subject matter differs, this volume is premised on *Toward
a Unified Ecology*. In that book, Allen and Hoekstra (1992) took pains to
separate questions of scale from those that turn on what we called criteria that define the type of system under observation. Because changes
in both scale and system type are responsible for changes in the appear-

ance of a system, these two concepts are easily confused. For this reason we again insist that scaling considerations be kept separate from criteria for class or type of system.

Scale governs material performance: With all else equal, doubling the size of a structure causes radical changes in its material performance. The reason is that even if proportions remain the same in the larger version of a system, certain underlying factors cannot be changed. These intransigent factors relate to the properties of the material from which the system is made. For example, surface tension properties of water remain the same in a larger system and therefore relate differently to the greater mass than they do to the mass of a smaller equivalent system. A pond skater insect the size of a dung beetle would sink (fig. P.1).

By contrast, observation criteria come from the observer's decision to focus on insects in the first place. The observation criteria provide the rules that give the system its type or class by recognizing some relationships as important while ignoring others. For example, considerations of pond skater insects tacitly use the concept *organism*, with its relationships between body parts and its physiological coherence taken as a given. In *Toward a Unified Ecology* we organized the chapters around particular criteria, *organism* being one of them, and then looked at scaling implications under those criteria.

Our criteria were *landscape, ecosystem, community, organism, population,* and *biome,* introduced in that order. The last two chapters were about applying the earlier criterion-specific notions to management and basic research, respectively. In this book, some of the chapters use a particular criterion, but the criteria are necessarily treated in a different order and with less even treatment across the full set of criteria.

For Tim Allen, this is the fifth book in a sequence that has moved from the almost completely abstract to the very concrete. In *Hierarchy: Perspectives for Ecological Complexity* (1982), he and Tom Starr laid out the general notion of hierarchy theory and an array of topics to which it might apply. Four years later he was the junior author behind Bob O'Neill, Don DeAngelis, and Jack Waide in *A Hierarchical Concept of Ecosystems* (1986). This monograph explicated the notion of scaling and hierarchy in the context of ecosystems, community ecology, and some aspects of population biology. Then, in 1992, Allen and Hoekstra pub-

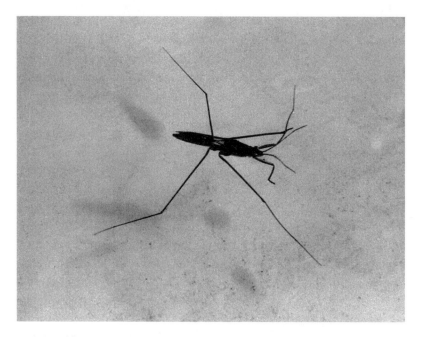

FIGURE P.I

A water strider. A larger animal would not be able to float on surface tension alone. There are indentations where the legs rest on the water surface, and through changes in refraction, they appear as shadows below.

Photograph by S. Dodson.

lished *Toward a Unified Ecology*. There, ecology in its many manifestations was the organizing force. The method of unification invoked notions of scaling and typing of systems through criteria for observation; hierarchy theory was merely the vehicle for the substance of the book. At the end of that book we took notions of basic ecology and pressed them into the service of ecological management. A fourth book by Ahl and Allen (1996) is theoretical and abstract, avoiding the confusion that often occurs at the base level of tangible things in material places.

In the present work we use the same vehicles of scale and criteria to address a central issue with concrete implications: ecological sustainability. This is a book about sustainability, not hierarchy theory. Ecological sustainability is not a matter of structure and function; it is a condition. This book is centrally about that condition in ecological and management terms. This is our least abstract treatment to date and represents a continuation of a trend from esoteric, academic discourse toward execution and action. This is our best effort at making ecological theory practical and useful. It is a translation of theory into a form in which we hope it will be useful in guiding policy and achieving ecological application.

This is Thomas Hoekstra's third book. After *Toward a Unified Ecology,* he co-edited his second book, *Arid Lands Management: Towards Ecological Sustainability* (1999), with Moshe Shachak. That book focused on experiences in the Middle East, North America, and Australia, recognizing that as living systems change, so research priorities and decisions about resource management must adapt. By relating various kinds of ecological systems to the question of sustainability in arid lands, *Arid Lands Management* offers new directions for research and management.

Sustainability has no inherent value; its value is purely instrumental. Social science is inextricable from ecological sustainability and central to the human future. Historical science reminds us that we are not the first to encounter a given kind of problem. Therefore the two ecologists, Allen and Hoekstra, teamed with social scientist Joseph Tainter for this and other recent work (Allen, Tainter, and Hoekstra 1999, 2001; Allen, Tainter, Pires, and Hoekstra 2001). Tainter's research on the relationship of sustainability to human problem solving began with *The Collapse of Complex Societies* (1988) and continued in his contributions to *Evolving Complexity and Environmental Risk in the Prehistoric Southwest* (Tainter and Tainter 1996a) and *The Way the Wind Blows: Climate, History, and Human Action* (McIntosh, Tainter, and McIntosh 2000). This work has aimed both at understanding historical cases of collapse, resiliency, and sustainability and at developing a general understanding of sustainability to apply to our contemporary situation.

A book always comes from more people than its listed authors. We are pleased to thank Joyce VanDeWater, Kandis Elliot, and Claudia Lipke for their excellent work in preparing many of our illustrations, Pamela Stoleson for her work on the index and references, copyeditor Carol Anne Peschke, and Robin Smith and Ron Harris of Columbia

SUPPLY-SIDE SUSTAINABILITY

1

The Nature of the Problem

A NEW GLOBAL SYSTEM

The issue of sustainability has emerged from problems that have become apparent on a global scale; some are new material configurations and others have emerged as phenomenal now that there are the tools to address them. The view of our planet in its entirety from space has dealt a death blow to flat Earth societies. The British Broadcasting Company had the wit to have a dignitary of a British flat Earth society on its panel of experts commenting on the live coverage of the first *Apollo* moonshot. It appeared disconcerting to that dignitary, but he was spirited in the explanations and reservations he expressed. Far more important was the effect of the moonshot on everyone who saw it at the time. It was a very moving image: Here we are, all alone, but at least we are alone all together. The image of an earthrise from the Moon is now part of popular culture, a wallpaper motif for a teenager's bedroom (fig. 1.1). The effect of that image and others like it on the global community has been enormous.

At the time of the first moonshot Allen was in Nigeria and saw a local artist's tie-dyed images of the *Apollo* lander, depicted as the same size as the Moon. On a spindly staircase an astronaut, the size of the Man in

FIGURE 1.1

The *Apollo* photograph of the earthrise.

Photograph by NASA.

the Moon, descended to the Moon's surface, looking back at Earth. A view of Earth from space makes a more powerful statement than anything coming from the writings of environmental professionals. Without it, one might suspect that Earth Day would never have never been the event that it was, and Rachel Carson and her *Silent Spring* would be of only academic interest, instead of Carson herself being an iconic figure. Allen recently asked his ecology students to raise their hands if they had heard of Marston Bates; nobody moved, but half the class knew of Rachel Carson. Carson and Bates wrote in a very similar vein, but Bates wrote before the space program whereas Carson wrote during it. The moonshot showed us that Earth is our one shot, and that is it.

Our new ethic recognizes physical limits much more explicitly. The system it replaces, that of the Industrial Revolution, was buoyant in its confidence that the application of more power would solve any problem. When Allen first started using a word processor to keep in touch with his family, his father, Frank Allen, wrote back, revealing himself as a child of the Industrial Revolution. He congratulated his expatriate son on his new communication device, saying it was a fine "engine." In German, the words for "electric motor" and "electronics" translate into English as *strong current* and *weak current*, respectively. Frank Allen apparently made no such distinction, for all machines were engines to him. Problems in the industrial age started at a modest size (fig. 1.2) and were attacked with the application of more size and more power. Industrial optimism, the notion that all important limits could be overcome, gave way to a greater realism. That more sober view recognizes that humans cannot predict the consequences of action on the global scale, at which modern problems reside.

Sustainability was not an issue in the nineteenth century, for in an expanding sphere of influence, Man was master of his destiny. There are still many holdovers of that view. A man of vision recently lost to us was Carl Sagan, the popularizer of physical science and a very good atmospheric physicist. Television tributes to him noted that his inspiration to become a scientist was the 1939 World's Fair. Sagan's vision is not ours. The 1939 World's Fair was the hurrah of the industrial age. We see satellites orbiting Earth as crucial for modern problem solving, and they will continue to influence our lives, but "To boldly go where no man has gone before" is nothing that matters. Bigger spaceships going greater distances is the Industrial Revolution model of bigger machines applying more power to overcome human problems of larger scale. But space travel is not just a matter of building a bigger iron bridge across a wider Victorian estuary. The notion of space travel as a way to relieve human crowding on this planet comes from the naiveté of the industrial age. Planetary limits, something the Victorians never encountered, now press issues of sustainability into the public consciousness and onto the agenda for environmental scientists.

The industrial model for our planet, though clearly outmoded, is remarkably persistent. Environmental scientist and conservationist Hugh Iltis, the plant systematist who discovered perennial corn,

3

FIGURE 1.2

Early industrial power:
The earliest phase of the
Industrial Revolution was
modest and used water
power. Here a spade mill
has two water wheels
(one for the bellows and
the other for the ham-
mer). The other structure
is a reconstructed skutch
mill for removing the soft
parts of flax plants to
make linen. Both build-
ings were moved to the
Ulster Folk Museum for
preservation.

Photographs by T. Allen.

informed an economist colleague that the nearest star was three and a half light years away. Somehow, a mere three and a half years seemed to reassure the economist as to the viability of space colonization. At 50,000 miles per hour, it would take more than a million years to get there. There is no prospect of colonizing other solar systems. There is nothing to be done about the hole in the ozone layer except to stop making chlorofluorocarbons and wait a century. If we can do nothing active to repair the hole in our own atmosphere, then other celestial bodies in our solar system cannot be made habitable by human manipulation of their entire atmospheres. For example, the Moon is too small even to hold an atmosphere, and planets much larger would crush human colonists. Without the surface of whole celestial bodies made Earth-like, colonization of space can have no direct effect on the sustainability of the human population. Sending a few astronauts off the planet cannot ease population pressure. Therefore, as a population, we are stuck on Earth. Our space exploration has had a significant effect on our views of ecological systems, but that is not the same thing as changing important material ecological flows here on Earth by removing human excess.

Modern technology allows scientists to see problems that are larger in scale than ever before, and that same technology promises solutions to some of those problems. The Victorians were so confident that they could not foresee that their great-grandchildren would be heading into unimaginable problems. For Victorians, things were either too large to consider or small enough to be challenged with the iron technology of industry. Perhaps we are prepared to struggle because we can imagine things that Lyle, Darwin, and Spencer could not. Ecologists and natural resource managers are being asked to address new problems that are of much larger scale than a coal mine that must be connected to a railway or a canal to a distant source of steel (fig. 1.3). Humanity is cognizant of global problems that invite a human-contrived solution. Although we may not be as confident as were the Victorians, modern humanity appears willing to engage an altogether larger set of issues.

Darwinian evolution is the model that has had the greatest influence on both lay and professional scientists' views of sustainability. Darwin was inspired by Lyle's geology, which implied a long time line along which biota have changed. So before the twentieth century, scientists

5

C

D

FIGURE 1.3

Industrial transportation. Now barely used for commercial transportation, the Lee Navigation Canal in east London flourished in Victorian times. (A) This granary and mill is typical of the industry that took advantage of canal transportation. (B–D) The scale of Victorian industrial technology remains human and now is viewed with some sentimentality.

Photographs by T. Allen.

were aware of the long-term and wide spatial scale over which some biological and ecological change occurs. Even so, Lyle's geology and Darwin's evolution both take so much time to unfold that they are very abstract. Certainly humans could not plan or influence happenings over such a long time. Notice how Darwinian evolution is cast in terms of "survival of the fittest," a Spencerian phrase, but one Darwin liked and quoted. The fittest are not those that are the most conditioned to play the most vigorously but those that fit the best into their environment. The prevailing paradigm of biology today is directly descended from Darwin's, and it is one of genotype interacting with a controlling environment to produce a phenotype. In that paradigm, environment is context, and contexts are unchanging. Although earlier observers might have cogitated about civilizations falling in the face of global events (as many people still do), there was never any thought of doing anything about it. Prior conceptions of human sustainability were of things happening to living and human systems, mere actors inside an inexorably large context. For the Victorian scientist, life fits in; it does not generally control its environment. The Victorian era, and the modern holdover views from the industrial age, use the Darwinian model of a huge, unchangeable environment for the individual.

Now we see things on a scale between the unimaginable size of Darwinian eons on one end and daily happenings on the other. The issue of sustainability was of no concern in the nineteenth century because there was no thought of running out of resources, and nothing was going to happen to change things much in the imaginable future (fig. 1.4). Now modern humans see the world as threatening change, but change that, through our best efforts, we might either blunt or ride. If our technology lets us see the world whole, then there is some thought that the same technology might allow humans to plan and influence even something as large as changing continents.

The present authors do not see a simple or permanent technological fix for sustainability, but we are not without hope. A technological fix smacks of an Industrial Revolution model. In fact, technology in itself is as much part of the problem as it is a solution. In this we do not refer only to big, smoky industry; we also see all technology as coming at a price, even green technology. The problem of sustainability is as much a matter of understanding social dynamics and human nature as it is an environmental crisis. The enemy we have seen is ourselves, and it is not just our polluting, consumer selves. The very act of solving environmental problems spends resources, and these resources are in themselves responsible for creating some other problem. It is not that we think solving environmental problems is a bad thing to do. But doing so always presents a dilemma at some level of analysis. The issue of sustainability turns on the nature of problem solving itself. So we begin here a strongly self-reflexive journey as we try to solve the larger problem of a society that engages in a self-defeating struggle. Overcoming immediate problems is part of a process that has brought other societies down, and there is no reason to suppose that modern society is involved in a different process. We are not sure that the indications we offer will work, but they are much better than doing what society does now. At worst, doing what we recommend anticipates retrenchment, so that it can be done with greater humanity. At best, we may have a solution that addresses the larger process, taking the self-defeat out of the struggle.

ECONOMICS, SOCIETY, AND ECOLOGY

Sustainability can be approached at several levels of generality. Our treatment hopes to be wide in its coverage. A narrow view of sustainability would stop at sustaining the biogeophysical system— the species, forests, and rivers. Certainly, without a viable biophysical component, wider views of sustainability cannot work. However, sustainability without a social justice component will not work either. Social justice addresses local considerations of individual sacrifice but in support of a larger system that offers real or perceived benefits for the individual. Along with all this, the whole system must be economically viable. Indeed, some instances of social injustice come from an inviable economy that consumes goodwill as it cuts corners or requires more work from citizens to make up the shortfall. Closing the loop, economic inadequacies take a toll on the biogeophysical systems as overcropping or pressing marginal ecological systems into service destroys soil and extirpates species. Across its widest purview, sustainability works with three major areas of dis-

FIGURE 1.4

A system despoiled. While Wordsworth wandered "lonely as a cloud" through the Lake District of England, he and the other Lakeland poets imagined that they were escaping the worst of the industrial blight on the country. They did not know that acid rain from Manchester was removing fish from the smaller lakes.

Photographs by T. Allen.

course (fig. 1.5). At this highest level, there are economic, social justice, and biophysical components, all of which are crucial. Various scholars have considered biogeophysical, social justice, and economic sustainability. Some have even started to explore the interface between two of the three areas, as in the emerging fields of environmental economics (Carpenter, Ludwig, and Brock 1999; Carpenter, Brock, and Hanson 1999) and rural sociology (Wolf and Allen 1995). However, we are aware of nobody who has worked all facets of sustainability together, and certainly not with a general model that applies across a range of societies of different sizes and degrees of complexity.

COMPREHENDING SUSTAINABILITY

We are accustomed to thinking of achieving sustainability by doing without—by consuming less and paying more for what we do consume. Regrettably, much of our national and international debate is phrased in such terms. The Kyoto agreement on greenhouse gases, for example, has been cast as a conflict between sustainability and growth. Depicted in this way, the actions needed to sustain the climate to which we are accustomed, for example, will always appear unfavorably. The public's choice is a foregone conclusion. The immediacy of quarterly balance sheets (for businesses), unemployment levels (for politicians and those unemployed), or poverty (for much of the world) commands attention far more readily than the threat that *someday* things may go quite wrong. Journalists amplify the problem by their tendency to present all policy disputes as combat between opposing champions. Even when nearly all scientists agree on a matter, journalists will always find one who doesn't (or perhaps not even a scientist), then present the dispute as a contest between arguments of equal merit. This distortion flows inevitably from allowing business and political leaders and journalists to define the terms of sustainability debate as consumption and employment versus sacrifice and unemployment.

We present here a more nuanced approach. We will show that sustainability is an active condition, not a passive consequence of doing less. One must work at being sustainable. It has costs and benefits and

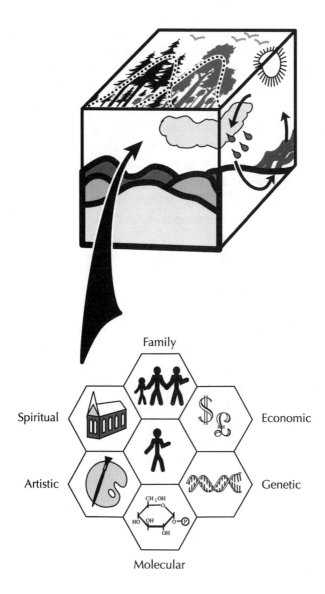

FIGURE 1.5

Humans in the ecological context. Inside the multifaceted ecological system, the humans are not just social creatures; they too are multifaceted, such that describing them in terms of just social justice and economics is a very austere description. In this book we mean to include all the facets of being as part of sustainability.

After Allen and Hoekstra (1994).

takes both knowledge and resources. Certainly our species cannot infinitely increase its consumption, but to concentrate narrowly on that alone is to miss much that is important. We focus in this book on the roles of hierarchy and complexity in sustaining ecological systems, human societies, and problem solving. We argue that being sustainable consists of such key approaches as the following:

- Manage for productive systems rather than for their outputs.
- Manage systems by managing their contexts.
- Identify what dysfunctional systems lack and supply only that.
- Deploy ecological processes to subsidize management efforts, rather than conversely.
- Understand the problem of diminishing returns to problem solving.

We sketch in these pages an understanding of sustainability that is more fundamental than mere exhortations to do such things as use public transportation and take colder showers. Sustainability entails *management* of systems and their contexts that is intensive and heavily knowledge based. We will achieve sustainability when it becomes a transparent outcome of managing the *contexts* of production and consumption rather than consumption itself. If we shift our management emphases to managing from the context for whole ecosystem functions, rather than for resources, the cost of problem solving will diminish and the effectiveness of management greatly increase. When a manager gets the context right, the ecosystem does the rest. Because the material ecosystem supplies renewable resources and makes them renewable, we call our approach *supply-side sustainability*.

A brief discussion of these five points gives a taste of the lessons this book conveys.

Manage Systems, Not Outputs

Managing to maintain the outputs of productive systems amounts to sticking a finger in the dike continually. Leaks spring inevitably and must be plugged, typically temporarily, ineffectually, and at great cost. The problem generating the leaks is never addressed, so the costs of repairing leaks can never be reduced. Managing for outputs is how we have typically practiced agriculture, forestry, and much else. In the social context, it pervades our approach to criminology (which, though not a focus of this book, provides some fine examples of our general points). The approach we recommend is to understand the productive system as fully as possible and manage for that. Sustainable outputs follow automatically, potentially at reduced management costs. In biological resource production this is exemplified in such approaches as ecosystem management (in forestry) and integrated pest management (in agriculture). In criminology it would consist of alleviating the factors that are thought to generate crime rather than trying to fortify every house and business and incarcerate every offender. In a matter such as illicit drugs, managing for systems would entail trying to understand and ameliorate the inducements to consume drugs rather than trying to seal national borders—the ultimate case of sticking fingers in the dike.

Consider how a supply-side solution might work in the case of population management. Programs of population control in Third World countries have emphasized the application of birth control technology combined with exhortations to use it. It is a case of managing for outputs, and it is predictably resented and ineffectual. A supply-side approach would emphasize changes to the whole system that produces so many deleterious outputs. A supply-side approach would address the systemic factors that influence peasants to reproduce so vigorously. Such an approach would address education (including greater emphasis on education for girls and women), high seasonal labor demands, childhood mortality, and fears of old-age poverty. Addressing these systemic problems would take high initial expenditures. Were the program to work, though, these expenditures would be less than the costs to battle excessive outputs indefinitely. The initial costs would be repaid many times in a population that is more productive and secure in an environment that deteriorates less rapidly. Properly managed, such a supply-side solution would generate positive feedback loops in which productivity growth that is greater than population growth would generate the wealth to pay for increased education, agricultural mechanization, and moderate old-age pensions.

Manage Contexts

Any system is controlled one level up: by its context (Allen and Starr 1982; Allen and Hoekstra 1992). Management efforts are most effectively focused not on the system of interest (e.g., pest outbreaks in forests or agricultural fields) but on the contexts that regulate such systems (perhaps landscapes consisting of overdense forests or continuous monocropping) (fig. 1.6). To continue our social example, crime is controlled in part by the number of men born 18 to 24 years ago. Demography is part of the context of crime. Although politicians proclaim their success in reducing crime or blame their opponents for failing, crime fluctuates in response to the birth rate two decades ago. We can do little to control demography (which is itself controlled by much broader contexts), but there is potential to change what children learn and the environments in which they grow. The factors at work here include socialization and education, which are best approached not directly through children themselves but through their contexts: parents and educators.

To manage one's context one must first know it (Tainter 1999). Tainter has worked among villages in southern Mali that are experiencing uncertain agricultural production in the face of climate change and population growth (fig. 1.7). The villagers understand that their situation is deteriorating, but some ascribe this decline to Allah. Although we respect their religious beliefs, we cite this as an example of not understanding one's context. Part of their context, climate change, is beyond their control, but another part, demography, they can affect. Their first task is to develop a new metaphor to describe their context.

Misunderstanding context may produce unwanted results. Jack Goldstone (1991) examined political revolutions from the seventeenth through nineteenth centuries. Revolutions typically promise such ideals as freedom, democracy, or other political or economic rights. Yet with a few notable exceptions (such as the American Revolution), the usual outcomes of a revolution are dictatorship and oppression. Goldstone shows that the revolutions that have produced the latter outcomes followed periods of rising population, which produced an imbalance between population and resources, inflation, fiscal distress among nearly all classes and the government, reduced opportunities,

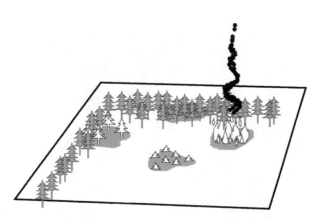

Mosaic of Patches in a Contextual Matrix

Humans Subsidize Local Unit

FIGURE 1.6

Manage from the context, not for resources. Most environmental management is done to make up for a lost context. Offer the things that the context would have offered, and the system takes care of itself.

After Allen and Hoekstra (1994).

FIGURE 1.7

Understanding context: environmental education in southern Mali. The Malian government sponsors environmental education in rural villages, with the environmental message delivered through professional entertainers. Many villagers understand that today's conditions result from increasing population and their own actions, but others ascribe environmental deterioration to Allah.

Photograph by J. Tainter.

and general discontent. The government makes an easy target for this discontent and, having been fiscally weakened, is overthrown. Yet merely changing the government does not address the underlying problems. Attempting to address the grievances behind such a revolution (e.g., inequity, lack of opportunity, high prices, and oppression) amounts to managing outputs. The context of demographic–resource imbalance that produces the grievances is not understood and cannot be addressed anyway in programs aimed at political issues. Revolutions therefore typically fail. Totalitarian control prevents the new rulers from being overthrown when grievances cannot be relieved.

Supply What Systems Need

This point hardly needs elaboration. Efficient management identifies what dysfunctional systems lack, which may be such things as nutrients, consumers, energy (e.g., fire), below-ground processes, or public understanding. This requires that management be focused and knowledge intensive. To know precisely what ecosystems lack and provide only that takes research and monitoring on a variety of processes. It also takes managers who can shift their focus between general and specific, who understand a broad array of ecological phenomena, and who can comprehend both social and biophysical processes.

This approach applies both to ecological systems managed just for biophysical processes and to those managed to meet human needs. In chapter 4 we discuss how Native Californians used fire to manage chaparral vegetation for early seral stages, which support higher densities of deer. Given a management objective of maximizing a resource useful to people, Native Californians acted as the context for that resource and supplied fire whenever needed.

Attempts to control acid rain are finding that this is done most cost-effectively not through traditional command and control but through a flexible system in which the government sets targets and grants flexibility to industries to decide how best to meet them (Kerr 1998). Pollution abatement efforts need flexibility and incentives, which the 1990 Clean Air Act Amendments provided.

Let the Ecological System Subsidize Management

In a world that will know fiscal constraints for the foreseeable future, management can accomplish most by using subsidies. In the vicinity of Albuquerque, New Mexico, for example, the historic cottonwood forest along the Rio Grande is dying (fig. 1.8). Most areas experience no regeneration, and detritus on the forest floor fails to decay. Saltcedar is displacing cottonwood as the major tree. The cottonwood forest can be sustained by planting poles to a depth of several feet, but over an area of tens of thousands of acres this is economically infeasible. Clifford Crawford and his colleagues (1999) have shown that overbank flooding during spring runoff does the job nicely: Detritus decays more rapidly, and a fresh generation of cottonwood seedlings will establish themselves if the site has open canopy. Of course, this is how the gallery forest functioned for millennia before the current era. In a time of managed ecosystems it is still how the forest functions best. Today's management objectives are best subsidized by services that are freely available whether we use them or not: winter precipitation and gravity.

Understand Problem Solving

Increasing the complexity of a management system is a common way to address problems. As challenges grow in scale and difficulty, which inevitably they seem to do, problem solving must adjust commensurately. Management systems thus seem inexorably to grow more complex. Complexity often is successful as a problem-solving tool (hence the great complexity of contemporary society and technology), but it is also costly. In southeast Alaska, for example, conflict between timber production and Native hunting generates growing levels of litigation, regulation, and legislation, which are all elements of complexity. Under this process, the costliness of the management system has grown to the point that the net economic value of subsistence harvesting has declined significantly (Tainter 1997). Historically, problem-solving systems that develop in this way either collapse, are terminated, or must be subsidized. We discuss this problem in depth in chapters 2 and 3. In our conclusions we argue that one approach to effective problem solving is to apply the first four points just discussed.

SUSTAINABILITY IN A SOCIAL CONTEXT

Many fine thinkers have addressed themselves to understanding and defining sustainability. A disciplinary approach is inherently incomplete. Each discipline defines sustainability so narrowly that not only is much of the matter excluded from consideration, but that which is included cannot be addressed successfully or discussed with others. Disciplinary views generate inherent narrowness of perspective that precludes discourse and promotes counterefforts. Thus the conserva-

FIGURE 1.8

Cottonwood regeneration in the Rio Grande gallery forest, central New Mexico. The older tree at left regenerated through ecosystem services: winter precipitation and gravity, which caused overbank flooding and moistened the soil. The smaller, planted tree reflects how reforestation is now accomplished: the management effort subsidizes the forest.

Photograph by J. Tainter.

tion biologist's concern for saving species generates reactions from business, from resource-dependent communities, and from politicians that make it extraordinarily difficult to meet the original objective. Similarly, the economist's concern for monetary valuation and markets ensures conflict with environmental advocates and with others who believe that not all values can be monetized. Such problems cannot be resolved within a disciplinary framework, for each discipline or point of view has its own vocabulary, assumptions, and values. Not only can the holders of different viewpoints not communicate, but they may not even share the same perception or experience of the phenomenon (cf. Feyerabend 1962).

We present here a collaboration between two ecologists and a social scientist. It is a product of many hours of discussion, which began metaphorically as ships passing in the night and resolved ultimately, in Sander van der Leeuw's words (1998b:25), to the point where our concepts and terms had been "negotiated to (relative) homogeneity." The effort was made easier by our use of hierarchy, systems, and complexity theory, for the abstract concepts of these approaches apply to systems of many kinds. Even so, it is difficult to fuse social and ecological science seamlessly. Recognizing this, we have tried to achieve (borrowing again from van der Leeuw 1998b:25) a "bee's-eye view": a multifaceted view of contiguous panes. One accepts the fracturing of such a view to gain the benefits of a multidisciplinary perspective.

The integration of social and ecological science is as necessary as it is obvious. Either approach, standing alone, will falter. We present here some elements of a social science perspective; ecological science is treated later in this chapter. Environmental advocates, and even some of our colleagues in the biological sciences, speak and write of "restoring" ecosystems to conditions described by such terms as "natural" or "pristine" or, more scientifically, within the range of "natural" or "historic" variation. Such terms often are undefined, but their common meaning seems to refer to conditions before human influence. When might that have been?

Much of the social science in this book is historical, for sustainability is by definition a long-term affair. A brief historical example deflates the notion of restoring ecosystems to conditions before human influence. In the first century B.C. (the age of Julius Caesar and many other well-known figures), the Romans financed their continuous foreign and civil wars with very high production of silver coins. The metal was produced from lead-bearing silver ores, and the smelting process caused atmospheric lead pollution that circled the northern hemisphere and left a signature in the Greenland ice (Hong et al. 1994). As Marc Antony and Cleopatra dallied in the eastern Mediterranean, the silver coins they minted to pay their troops caused lead to be deposited throughout the northern hemisphere at levels not to be seen again until the Industrial Revolution.

For at least this long, then, humanity has affected biophysical processes at scales ranging from hemispheric to global. Of course, human influence at the local level is far older—as old as humanity itself. Even in North America, only recently settled by people, restoring ecosystems to conditions before human influence takes us back to the Pleistocene. Even if we could bring back Pleistocene-like conditions, Canadians would surely object to being covered in ice. Sustainability calls for broader disciplinary integration and subtler conceptualizations than are offered by current efforts.

Ecosystems clearly cannot care whether they lose species, leak nutrients, or have their processes degrade. Such things matter only because people worry about them. To rephrase an old philosophical conundrum, if a nutrient leaks from a forest and no one knows, can it matter? Sustainability is a topic of human values. Once this simple point is understood, dilemmas imposed by simple biological or economic conceptions diminish. In the gallery forest of the Rio Grande, as we have just discussed, the native cottonwoods are being displaced by saltcedar, an introduced species with a shrubby appearance and wispy, pale-green foliage. Responding to public wishes, much effort and money are expended to maintain cottonwood as the primary tree of the forest canopy. From the perspective of management efficiency, this is a fine dilemma. Saltcedar is the more sustainable species. It takes care of itself, and left alone it outcompetes cottonwood under today's conditions. Yet we try to suppress it. Cottonwood apparently is unsustainable today, yet we encourage it. Such a conundrum forces us to realize that sustainability, as Stephen Pyne (1998:98) might say, "is not an ecological condition so much as it is the interplay between a continuously evolving state of nature and a constantly changing state of mind."

Degradation often is taken to be the opposite of sustainability. Yet it manifests itself in counterintuitive ways. Sander van der Leeuw and his colleagues have studied degradation across parts of Europe and the Mediterranean Basin. Van der Leeuw pointed out that degradation is a social construct. It has no absolute references in biophysical processes. Some popular conceptions of degradation cannot stand close scrutiny. In the Vera Basin of Spain degradation manifests itself in erosion—a common understanding of the term. In Epirus, in the northwest of Greece, however, degradation appears otherwise: as an increase of scrub vegetation that chokes off a formerly open landscape. A centuries-old pastoral life, in which local villages were sustainably self-sufficient, is now impossible. To urban residents the landscape now appears "natural," but to Epirotes it has been degraded. Moreover, the spread of shrub and tree cover has reduced the supply of groundwater and the flow of springs. As mountain vegetation thrives, that lower down declines. When soil is eroded from the Epirote mountains it forms rich deposits in valleys that have sustained agriculture for millennia (van der Leeuw 1998a; Bailey et al. 1998; Bailiff et al. 1998; Green et al. 1998). Here mountaintop species have suffered while agroecosystems thrived, and the contrast with erosion in the Vera Basin is profound. In the realm of sustainability and degradation there are winners and losers, and the only constant is that these terms can mean whatever people want them to mean in specific circumstances.

Clearly we must define our conception of sustainability. The difficulty in defining sustainability is precisely that it is a matter of values, which vary between individuals, groups, and societies and change over time. Before the nineteenth century, for example, the ideal landscape was an agricultural one. This was the landscape to be sustained, and early in American history it was considered the basis of Jeffersonian democracy. Today a landscape of small farmers is largely a quaint remembrance, valued more for nostalgia than for political economy. Many urban residents today value wilderness (or at least their conception of wilderness), preferring recreation to commodity production. That value too may change. We might wonder why we struggle to sustain forests that take centuries to mature when centuries from now no one may care.

Similarly, the impact of foot-and-mouth disease in Britain at the time of writing is large for the farmers, wrenching away a lifetime of careful breeding of a herd. However, of more economic importance is tourism in England's green and pleasant land. It is distinctly possible that farming in Britain may become a quaint occupation of grooming the landscape for its visual attributes. The profit from produce sent to overseas markets collapses under a policy of vaccination, for infected and immune animals are indistinguishable; British beef is quarantined indefinitely. Tourism and its economic clout may turn Britain into an agricultural museum.

Carried too far, such reasoning veers close to nihilism: Why sustain anything when we can't know what future people will need or value? On the other hand, the problem of valuation could equally lead to the opposite reasoning: Because we can't foresee future values we must preserve everything. We advocate neither view and raise the dilemma merely to emphasize the value-laden nature of sustainability decisions. Although many advocates of sustainability feel that the things they seek to protect have intrinsic value, we argue the contrary. The things we want to sustain have only the values we assign to them (Tainter and Lucas 1983; Allen and Hoekstra 1994), which are transient, variable, and mutable. Only when this is recognized can we expect to diminish the political invective that infuses sustainability debates. Deciding what to sustain and how to accomplish it are matters for negotiation and consensus (Tainter 2001).

Recognizing the idealistic or value-laden nature of sustainability, the best definitions have been general. The one most widely cited was offered in 1987 by Gro Harlem Bruntland, then prime minister of Norway: "Sustainable development is development that meets the needs of the present without compromising the ability of future generations to meet their own needs" (World Commission on Environment and Development 1987:43). Although this definition will no doubt continue to be widely cited (almost as a totemic ancestor), we do not find it useful. It borders on tautology: *Of course* sustainable development concerns tending to the future. The word *needs* suggests material requirements, although it could be stretched to cover the intangibles that many people value, such as ecological processes, endangered species, or uncut forests. The definition is too general, however, to guide decisions, which perhaps befits a political leader. It is so vague "as to be consistent with almost any form of action (or inaction)" (Pearce, Atkinson, and Dubourg 1994:457).

Writing from the perspective of economics, Pearce, Atkinson, and Dubourg modified and condensed Bruntland's definition to mean non-declining human well-being (1994:458), *well-being* bearing a meaning that economists prefer to *needs*. If human well-being is considered to include intangible values (ecological processes, species, pleasing landscapes), this approach seems to improve on Bruntland's. For our tastes, though, both definitions focus too much on outputs (the things that satisfy needs or promote well-being) rather than on the contexts of production. We need a definition better suited to our approach.

We suggest that when confronted with the term *sustainability* one should always ask, "Of what, for whom, for how long, and at what cost?" Incorporating these questions, we define sustainability as *maintaining, or fostering the development of, the systemic contexts that produce the goods, services, and amenities that people need or value, at an acceptable cost, for as long as they are needed or valued.* Our concern, as we emphasize throughout, is context, not outputs.

It is important to distinguish *sustainability* from *resiliency*. Sustainability is the capacity to continue a desired condition or process, social or ecological. Resiliency is the ability of a system to adjust its configuration and function under disturbance. In social systems, resiliency can mean abandoning sustainability goals and the values that underlie them. Sustainability and resiliency can conflict.

Social systems can be sustainable and resilient (social goals are flexible and in harmony with underlying ecological processes), unsustainable but resilient (the system adjusts to perturbations but not as people wish), or sustainable but not resilient (sustainability goals are feasible within narrow parameters but inflexible). The first condition (sustainable and resilient) represents the mythical harmony that many writers impute (idealistically and often unrealistically) to tribal and peasant societies. Because such societies have dominated nearly all of human history, they may indeed, as a general class, be sustainable and resilient. The second type (unsustainable but resilient) might describe contemporary industries that must reorganize to survive in a global economy. The third case (sustainable but not resilient) describes societies that have diminished their own adaptability, which can arise from over-centralization. Today's classic example is the former Soviet Union, which could only expire rather than adapt. An especially telling historical example is the overirrigation of agricultural lands in ancient Mesopotamia, which we relate in chapter 3.

We focus here primarily on sustainability; understanding what makes societies resilient (or not) is another topic altogether. Most of us prefer the comfort of an accustomed life (sustainability) to the adventure of dramatic change (resiliency). We find it difficult to recognize, let alone alter, the ingrained values that underlie our sustainability goals. A fully resilient society would be a valueless one, which by definition cannot exist. Accordingly this is not a book about how to change sustainability goals, the essence of resiliency. We take a desire for sustainability and accompanying values and goals as given and address ourselves to discussing how sustainability is to be achieved.

PAYING FOR SUSTAINABILITY

Accepting that sustainability does not become an issue until human involvement is significant, little can be achieved without economic sustainability to support the activities that pay for action. Dan Janzen (1999) eloquently argued that any tropical forest that cannot support itself economically will be gone in short order. He recommended harvesting a small amount of material for genetic resources or certain exotic pharmaceuticals that are worth literally more than their weight in gold. Sometimes the links between biogeophysical systems and economics are simple and direct, as in Janzen's forests, but the links may equally be more general, as in a society being able to afford to set green areas aside.

Proper attention to the economics in ecological sustainability lies behind several of the new concepts that arise from the move upscale to larger natural resource issues. For example, the term *Ecosystem Approach* explicitly includes an economic component. The term is part of the vernacular of the environmental scientists who focus on the Great Lakes Basin ecosystem and is central to the work of the International Joint Commission for the U.S.–Canadian border. In the Great Lakes region, *Ecosystem Approach* implies a particularly inclusive ecosystem concept, which is outward looking and active in its management style. The ecosystem approach explicitly puts the human creature, with its institutions, into the ecosystem as a working component (Allen, Bandurski, and King 1993). In an early statement of the ecosystem approach Jack Vallentyne (1983) and others divided the ecosystem into three components that correspond exactly to the economic, social justice, and biophysical parts that we identified earlier as aspects of sustainability. Business is likely to be one of the major institutions that foots the bill for remedial action directed at ecological distress. Therefore, it only makes sense to include commercial forces as working components of the ecosystem.

MAINTAINING THE POLITICAL CONTEXT

In matters of sustainability, social justice is a critical sector of concern because without some sort of workable social contract, the people in the system will not cooperate. Under the worst scenario there is revolt into anarchy. Warlords make poor conservationists. The term *ecocide* was popularized in the news reports of the wanton destruction of oil wells by the retreating Iraqi troops in the Gulf War. Without social justice, both the economic and the biophysical system are not sustainable.

In Dan Janzen's tropical forest, local people are employed as taxonomists and wardens, thus sustaining the local economy and engendering a pride and commitment to sustain the whole enterprise. In paying tropical societies a fair price for preserving forests, there is a greater measure of social justice in the North–South tension. That makes the politics of Janzen's forest easier to sustain. On the other hand, merely dumping money or resources into the system does not in itself lead to benefit. For instance, the large commitment of aid into Somalia gave strength to military leaders who organized themselves around the resource (Besteman 2002). Social injustices such as nineteenth-century gunboat diplomacy or twentieth-century economic imperialism in the politics of bananas are expensive to maintain because of local nationalist forces. Furthermore, locally maintained tropical forest reserves present humanity at large with the materials for cancer drugs and pest resistance for field crops instead of merely tea or cheap fruit. Social justice is a prerequisite for sustainability.

THE ECOLOGY OF SUSTAINABILITY

We consider economic and social justice as necessary, but the biophysical aspects of sustainability are central. Without a material system capable of functioning for a long time, there is nothing to sustain. Furthermore, inviable material systems deny economic sustainability. The economic hardships of the Great Depression were made worse by the failure of the agroecosystem in the Dust Bowl. It is hard to conceive of social justice without a material base on which the luxury of justice can rest. In the absence of a sustainable agroecosystem, social justice is liable to disintegrate. The drought that stressed the biophysiology of the horn of east Africa was a prelude to a disintegration of social justice in Somalia, leading to the collapse of the agricultural system. Even as the drought was relenting in its physical effects, farmers were disinclined to plant because of the catastrophe in social justice coming from the drought. Looting makes harvest not worthwhile. Even as the environment of social injustice relented, United Nations rations had depressed the price of food so that Somalian farmers could not obtain a return on their efforts. The repercussions of the prior failure of the biophysical system in drought took an economic turn, even as the drought eased. Biogeophysical systems depend on and cause the condition of economic systems and social justice, but the most tangible member of the trio is the biogeophysical system.

DRIVEN BETWEEN DISCIPLINES BY TECHNOLOGY

More than ever before, the modern world is self-conscious about being technological. Note, however, that the industrial age was different from its forerunners by virtue of its technology, and other ages have been named for the metal technology that started them. Even so, the self-consciousness we moderns have about being technological is undeniable, whether or not those attitudes are well placed.

The explanation for the modern preoccupation with technology is that technical advances impinge on us in a very personal way. The common experience is that technology makes things more convenient, provides us with entertainment, and affects many daily routines. The price of mechanical goods with moving parts, such as garden tillers, increases with inflation, while the technology associated with many of them remains essentially the same. Engines belong to the industrial age, and in these days they do not become easier to make, nor are they designed on new principles. Electronic devices show an entirely different pattern. Televisions remain the same nominal price despite inflation, and cutting-edge electronic devices, such as computers, remain the same price while doubling in capacity every 18 months. Consumer goods that use new technology are readily available to the public, and improvements are taken for granted. By contrast, in Victorian times technology was slow to change and impressive. Members of the Victorian public experienced industrial changes only infrequently, as when a journey took the traveler over some large steel trestle hundreds of feet in the air. Whereas the technology that defined the Industrial Revolution was impersonal and distant, the modern world is cast as technological because the new technology is more intimate.

In some parts of the ivory tower there are still attitudes that carry over from the industrial age. The confidence of Victorian Britain survives in some who still expect that our technology will bail us out of most of our large problems. However, many parts of academe are less optimistic. Perhaps some energy engineers still think utopia will come in the form of clean, abundant energy, but large sectors of society are far less optimistic about our modern ability to solve problems. It is telling that in 2001 a new society of ecological engineers was established. At its second meeting in 2001, it saw itself as clearly distinct from environmental engineering, which is really just a branch of conventional civil engineering. So sensitivity is on the rise. In academe, women's studies, psychology, and urban and regional planning faculty must surely be less than fully enamored of a technological fix to societal ills. There were naysayers in the eighteenth and nineteenth centuries who doubted the unbridled success of an industrial fix, of course. The Lakeland poets, for instance, actively escaped the city and all its promise. These days, doubts about technology being on balance worth the cost are not confined to an aesthetic fringe. Rather, they are held by large sectors of the public suspicious of everything from genetic engineering to simple irradiation of food. If anything, technology is more commonly viewed as being a large part of the problem for society, a very different attitude from that of the nineteenth century.

Some modern problems do indeed have a recent material origin with a man-made cause. The thinning of the ozone layer is a recent phenomenon that appears not to have existed before the mid-1970s. Global issues and climate change have become front-line concerns in the scientific community as a result of this material change. But the fact is that most of the modern natural resource and environmental dilemmas are not new; they are merely receiving attention when before they were either unseen or ignored. For example, acid rain is nothing new. The supposedly unspoiled Lake District, where the poet wandered lonely as a cloud, was already devastated in Wordsworth's day by sulfurous smoke of the Industrial Revolution. Only recently has it become clear that the mountain lakes with no fish in them are acid not because of natural processes of leaching and acidification by sphagnum moss but because of acid rain from the satanic mills. Some global problems are new, but many others are old and only just recognized.

The heightened awareness of professional ecologists to global problems has a base in technological change. Technology has delivered tools that allow anyone to see problems that were heretofore so large in scope that they were taken for granted as just context. If fish were scientists, the last thing they would find would be water. Only with new tools thrust upon them do environmental scientists see global problems clearly. One of our theses is that tools are the drivers of most new scientific frameworks. For example, not until computers were generally available to scientists did the new science of complexity emerge in the form of fractals and chaos. In ecology and whole organism biology, examples might be cinematography in animal behavior or videotape in cognitive development. Allen even had a question on one of his undergraduate examinations 35 years ago that asked how polyethylene bags had changed plant taxonomic field collection. The new measuring devices and data collection methods have redefined what is ecologically possible and interesting.

31

The critical effect of tools is that they rescale the activities of the scientist. Even in the case of polyethylene and plant taxonomy, the plastic bag rescales the decay rates of specimens relative to the time taken to get back to the herbarium. More obvious examples of rescaling human activity are the telescope and microscope, but even the quadrat pressed a rescaling on Pound and Clements (1898) in their vegetation studies. The quadrat is a bounded unit of area, often square, in which plants are counted or assessed in some other way. It was too small to do biogeography in the tradition of von Humbolt and Bonpland (1809) early in the nineteenth century, but it was too heterogeneous and far from the laboratory to use the adaptationist arguments of the German plant physiologists to explain what one found in quadrats. From that mismatch, Clements (Pound and Clements 1900) was forced to organize his work around the plant association, which became the cornerstone of a self-conscious new discipline, ecology. Thus even in less obvious cases technical methods rescale the observer to the material functioning of what is observed.

New tools have rescaled the issues that can be addressed, and some of them fall outside the traditional disciplines. This appears to be a normal part of the change in scientific culture. The large-scale problems pressed on contemporary scientists by the new technologies often fall between the standard disciplines such as botany, geology, or meteorology, challenging the standard disciplinary structure of the scientific community. It can be argued that ecology is still emerging as a quasi-discipline exactly a century after the quadrat imposed the critical change of scale on Pound and Clements. Not yet a discipline in its own right, ecology is rather a collection of variously interrelated activities from soil science to meteorology. Modern issues that face humanity demand new aggregations of classes of ideas linking economics to history and natural resources to political science.

The critical character of our age is its rapid pace of technological advance. This is spawning a host of interdisciplinary activities, some of which may eventually become the scientific disciplines of the twenty-second century. As with all tools, modern technology is rescaling the human creature's relationship to the world, this time in very literal terms.

PREDICTION IN LARGE SYSTEMS

In this book we propose a redefinition of the problem of ecological and social sustainability that pulls together academic disciplines that have heretofore moved at best in parallel. The problem of global and continental sustainability is so large that standard disciplinary approaches cannot hope to address the central issues adequately. Lines have been thrown between ecology and economics so that ecology has some connection to human functioning. But a few loose pairwise ties will not do. The biogeophysical sciences have not yet pulled together with the social sciences at large to form a coherent unit to engage the pressing problems of large-scale sustainability. Consider first the limitations of ecology when it is deployed in isolation against the challenges of sustainability.

The view back to our planet from our local sector of space has allowed humanity to recognize the constraints of a closed system. Of course Earth is not completely closed, but it is as closed as any ecological system that we are likely to encounter. Beyond this, the mechanics of space capsules rescales humans to press against the limits of being in a substantially closed system (fig. 1.9). A new study of the ecology of biospherics gives a workable model for some of the limits recognized as important in the contemporary view of our planet.

Constraints are the stuff of prediction; all predictions are that the constraints remain in place to give the predicted effect (fig. 1.10). In a closed biosphere, be it a space capsule, some larger structure such as Biosphere II, or the entire planet, the system is constrained in a mechanical way that gives insights into the functioning of ecological process within. Enclosed biospheres are such a challenge that they not only press upon prevailing theory as to ecological processes but also generate new hypotheses.

The essentially closed nature of our planet allows for powerful prediction at a global scale. These predictions are so reliable that atmospheric scientists take them for granted in the day-to-day conduct of their science. However, aspects of atmospheric sciences that apply at only slightly less than the scale of the entire globe are very difficult to predict.

FIGURE 1.9

Space travel imposes deep constraints on human ecology.

Photographs by NASA.

Whereas the heat budget of the globe as a whole is well described, the likelihood of global warming and, even more so, its effects on regional climates are poorly understood. The reason is that at a regional scale the system is still large but does not press against unambiguously defined constraints on the whole planet. Peter Greig-Smith (1971) once conceded that tropical rain forest may not be the most diverse vegetation on the planet, but he went on to argue that it is the most diverse that is physically much larger than the field botanist. The difficulty with conducting science at a scale significantly larger than ourselves is that the entities posited are the context of the investigators, who are for the most part nested inside that which they study. Except for the whole planet, bounding entities larger than ourselves is difficult, and our capacity to recognize structure and change at that scale is limited.

PEACE AND FUTURE CANNON FODDER

The Tiger : " Curious ! I seem to hear a child weeping ! "

FIGURE 1.10

An unlikely prediction. In a 1919 cartoon in the *Daily Telegraph,* Wil Dyson made a remarkable prediction of World War II. The constraints were that an industrious people treated harshly in war reparations will continue to be industrious, and the harsh treatment will set the stage for a reaction. They will be back in a generation, and so the next conflict will occur in 20 years or so.

Redrawn in the same style as the original by Kandis Elliot in Ahl and Allen 1996, reproduced with permission of Columbia University Press.

Reductionist methods generally fail in large-scale systems where the constraints are less than obvious. Reductionist science is so sure that the phenomenon is self-evidently interesting and worthwhile that it proceeds immediately to data collection. The measurements are performed at some prescribed lower level at which the explanatory principles are confidently asserted to reside. Reductionist self-assurance comes from past successes using certain techniques where the assumptions are workable. In situations in which reductionism does work, there is general agreement as to the critical questions, and the entities in question are bounded on well-known principles. The analysis of patterns to define the problem has already been done, and that means that the confidence in the opening assumptions of reductionism is well placed in that context. However, in large-scale ecological systems reductionism is not successful because description of the organizational structure is primitive; there are no obvious bounds, so organizing principles lack an agreed currency.

With limited success, ecology has been on a reductionist binge for a quarter of a century, where mechanism is sought inside a simplistic system characterization. There was a full decade of experimentation elaborating competition as an explanatory principle, until Simberloff pointed out that competition had not yet been demonstrated. In the context of ecological sustainability, Dan Simberloff (Simberloff et al. 1992) argued very effectively that austere ecological theory is well and good, but until it is tested on the specific situation it is unscientific. Science, he argued, is a matter of creating theory and then testing it on material systems to see under which conditions its assumptions hold. If they do indeed hold under the conditions where one wants to apply them, then one is justified by a scientific approach in imposing a certain management regime. Unfortunately, the scientific culture of ecology does not regularly use a cycle of (1) find a phenomenon, (2) propose a theory, and (3) test it by looking for disconfirming evidence.

The minimal ability of ecological investigations to predict has one notable exception: island biogeography. The tenets of island biogeography are that the number of species on an island is the product of the balance between invasion and extinction. This is largely tautologically true, for there is no other way for a species to be either absent once

37

present or present except for endemic evolution. The model is extremely successful in predicting patterns of species number by introducing surrogates for invasion and extinction rates. If the members of an archipelago are all approximately the same distance from a mainland source of invasion, the number of species of a given general type of organism, such as birds, is related logarithmically to the area of the island. On the other side of the coin, the line plotted logarithmically of species number against island area is straight and of a given slope for birds but another slope for insects. The extinction argument says that, with all else equal, a species on a large island is more resilient than one on a small island. The converse invasion argument says that with islands of the same order of size, islands closer to the mainland source have greater invasion and so have more species.

With such robust quantitative predictions one might have confidence in using island biogeography theory as the basis for remedial action or conservation. Indeed, it is much more promising than most theories, and it does suggest very reasonable hypotheses for design of nature reserves. Of course there are assumptions, albeit reasonable assumptions, such as that actual oceanic islands are good models for nature reserves isolated in an agricultural landscape. The theory suggests that, with all else equal, reserves that are more sustainable are large, roughly circular, and connected to other reserves by corridors (fig. 1.11).

With an appreciation for scale finally permeating the population literature, the concept of metapopulations has emerged. The model in the metapopulation approach is that populations regularly go extinct, so the sustainability of a species does not depend on the sustainability of individual populations. The unit of sustainability in a region would be better modeled as a set of populations winking off and on. There is supposedly a good chance that a population will act as a source of a new population before it goes extinct. Simberloff et al. (1992) suggested that this model replaces island biology as a model for rare species preservation in nature reserves.

For both island biogeography and metapopulation models, Simberloff (1988) argued that they have not been sufficiently tested empirically to see whether they apply to practical situations of restoration or preservation. Although metapopulation models sound fine in theory,

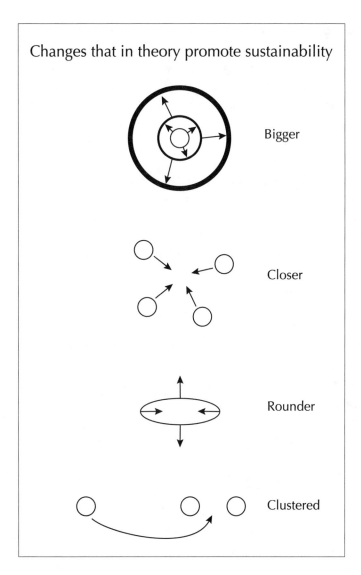

FIGURE 1.11

Characteristics that, in theory, improve reserves. The principles of island biogeography suggest that several large reserves connected by corridors are better than a small number of isolated, small reserves. Caughley (1994) found very limited support for this view in empirical evidence, and Simberloff et al. (1992) concurred and offered many reasons why corridors may expose endangered species to danger.

most populations that wink in and out between invasion and extinction do so in the context of a large permanent population from which almost all the invasion comes. Clearly, when such source-to-sink rigidity applies, metapopulation models offer no insight into proper remedial action to achieve sustainability. The metapopulation is held in the vise of that large permanent population, and that source population is the only real concern when action is planned to sustain the species in the region in question (fig. 1.12).

Simberloff indicated that working reserves and conservation programs in progress have been abandoned because they did not conform to island biogeography theory, even though there has been no empirical test of the relevance of the theory in those cases or a refutation of the approaches that were abandoned. In a lecture that he gives on the topic, he shows a slide of a corridor that has been left not as a remedy to help stave off extinction but rather as a palliative to political sensitivities so that logging can continue. Corridor theory is much more complicated than first-order theory suggests. Linda Puth (1997) has shown that it is worthwhile making the corridor concept more sophisticated by identifying that the reduction of dimensionality of corridors need not only be from the two dimensions of the plane of the landscape to the one dimension of the corridor. Some corridors appear to exist at a fractal noninteger dimension and so apply as a probability function rather than as a thing in a place. Some animals view another species' corridor as a boundary. Simberloff et al. (1992) suggested that there are few data to indicate that the corridor system around the Everglades helps sustain anything. Simberloff (1983) certainly managed to put a stop to most of the pointless work premised on simplistic notions of competition, and perhaps he can help landscape ecology by insisting on refutation in a material setting for the value of corridors for sustainability.

Simberloff went on to argue that essentially all the emerging ideas on sustainability and whole ecosystem management are also based as unscientifically as management based on population theory. While encouraging his skepticism, we do not go all the way with him. Note that unlike narrow population biological theory, ideas in whole ecosystem management are new in style because they are expansive and represent a movement of theory upscale and toward realistically hetero-

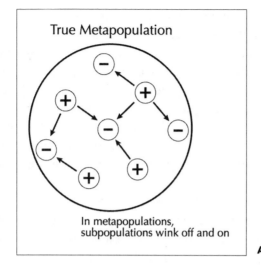

True Metapopulation

In metapopulations,
subpopulations wink off and on

A

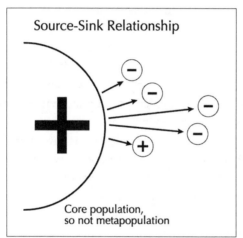

Source-Sink Relationship

Core population,
so not metapopulation

B

FIGURE 1.12

Theory and nature reserves. (A) Metapopulations work on the principle that when a population goes extinct, some other local population will supply invaders. It is really a metapopulation only if there is likely to be reciprocity in that a local site that has offered invaders itself accepts invaders when its population has become extinct. (B) More often there is a source–sink relationship, in which a large population is the source for local populations that rest in its context.

geneous settings. Simberloff's reasoning is that there is little empirical data to suggest that whole ecosystem management models are applicable. His critique may be valid at the level of the field site and the specific activity of the individual manager. Indeed, there is some urgency that rigor be applied and empirical tests of management practice based on the new ecosystem management be undertaken. Even so, the critique is premature. At that higher level we have found that policy statements from the U.S. Forest Service about reordering priorities toward whole ecosystem management are strictly scientific, at least in their idealized formulation.

Let us explain where the rigor lies. The U.S. Forest Service has found that managing narrowly for commodities has failed to meet expectations empirically, and managing for commodities, as done to date, is clearly not sustainable. As a result of this empirical finding, the agency proposed turning away from a rejected hypothesis: that managing for commodities would produce sustained yields. The new hypotheses revolve around sustainability and whole ecosystem management. True, at this point we do not know whether it will meet suitable goals for natural resource management, but now is the time to elaborate alternative theories. If whole ecosystem management does not offer better results than managing for the entire resource, then we need to rethink the issue. This is the principle of adaptive management proposed by Carl Walters (1986). Either way, once there are data to suggest that whole ecosystem management is or is not a good model, then that is the time to present alternative models so that the best versions of it or alternative models are presented for empirical testing.

In the use of scientific models, the truth of assumptions is not the point. Mapping back to the material system is not to verify assumptions in a narrow sense but is rather to see the extent and place to which the formal model maps. Does it map to a sufficient extent to the type of system of interest? In physics we all know that there is no such thing as a frictionless pendulum, but this is nevertheless the basis of a surprisingly large sector of physics applications to engineering. The ultimate set of principles for achieving sustainability we will propose embodies a set of assumptions that will guide management action that resonates with the larger societal context. In developing models for ecological sustainability, because we cannot control for every detailed eventuality, we want to find which assumptions can we afford to make and still have the predicted, desirable management outcome.

Prediction will at first be very difficult in achieving sustainability in large systems because we have little experience working at that level. We therefore need an effective exploratory period using the methods of holism. In such preliminary stages of the investigation of a new idea, a holistic approach generally works better than reductionist prescriptions. We should hasten to add that, according to our definition of holism, the holist does perform reductions. Because all explanations are a matter of reducing the system to a set of lower-level explanatory principles, when a holist offers an explanation, it is a matter of reduction. For our purposes, the difference between reductionism and holism turns on the manner of the search for the explanatory principles that will be used in the reduction. The holist is much less confident than the reductionist and so uses a more exploratory style. By looking for a repeated pattern, the holist seeks reassurance that the original observation that started the line of investigation is more than a local quirk of some observational procedure. Thus reassured that there is a reliable phenomenon at the base of it all, the holist analyzes the pattern of behavior surrounding the phenomenon to see what properties one might expect the explanatory principles to have. Thus informed, the holist says that he does not know what the explanatory principles are, but they should have such-and-such properties. When something fitting the bill is discovered, it usually seems painfully obvious. Holism is the strategy that is more likely to reveal predictive principles when the endeavor is new, as is the search for principles for ecological sustainability. Reductionism emerges for us as a protocol, not a belief system, and is what follows when holistic approaches have delivered a set of reliable principles that can for the moment be taken for granted.

STANDARD PRACTICE FOR DIFFERENT REASONS

Although the challenges of the modern condition are great and they demand a new approach, the change we recommend is large but tractable. It is a mistake to imagine that all standard procedures of environmental scientists must be rejected. Environmental scientists have achieved a lot in standardizing protocols and developing methods, and it would be foolish to cast all that aside. To do so would make the goal unachievable. Many of the established protocols in environmental science will continue to be used. Thus what will be done will be quite similar to what has been done in the past, but it will be played out in a new setting with a new agenda. Although the various local actions will be the same, important but subtle differences will arise in the order in which things are done or the combinations of them that are used. These subtle changes will come from operating in a different context. The reasons that guide action will be different.

Those concerned with biophysical sustainability are well aware that the problems are very complicated, but before we offer any solution, we will suggest that the situation is even more complicated than standard practice indicates. The added complications come from two directions. First, a proper appreciation of scaling effects inside a well-defined subdiscipline leads to a set of complications that are not part of the conventional wisdom. Second, specialists in one area of discourse are required to take other specialties into account. Specialists will be forced to recognize that the material system must be cast in terms foreign to them. For example, population biologists intent on sustaining a local species will be forced into a political arena that is the context of the local landscape ecology. In these unfamiliar arenas of discourse, the specialist will encounter a different version of the scaling effect crisis that challenges their own specialty, this time away from home. Unavoidable issues of scale and type impose a new order of complexity.

Thirty years ago, John Neese, a seasoned teaching colleague, told Allen that he had just offered his students an elegant solution to a problem they did not have. The lecture on multivariate analysis of vegetation had been a failure. To avoid that problem again, let us now lay out the problem of just how complicated is management based on a rational ecological analysis with a case history. The example unravels some of the complications that arise inside a study that remains strictly within a subdiscipline once scaling effects are considered. It unfolds a scenario in wildlife ecology, Hoekstra's specialty. The intent is to ensure that the reader recognizes the problem of complexity.

As a population biologist studying pronghorn antelope changes the scale of the object of study, the explanatory principles change type.

Allen and Hoekstra considered this at some length (1992), but it provides a succinct example here. At the scale of a doe and her fawn, the relationship is genetic and reproductive. At the next scalar level, herds of about a dozen individuals exchange members quite often when the groups meet, so membership has no genetic basis. The herd is for avoiding predation. At the next higher scalar level, that of the local population of herds, there is a high degree of relatedness within the group, for there is little exchange across the unfavorable terrain that separates groups at this larger spatial scale. Local habitat limitations circumscribe the local population. At the level of the entire species, the range is defined by invasion balancing local extinction across a region of pockets of favorable habitat at the species' limit.

At each of these spatiotemporal scales, the units to which the individual belongs are not just successively larger but are of a different type, held together by unrelated processes. Add to this the fact that the relationship of the successive groups is to a differently defined habitat. The habitat changes from local feeding and nurturing strategies, to a landscape full of predators, to a region that is homogeneously favorable, to a large landscape that is heterogeneous, with only islands of favorable habitat across a range of scales. A further complication is that for each of the groups there is a specific relationship of density-dependent to density-independent factors. Thus even within a remarkably simple and well-defined system, there are untold complications. And we did not even consider issues of land ownership and politics of local versus interstate issues in the context of national agencies and policies. Every individual belongs at each of those four scalar levels, and each level raises remarkably different management criteria within the domain of a wildlife population biologist. No wonder population biologists have their work cut out for them, and they should command our respect for their management achievements, even though we are not satisfied with the status quo.

Environmental scientists, managers, and administrators often view themselves as broad in their scope of activity, interest, and competence. Such a self-image might characterize our population biologist concerned with the issue just described. Though specializing in antelope, the scientist might easily be versed in plant and animal studies, in manipulating population equations, and in various observational protocols. Therefore many population biologists might view themselves as fully competent to deal with community issues with regard to sustainability. After all, so the reasoning might go, if all the populations are in place and fully functional, then the biotic community will take care of itself. On the face of it, the self-image of broad, fully adequate competence might seem well founded.

Now we need to disabuse such a scientist as to the general utility of such a scope of activity and breadth of competence. Note that as we proceed, we will not introduce complications that come from the richness of the setting. Plants grow in, animals live on, and both contribute to soil, as the population biologist knows. We will not even introduce such elements into the equation; let us assume there is a fully staffed department of soil science at hand, and the population biologist will know when to consult it. That is what most of us would do. Let us also presume that there is an equally well-qualified nutrient-cycling expert and meteorologist, so that any related ecosystem problems can be sent to them and solved in collaboration.

Even so, population problems are much more complicated in themselves than the state of the art can handle. More than that, populations seen in a community setting are much more complicated again. Population biologists use various powerful principles such as competitive exclusion. That principle says that in the same niche, two species cannot coexist, and one will be driven to extinction. True, there is equivocation as to sameness of niche, but there is abundant experimental evidence to show that the principle holds beyond the tautology that if they coexist, then they cannot be in the same niche. The trouble is that in a community, competitive exclusion is precisely what is not interesting. One is not interested in the order of winners in a competitive hierarchy but in how the losers do not seem to lose with the regularity that competitive exclusion suggests. In a community setting, one is interested in how competitive exclusion is generally withheld. In a sense, the community is exactly the opposite of what a population biologist would study. If a community consists of populations, the question becomes how is it that coexistence is the overwhelming rule. The answer is that aggregate populations constitute communities only in the rudest bookkeeping terms, not in a manner that lends community predictions.

Worse still, the things that make an organism a member of a population are not the same facets that are important in placing that same organism in its community. Population members breed with each other, whereas community neighbors of different species do not. Density dependence as a controlling factor can be calculated inside the homogeneous population, but most such relationships between members of different species are hideously complicated, even considering only two species at a time. Interaction between individuals of different species often is the default relationship in a community. Furthermore, relationships between populations usually are known only for a very small set of environmental conditions. The richness of communities indicates that most relationships between community members importantly do not occur close to the narrowly specified conditions derived from convenience of experimental protocol.

The strength of a population approach is the ability it confers to write equations. This is possible only by assuming remarkable homogeneity within populations with regard to scaling of individuals and spatial circumstances. Meanwhile, community ecologists who focus on species do not even attempt to write such equations. Instead they must resort to other principles such as succession, climax, and disturbance regimes. If two populations can be modeled together, three under the same approach represent a real problem. With more than three, even a very simply specified system often exhibits chaotic behavior. Chaos means that the research cannot be replicated. It dies in the rounding error in initial conditions. Note in the aforementioned population biologist's community nightmare, there was little mention of spatial subtleties. Some population biologists do specialize in spatial statistics, but those are gross generalizations for a landscape ecologist. Once a landscape ecologist insisted on things that are mundane but important in the material functioning of the system from the landscape perspective, the population ecologist probably would be overwhelmed. Two whole volumes by Hof and Bevers in the present series of books are devoted to separate aspects of introducing a spatial component into population management.

The message here is not about the limits of population biology, for a horror story just as grotesque could have been concocted for community, landscape, or ecosystem ecologists. The message rather is that standard subdisciplinary approaches of all sorts bog down when scaling effects are taken seriously, and scale is introduced as a formal consideration. Clearly sustainability must involve action at different scales that relate to systems of different types. A community cannot be sustained unless its associated populations are healthy. A set of recommendations as to sustaining even a single population that did not consider the community would be unreliable. Even if the short run worked out well, sustainability by definition involves a long time line. One would have to wait only a short time for community considerations to spoil the plans derived from population models. Notice that there are no esoteric arguments involving speculative theory here. Common sense says that sustaining a material ecological system must involve a rich set of types of description and a wide array of scales. Given the complications of system scale and type, the problem of confidently prescribed sustainability might appear insurmountable from the vantage point of the conventional subdisciplines inside ecology.

As we noted earlier, we have enough problems without having to reinvent the various subdisciplines. Fortunately, we do not have to do that. In fact, it would be worse than a waste of effort. We have found that when a problem is dissected into differences in the type of system and differences of scale, what seems impossible at first becomes tractable. The overwhelming confusion often arises precisely from mixing scale and criterion informally and unwittingly. When the scaling and typing of the system are performed adeptly, as prescribed in *Toward a Unified Ecology*, the tried-and-true standard methods, such as mark and recapture, suddenly become remarkably useful and effective.

Notice that the approach of Allen and Hoekstra (1992) does not recommend a slovenly holism that tosses everything into consideration all at once. Not only is the adage that everything is connected to everything else untrue (most terms in an interaction matrix are most sensibly set to zero), but surrendering to such a notion is intellectually paralyzing. If everything is indeed importantly connected to everything else at every scale according to every criterion, then there are no workable simplifying assumptions. Vague hand-waving about the whole system would indicate that there is nothing that can be done, and so it would follow that we should stop worrying about sustainability and brace ourselves for the collapse. By contrast, the approach we use is disciplined. It is holistic in that it looks outward to other scales and other ways of looking at the system, but it does so in an orderly fashion. The responsible thing to do is studiously ignore most of the marginal connections. It is imperative to work out what are the significant connections to be designated as important and deal only with them. Then one will discover the subtleties of their scaling properties on the material system at hand. The problem is tractable, and the good news is that it includes a lot of what competent specialists know already. However, the formal treatment of scale and criteria for observation and importance is an essential, and so those are the ordering principles on which we insist, and that is why we use *Toward a Unified Ecology* as a point of departure.

SOCIAL AND BIOGEOPHYSICAL INTEGRATION

As far as we know, there is no other treatment of a rich description of the biogeophysical system as the warp and a rich sociopolitical system as the weft fashioned into whole cloth (although note the edited volume of Gunderson, Holling, and Light 1995, whose title includes "renewal of ecosystems and institutions"). Although there are whole journals devoted to subjects such as ecological economics, even these outreaching missives are fairly narrow in the connections they make. In this book we describe an emergent structure that is large enough to characterize the essential problem of ecological sustainability.

We will look at both ecological systems and social systems using the same metric so that the two can be linked in a consistent fashion. There have been other efforts to use common metrics, such as dollar values of environmental degradation. Though useful enough, the metric of monetary currency is a capricious link, chosen for its convenience in economic terms and the tangibility of a dollar in your pocket. As might be expected, a metric that is fashioned for comparing societal change and environmental change is unlikely to be in common use in either of the traditional accounts in sociology, economics, and ecology.

The unifying principle is organizational. If the biogeophysical environment in a world full of more than 6 billion people is to be made sustainable it must be predictable, and prediction is predicated on organization. We will look to see how the organization of the biogeophysical world changes with human interference. Human societies too are organized, and we will address how that organization changes as it encounters problems in the biogeophysical and social worlds. The connection is through resource exploitation. A most important class of resources are renewable, and their renewability turns on the working of whole ecosystem function. An important organization that we will address looks at changes in societal organization in relation to biogeophysical changes.

Our treatment will be different from conventional scientific discourse because we will mediate the environment–society relationship through subjective values. Value occurs in the value of resources as they are exploited and the value that is preserved by not exploiting the environment. On the societal side, there is value perceived in sacrifice for society, value in the benefits society offers various individuals, and the values that characterize the cultural aspects of societies. As time passes and material processes unfold, changes occur as some values override others to redefine the situation for the actors. For example, when the value of a resource falls below the value of the human effort to achieve that resource, then the value of sacrifice exceeds the value of the benefit to individuals in society. Such changes in value may run through ecological and societal observables, such as environmental degradation and inflation, so the orthodox measures within conventional disciplines still pertain. Attempts to be value-free in conventional scientific approaches to parts of the problem of sustainability are still valid. That quasiobjectivity is the means whereby scientists achieve a modicum of even-handedness in, say, statistical sampling. Even so, larger values at the level of context of the discourse remain intact despite removal of overt bias at the technical level of the day-to-day practice of scientific calibration. Therefore, in contrast to conventional normal science discourse, we will attempt to give disciplinary science more power by offering a new context in which values are an explicit part of the problem as well as a means to a solution. Because we cannot get rid of values at the contextual level, it is best to use them rather than pretend they are not there.

Technology is particularly value laden because it is created to overcome undesirable limits. Beyond this, technology has negative values ascribed to it because it comes at the cost of some good things that existed before the technological change. There is a bit of a Luddite in most of us. For example, Allen began to use a graphic interface on his computer reluctantly because the icons are less than transparent and diagnosis of malfunctions appears harder. If alphabetic description was good enough for the Greeks, it was good enough for him as he typed "cd/" to change directories on his computer. In the end, attitudes toward technology turn on the value of what is familiar—either positive value for familiarity with prevailing technology or negative value associated with the loss of what was familiar. Technology gives benefits and comes at a cost, for it is both a solution and a problem at the same time.

The modern dilemma includes the costs and benefits of solutions such as technology. Significantly, it has always been a dilemma. In the beginning the benefits of problem solving greatly outweigh the cost of finding and implementing the solution. But quickly there is a diminishing return, until the negative outweighs the positive. The value of the familiar can skew perceptions in favor of prevailing solutions for quite some time, so human societies usually push on with old solutions well beyond the breakeven point. Sometimes the excess is achieved through coercion from elites. Sometimes it is just common traditional values held in the face of difficulty. As a result, environmental degradation has been the norm through history. It may be our downfall, too, unless society can find a way to discount the cost of elaboration of society and technology as it happens. We are not sure we can wholly overcome the dilemma in a new synthesis, but without synthesis to a new level of discourse and management strategy, our way of life may change unpleasantly.

I

COMPLEXITY,

PROBLEM SOLVING,

AND SOCIAL SUSTAINABILITY

2

Complexity and Social
Sustainability: Framework

S ocial sustainability depends ultimately on ecological sustainability, but as suggested in chapter 1, it cannot be reduced to biology in a direct or simple manner. Similarly, undegraded ecosystems do not guarantee sustainable societies, for social sustainability depends greatly on what happens within and between societies. Although we know that the soil, water, and air of Russia are badly polluted, for example, no one suggests that this caused the demise of the government of the former Soviet Union. While we see social and ecological sustainability as intertwined, we can achieve clarity by analyzing them in part separately. This chapter and the next establish a framework for understanding what allows societies to be sustainable or causes them not to be. Because societies manage ecosystems, this framework will merge easily with the discussion of ecological hierarchy in chapters 4 through 6.

Explicit sustainability programs are something fairly new. We envision that such a program would involve several steps, notably the following:

1. Establish sustainability goals, which specify the things to be sustained and the conditions in which they are desired. Fol-

lowing from chapter 1, the sustainability goals preferably concern system contexts rather than outputs.

2. Identify the phenomena that are important to monitor so that one may know whether the sustainability goal is being met.
3. From a knowledge of constraints and from the data derived from monitoring, predict the trend of the contexts to be sustained.
4. Manage systemic contexts so that they stay within the state or range specified in the sustainability goals.

As easily as these steps can be stated, each will be very difficult to accomplish. We do not concern ourselves with establishing sustainability goals. These are idiosyncratic to specific groups or communities and must be established in each social situation through a process of negotiation and consensus building. Our focus is on understanding contexts so that one can know what to monitor and how to do so reliably. In both social and ecological systems, monitoring and predicting from misunderstood contexts (as the Malians in chapter 1 seem to do) are clearly to be avoided yet are disturbingly common. We begin with a simple example of how misunderstanding context can pervade an entire society.

MONITORING, PREDICTING, AND PROBLEM SOLVING

Sustainability is bound intrinsically to both monitoring and predicting. To assess whether we will exist tomorrow in the condition we choose, we monitor conditions today and make predictions. There is a paradox here, for to sustain something is to predict a future condition that derives unambiguously from today's knowledge and values. Yet *un*sustainability is substantially beyond our direct experience: Most people in industrial democracies lead lives that allow or require them to think little further than the inflation rate, the unemployment trend, or the monthly housing payment. Our societies have avoided genuine collapses (sensu Tainter 1988) for so long that the possibility does not enter day-to-day calculations. We are condemned to predict without experience what unsustainability would be like or what might bring it about. Yet many societies in the more distant past did prove not to be sustainable. Their misfortune is our potential luck: The records of these societies reveal the constraints that generate sustainability and collapse. Knowing these constraints allows us better to assess our present and future.

An especially interesting prediction came from the Greek historian Polybius. Taken to Italy during Rome's ascendancy in the second century B.C., he was befriended by many of the influential Romans of his day. He set out to explain to his fellow Greeks how Rome had come to dominate the Mediterranean Basin and its peoples, including themselves. His explanation was an exercise in understanding sustainability, which ancient writers thought of as the independence and survival of individual polities. Although Rome had come to dominate the Greeks and many other peoples, Polybius assured his readers that happier times were sure to come. In what may have been history's most remarkable prediction, Polybius foresaw the end of the Roman Empire before it had even reached its height. Unhappily for his readers, the prediction took 600 years to come true.

Although Polybius's prediction was perhaps comforting and uncannily accurate, its conceptual foundation was flawed. Along with the other intellectuals of his day he assumed a cyclical view of history, in which the development of human societies was likened to the biological cycle of birth, growth, senescence, and death. "Every organism, every state and every activity," he asserted, "passes through a natural cycle, first of growth, then of maturity, then of decay" (Polybius 1979:345). Polities, like organisms, eventually die. Foreseeing the fall of Rome therefore was like predicting that an ox will expire. Rome, Polybius (1979:310) assured the Greeks, "will pass through a natural cycle to its decay."

Given the automatic nature of this process, what was a knowledgeable person to do? Nothing could be done to avert a foreordained disaster, but the process could be monitored by those with the comprehension to do so. The unfolding of the prediction disclosed clues to its progress. These clues came from the fact that in the ancient Classical view, the social and biological worlds were interlinked: As one failed, so did the other. So to Seneca the Elder, writing in the first century A.D., the decline of Italian agriculture was a symptom of the aging of the Roman state (Mazzarino 1966:32–33). In the second century A.D. Pausanias, a geographer and traveler, described a landscape in Greece that was both deforested and depopulated. To ancient writers the ideal landscape was an agricultural one, filled with peasants producing food for the cities, taxes for the state, and sons for the army. When Pausanias described a Greek landscape that was no longer tilled as it once had been, he signaled to his readers a sophisticated allusion to a society in an advanced state of decline (Alcock 1993). The third-century Bishop of Carthage, Cyprian, was somewhat clearer in linking social and natural decay. He asserted

that the age is now senile. . . . The World itself . . . testifies to its own decline by giving manifold concrete evidences of the process of decay. There is a diminution in the winter rains that give nourishment to the seeds in the earth, and in the summer heats that ripen the harvests. The springs have less freshness and the autumns less fecundity. The mountains, disemboweled and worn out, yield a lower output of marble; the mines, exhausted, furnish a smaller stock of the precious metals: the veins are impoverished, and they shrink daily. There is a decrease and deficiency of farmers in the field, of sailors on the sea, of soldiers in the barracks, of honesty in the marketplace, of justice in court, of concord in friendship, of skill in technique, of strictness in morals. . . . Anything that is near its end, and is verging towards its decline and fall is bound to dwindle. . . . This is the sentence that has been passed upon the World. . . . This loss of strength and loss of stature must end, at last, in annihilation.

Quoted in Toynbee (1962 IV:8)

To assess political sustainability one had only to monitor the decline of the natural world and its products.

Despite his questionable reasoning, Polybius's prediction was innocuous. To proclaim unsustainability based on faulty reasoning and spurious monitoring usually is a harmless exercise. Most people ignore such a prediction, and however it turns out, life goes on. Regardless of the musings of Seneca or Cyprian, peasants still tilled their fields, soldiers guarded the frontiers, and the Roman government made decisions that guaranteed its own sustainability for several centuries. The great advantage to a defective prediction of doom is that one's reasons for the prediction and criteria for monitoring its progress do not matter. When a faulty analysis led zoologist Iben Browning to predict that New Madrid, Missouri, would experience a catastrophic earthquake on December 3, 1990, neither the prediction nor the geological tranquility of the day had any lasting consequence except in the public opinion of science (Toumey 1996:3–4). Provided that they do not generate prolonged mass hysteria or great diversions of resources, predictions of disaster probably will not affect the outcome.

By contrast, the stakes are entirely different with a prediction of sustainability and with monitoring to ensure that it comes true. Predicting sustainability does affect the outcome, becoming either a self-fulfilling prophesy or a self-defeating one. If one predicts that certain actions will lead to sustainability and that specific variables should be monitored, the foundations of one's reasoning matter a great deal. One had better be right. An error in deciding which actions will produce sustainability or which criteria of sustainability to monitor has the potential to produce a tragedy. The efforts of the former Soviet Union to plan, predict, and monitor do not need to be recounted. In their wake lie human and environmental tragedies from which it will take decades to centuries to recover. This is one outcome of predicting and monitoring.

Two principles developed in this book point to an enhanced comprehension of what is needed for a society to be sustainable. These same principles also apply to any of the institutions or activities that make up a society. The first, introduced in chapter 1, is to understand the difference between managing for the outputs of a productive system and managing for the integrity of the system. In part, this issue of outputs is a matter of what to predict and what to monitor. The second principle, developed in this chapter and the next, is a historical problem that turns on complexity in problem solving (Tainter 1995, 1996a, 2000). As we shall show, the two are related: Whether one manages for outputs or for the system that produces them has much to do with how one addresses the matter of complexity.

Human societies and their institutions are problem-solving systems. Among social activities, problem solving is central. Societies respond to stresses and opportunities that range from petty quarrels and local environmental problems to international crises and global climate change. Moreover, human societies are problem-solving systems at all hierarchical levels of institution and scale, from families at one end to international bodies at the other. Firms, nonprofit organizations, government and its agencies—each exists to solve problems, and to continue to exist each must show some success at doing so. Institutions that fail to solve problems, or to solve problems perceived to be important, lose legitimacy and support, as many governments have learned. Our persistence as a species results primarily from our ability to create problem-solving institutions. Organizations that study or manage environmental affairs, whether university departments or land-managing agencies, are problem-solving institutions no less than the Supreme Court, the bewigged lords of the Old Bailey, the military, or the Federal Emergency Management Agency.

Although flexibility in problem solving is a vital asset, it is also incompletely understood. As we discuss, sustainability or collapse follows from the success or failure of problem-solving institutions. Problem solving can have pernicious effects, for a solution that is successful now can contribute to future failure. Societal problem-solving systems often develop over periods stretching to decades, generations, or centuries. Thus their current condition rarely can be understood by examining only their recent past. Today's managers urgently need a science of problem-solving institutions, including a historical understanding of how they develop (Tainter 2000). The need is urgent because complexity in problem solving is one of the primary limitations to its effectiveness, and our problem-solving institutions today are the most complex that have ever existed.

The members of our optimistic, impatient, and technologically creative society will hesitate at the notion that there might be constraints to our creativity in problem solving. Our unique combination of historical ignorance and historical arrogance joins with the pervasive influence of neoclassical economic theory to perpetuate the notion that all problems are solvable if only there is resolution and incentive to do so. Yet the experiences of the Romans or any of the other societies that have collapsed suggests that there are conditions under which problem solving breaks down. The study of the history of problem solving, a study truly in its first stages, is producing new understandings of the relationship of problem solving to sustainability.

COMPLEXITY AND PROBLEM SOLVING

Over the past 12,000 years or so human societies have shown a tendency to become progressively more complex. Except for episodes such as the European Dark Ages, which punctuate the trend, the development of complexity has become so common that we regard it as normal and inevitable. This is a curious development, for truly complex societies are quite recent in our history and statistically rare. Social complexity is a highly unusual mode of human organization. As a species we may actually be complexity-averse in some respects, which suggests that our recent history is an anomaly needing to be explained.

At one time it was felt that social and cultural complexity were so clearly desirable that their development could be explained simply by the *opportunity* to become more complex. Complexity was thought solely to be a function of human creativity, surplus food, and leisure time. In many older reconstructions, the development of agriculture was thought to have provided food surpluses and leisure time to allow people to develop such things as arts, writing, and complex institutions. In other words, complexity was intrinsic to our species, lying dormant for the right circumstances to emerge.

We now know that this view was erroneous in nearly all particulars. People who live by hunting and gathering, for example, often are able to produce food surpluses but for various reasons don't do so. In some cases the technology to store food doesn't exist or is labor-intensive. In other cases food stores would cause people to be tethered to a place when other aspects of a land use system call for mobility. Food surpluses are not unique to agricultural production. Nor is leisure time. Agriculture is actually quite labor-intensive, and recent research among hunter-gatherers showed that they may work very little for their livelihood. When they were studied by Richard Lee in the 1960s, for example, the !Kung Bushmen of the Kalahari were found to work only about 2.5 days per week (Lee 1968). As anthropologist Marshall Sahlins observed, hunting and gathering produced the first leisure society.

In the world of complex systems, there is, to use a colloquial expression, no free lunch. Any society that is more complex—whether human, other primate, or ant—has greater structural differentiation

61

and higher levels of integration. A more complex human society has more institutions, more subgroups and other parts, more social roles, greater specialization, and more networks between its parts. It also has more vertical and horizontal controls and a greater interdependence of parts. The organization of such a differentiated structure is achieved through a greatly enhanced flow of information, which must be directed, centralized, filtered, and synthesized. Much social structure develops solely to process information. Hunting and gathering societies, for example, may have no more than a few dozen distinct types of social personalities, whereas contemporary European censuses recognize 10,000 to 20,000 unique occupational roles, and industrial societies may encompass more than 1 million types of social personalities (McGuire 1983; Tainter 1988). Similarly, early anthropologists studying the native peoples of western North America found 3000 to 6000 cultural elements among each group. In contrast, during the early days of World War II the U.S. Army landed more than 500,000 types of artifact at Casablanca and a corresponding logistical service (Steward 1955).

Complexity in social systems refers to differentiation and organization or to increasing organization. Figure 2.1 shows a clear contrast between a social unit that was simple, in an anthropological sense, and one that is much more complex. In the lower left quadrant of the photograph is the Dominguez Ruin, a small pueblo ruin of the twelfth century A.D. in what is now southwestern Colorado. The structure is small, simple, and undifferentiated, reflecting the group that produced it. The edifice rising behind it is the Anasazi Heritage Center, where the remains of the prehistoric people are stored and studied. It is many times the size of the small pueblo and requires a permanent staff and a fleet of vehicles. The staff is hierarchically organized and differentiated by specialization. The center's existence was authorized by the federal government, which provides the funds that it needs. The energy needed to heat and cool the building may well exceed what the entire prehistoric community consumed when the Dominguez Ruin was occupied. We undertake these extraordinary expenditures merely to interpret the small pueblo and others nearby. The two structures reflect societies that are vastly different not only in scale but also in complexity.

The greatest problem with the view that complexity is intrinsic is simply that complexity costs. Every element of increasing complexity

FIGURE 2.1

The simple and the complex: Dominguez Ruin and the Anasazi Heritage Center. Dominguez Ruin is in the left foreground.

Photograph by J. Fleetman, courtesy of Victoria Atkins and the U.S. Department of the Interior, Bureau of Land Management, and Bureau of Reclamation.

has an energy cost, which is the cost to create, maintain, and replace the parts of the social system, to support specialists, to regulate behavior, and to produce and control information. In a cultural system activated primarily by human energy this may amount to only about 1/20 horsepower per capita per year (White 1949:369, 1959:41–42). Today such an amount of energy can produce only a fleeting moment of industrial life. No society can become more complex without increasing its consumption of high-quality energy (Hall, Cleveland, and Kaufmann 1992). Moreover, before we discovered how to subsidize our activities with fossil fuels, increasing the complexity of a society typically meant that *people* worked harder.

So much of our behavior is complexity averse that it may be a fundamental human characteristic (Tainter 1996b). This is reflected in many sayings, a common one of which commands, "Keep it simple." Journalists find receptive audiences when they expose the "complexity of modern life." Some analysts suggest that participation in our national elections is so low because the value of one vote does not offset the cost of mastering complex issues. Politicians respond with simple slogans that suggest complex issues but do not require that one comprehend them. Much of the popular discontent with government arises from the fact that governments increase the complexity of people's lives. Governments add time-consuming activities that people would otherwise avoid (such as standing in lines or filling out forms), force increases in the intensity of labor (to pay taxes), and regulate behavior. So strongly do people resent government-imposed complexity that politicians build successful careers opposing it, journalists win prizes for exposing it, and separatist movements emerge in response to it. We are complexity averse when it is we who must bear the costs.

In science, the principle of Occam's razor has long been an accepted axiom. It states that simplicity in explanation is preferable to complexity. Christopher Toumey (1996:11–26) has shown that the incorporation of scientific values into American popular culture has varied inversely with the complexity of science. When science was simple enough that its main requirement was an inquisitive spirit, pursuing knowledge of the natural world was considered a worthy thing, even divinely sanctioned. As science grew complex and specialized, its goals diverged so sharply from American values that public attitudes began to range from disinterest to hostility. Alexis de Tocqueville commented on this trend as early as 1834. As we will discuss, the growing complexity of science as a problem-solving system has much to do with sustainability.

Thus the development of sociocultural complexity is one of the wonderful dilemmas of human history. Why have we consistently adopted courses of action that cost more of our energy, time, and money and that go against our deep inclinations? The reason is simple: Most of the time complexity works. It is a basic human problem-solving tool. Confronted with challenges, human societies often respond by developing differentiation of structure, increasing organization or developing new elements of organization (e.g., status levels, bureaucratic positions), and gathering and processing more information. Each is an element of increasing complexity and a type of change that we can implement rapidly. As a society differentiates, complexity makes the behavior of individuals simpler, more predictable, and more efficient while also providing integration to specialization.

Our success as a species typically is attributed to a large and richly networked brain, upright posture, and opposable thumb. In fact, these are only part of the matter. It is of equal importance that these qualities allow us to increase rapidly the complexity of our social behavior. Thus, although humans may be complexity averse, contemporary social institutions seem to be powerful complexity generators. All that is needed for growth in complexity is a problem that requires it and resources to fund the process. Because problems are inevitable, complexity seems inexorably to grow.

Growth in complexity generates tensions within societies and no doubt has always done so. Although on one level we value complex societies (commonly labeling them "civilizations"), they are something for which people prefer someone else to pay. The consequences of complexity aversion should never be underestimated, but the most immediate effect of complexity on sustainability comes from the economic realm. Complexity costs, as we have seen, and if complexity is a problem-solving response, then it can be viewed as an economic function. Societies invest in complexity. Often this is explicit, such as when the director of the U.S. Environmental Protection Agency tells a congressional appropriating committee that an expenditure of x dollars to enforce a regulation will ensure safe drinking water for y people. More often it is implicit, as in the case of today's automobiles, which get better fuel economy and pollute less but at a cost of complexity in design and assembly that has forced many backyard mechanics to find other ways to occupy their weekends. Implicit investments in complexity can be sinister, for their costs creep upward in nearly unnoticeable increments until a discontinuity is reached. At its worst, such a discontinuity can be a catastrophe, in both the common meaning of the term (Tainter 1988) and its meaning in René Thom's catastrophe theory (Renfrew 1979).

In any system of problem solving, the initial investments often are cost-effective and may give very high returns. Positive feedback between differentiation, integration, and economic and information flows enhances this effect. This is a normal economic process: Simple, inexpensive solutions are adopted before more complex and expensive ones. So in the history of human subsistence efforts, as we will shortly discuss, hunting and gathering has given way to agriculture, which in turn became more labor-intensive as populations grew. In some places subsistence agriculture gave way to industrialized food production that uses such high energy subsidies that more calories are consumed than produced (Boserup 1965; Clark and Haswell 1966; Cohen 1977; Hall, Cleveland, and Kaufmann 1992). Whenever possible we produce minerals and energy from sources that are most economical to exploit—ones that are least costly to find, extract, process, and distribute. In our social institutions we have changed from egalitarian organization, economic reciprocity, ad hoc leadership, and generalized social roles to increasing specialization of roles, economic inequality, and differentiated, full-time leadership. This is the essence of the development of social complexity.

No society can forever enjoy stable or increasing returns to investment in complexity. As the highest-return solutions in organization or resource use are exhausted, only more costly approaches remain to be adopted. Should these yield a disproportionate return, such that the society gains more than it invests, then complexity yields increasing returns. Increasing returns to complexity yield wealth, as happened when ancient centralized societies in Peru and north China began to practice state-directed irrigation. The surpluses thus produced were a key ingredient in the development of expansive empires, which at least in their initial phases yielded further positive returns—at least for the conquerors. Expanded territorial control in turn allowed irrigation systems to be extended still further, producing economies of scale. These are the synergistic effects called positive feedback.

Regrettably, it seems that increasing returns to complexity cannot go on forever. The highest-return ways to produce resources, process information, and organize and defend a society are progressively exhausted. Then, further needs to increase complexity must be met by more costly and less effective responses. As the cost of organizational solutions grows, the point is reached at which continued investments in complexity do not give a proportionate yield. Increments of investment begin to yield smaller and smaller increments of return. This is a decline in the marginal return (that is, the return per extra unit of investment). We characterize it as the problem of diminishing returns to complexity. Carried far enough it brings economic stagnation and ineffectiveness in problem solving. In its most severe form it makes societies liable to collapse and historically has led to conditions that are commonly called dark ages (Tainter 1988, 1999). A prolonged condition of diminishing returns to complexity in problem solving is a major part of what makes a society *un*sustainable and is a key element of our assessment of sustainability.

Two activities fundamental to being a sustainable society are producing resources and producing knowledge. Both are key elements of monitoring, predicting, and problem solving. Many people assume that sustainability is only a matter of producing resources. Yet in a world where all landscapes are affected by human activities, many are anthropogenically created, and most are managed to one degree or another, information becomes a key ingredient of resource production. The more difficult it becomes to manage in an environment of uncontrollable influences and conflicting demands, the more important becomes the role of knowledge. The more ecosystems lack key ingredients and develop states of structure and function that are no longer self-regulating, the more vital it is to understand how the system *should* work, to understand what has changed or gone missing, and to determine how structure and function are best restored or replaced.

In the next sections we examine the economics of problem solving in systems of producing resources and producing knowledge. In the cases we discuss, people obtain resources and information in economically rational ways. They prefer behavior and institutions that are less complex to those that are more. They prefer conservation of labor and other types of energy to uneconomical expenditures. When they are forced to adopt new institutions or ways of obtaining resources, they often experience diminishing returns. We discuss here examples of much of the spectrum of cultural complexity, from foragers to agriculturalists, ancient empires, and the emergence of industrialism. In each case across this spectrum, people struggled with complexity and costliness in problem solving, each time devising a solution to sustainability or accepting the need to change. The following discussion of the economics of problem solving sets the context for understanding sustainability and collapse.

PRODUCING RESOURCES

As illustrated in the following examples, rational human populations always pluck the lowest apple. Phrased more formally, provided that they have the knowledge necessary to do so, people first use sources of food, raw materials, and energy that are easiest to discover, acquire, process, distribute, and consume while providing the desired level of consumption. As consumption demands grow or inexpensive resources are exhausted, resources are used that are more costly to discover, acquire, and so forth while yielding no greater return. The marginal return to intensifying production declines.

The life of hunter-gatherers was once thought to be, in Hobbes's terms, "nasty, brutish, and short," an unending scramble for those scarce resources that an uncultivated Earth yields only grudgingly. We now know how incorrect Hobbes's speculation was. Hunter-gatherers may know and use their resources so efficiently that they have been evaluated for how well they exemplify optimal foraging theory (Winterhalder and Smith 1981). When hunter-gatherers have access to a resource that is nutritious and seasonally abundant, they typically use it to the utmost. If the resource can be stored, so much the better. Such resources reduce the cost of mobility. The !Kung San (Bushmen) of the Kalahari, as described by Richard Lee (1968, 1969), will settle into a grove of nut-bearing mongongo trees and literally eat their way outward. The nuts are so abundant that many San families can aggregate at a single location, something otherwise impossible in an environment that demands mobility and flexibility.

The Aborigines of Australia illustrate much about foraging economics. Australia was the only occupied continent in which, until recent times, no one practiced agriculture. The settling of Australia took place tens of thousands of years before the peopling of the Americas. The ancestors of the historic Aborigines settled the continent in a logical and efficient manner. The first areas occupied gave high returns to foraging efforts. These tended to be coasts and rivers and perhaps a few other areas. It took the filling of such first choices to force Aborigines into secondary habitats and ultimately into the central desert, where the energy return on labor investments progressively fell. Particularly harsh sections of the interior were not occupied until grinding slabs and other technologies were developed to process seeds. Seed collecting and processing can be a laborious way to make a living. Among groups such as the Alyawara, once welfare rations became available from the Australian government, the use of low-return seed-bearing plants was dropped, although other portions of the native foraging system are still pursued (O'Connell and Hawkes 1981).

The Cree of the Canadian boreal forest use a set of resources as different from those of the Alyawara as one could imagine, yet they make decisions about how to pursue them with similar logic. Because most of the energy in boreal forests is locked into vegetation that people cannot eat, to live there human foragers must hunt. Although large mammals, such as moose and caribou, are available, most of the animal biomass occurs as smaller-bodied fauna, especially hare. About half the area is covered by muskeg or peat bogs, which hunters find difficult to traverse in summer. Another 18 percent of the area is covered by water.

Cree hunters studied by Bruce Winterhalder (1981) throughout their yearly cycle pursue a set of species that yield a net gain of roughly 1500 kilocalories per hour or more. In the early winter months of November to December, for example, Cree foragers trap beaver and hunt moose, grouse, and hare (in addition to marketable fur-bearers). After break-up (late May to early June), hunters pursue waterfowl, muskrats, and beaver, as well as moose should they come upon one. When the net acquisition drops below about 1500 kcal/hour, hunters express dissatisfaction and stop pursuing a species. A hare snare line that had yielded about 1900 kcal/hour for several weeks was abandoned once hare became too scarce for a suitable yield. The availability of external subsidies today (snowmobiles, outboard motors, and fuel) reduces search time and allows the Cree to bypass low-return species (much as the Alyawara now do). Cree hunters now travel far to a place where a high-value animal such as a moose might be found, passing up evidence of lower-return species along the way.

Cultural behavior is so varied, complex, and interconnected that it is doubtful that any human group has ever behaved as perfectly optimal foragers. When people such as the !Kung need to work only 2.5 days per week, there is little selection to behave optimally and much opportunity to do otherwise. Yet most hunting and gathering groups described in recent centuries share one characteristic that is perfectly sensible from an efficiency perspective: They did not practice agriculture (Sahlins 1972; Cohen 1977), for it is often a laborious way to subsist.

Harold Conklin (1957) described shifting (or swidden) cultivation among the Hanunóo of Mindoro Island, Philippines. Shifting cultivation for these people is not just a way to subsist; it is part of their way of life. There are rituals and incantations for every stage of agricultural production, and the social organization as a whole facilitates the swidden cycle. The Hanunóo have a rich semantic lexicon describing their natural environment: soils, plants, and animals within it; types of swidden sites; the activities and stages of production; types of labor; crops; the stages of forest growth; and the feasts that culminate a harvest.

A swidden plot is selected from a complex set of criteria, including religious taboos, distance from the settlement and the locations of other settlements, previous uses, climate, and edaphic factors. The preferred swidden plot is old forest, for it is easiest to cut and will burn in a single firing. Among old forests the Hanunóo prefer secondary growth to forest never cut, for labor is an important consideration. Uncut forest is harder to cut than secondary, takes longer to dry, and rarely burns completely. The total labor cost for all phases of a swidden plot in primary forest averages 3180 hours per hectare. This drops to 2975 hours for secondary forest and 2865 hours for secondary growth in bamboo.

Once cut and burned the plot is typically planted to a seed mix dominated by rice, which remains the primary crop throughout the first year. For the rest of its life, which may be 5 to 10 years, a plot is devoted mainly to root and tree crops. The Hanunóo differentiate between 87 varieties of cultivated plants, each of which has its own time of planting, location in the plot, technique of planting, and characteristic spacing. Intercropping is the norm after the first year; many crops are cultivated simultaneously, within one swidden, in an overlapping fashion. The result is a diverse and constantly changing mosaic of intercropped cultigens. By April of the first year a swidden plot has five or six basic crop types. From May through August 20 to 25 more are added, and from September to January, 15 to 20 more. By the end of the first year a swidden plot contains 40 to 50 crop types, which means 100 to 125 different cultigens. Still other cultigens are planted in older swiddens and in swiddens reverting to forest. In times when the primary swidden crops fail, crops of tubers, fruits, and palm starch, planted in forest at various successional stages, provide a buffer. Two-thirds to three-fourths of an average swidden cycle consists of controlled fallowing. Banana trees may be productive for up to 20 years. By 30 years after cutting the swidden plot becomes indistinguishable, to the outsider or untrained eye, from primary forest. Much of the Hanunóo floral landscape, which appears to outsiders as second-growth forest, actually consists of older swiddens still producing crops.

Hanunóo agriculture is a complex system of landscape management that maintains high diversity, eases labor, and either mimics or uses the forest's natural processes. As Clifford Geertz (1963) described so well, swiddening at its best simulates the natural forest in both structure and function. Whereas much agricultural production consists of changing diverse communities into specialized ones, swidden is most successful where it maintains diversity. In both the primary tropical forest and aging swiddens, there is a very high ratio of nutrients locked in plant matter to those in soil. Burning of vegetation in a new swidden releases the organically stored nutrients faster than does the process of natural decomposition (although nutrients do leak during burning). As cultivated trees mature and the forest invades, a forest structure is restored that deflects potential disturbances. In an older swidden, as in a tropical forest, little energy escapes except that which people intend.

In Hanunóo swiddens, mature trees that do not burn shade the early crops and break wind and rain. Within the plot, after the initial wet season rice harvest, there is high diversity. More than 100 cultigens are planted, each with peculiar growth habits: trees, root crops, pole-climbing vines, ground-hugging vines, and shrubs. In the first 2 years a new swidden produces a steady supply of seed grains, pulses, tubers, underground stems, and bananas, ranging from 1 meter below ground to 2 meters above. Thereafter it continues to produce root and tree crops, the latter for many years. A swidden reverting to forest is allowed to go back to natural successional patterns, except that a small part of its energy flow is diverted to crops that nurture the human population without disturbing the forest itself. The Hanunóo let the forest successional pattern sustain them rather than taxing themselves to sustain the forest. This illustrates what we call the supply-side approach: using natural subsidies to minimize human costs and to give high and long-lasting returns on efforts.

Although subsistence agriculture may be more laborious, averaged over several seasons, than foraging, labor is not a limiting factor in most peasant societies. On the contrary, labor is often abundant,

71

underused, and inefficiently deployed. Labor demands for subsistence agriculture are subject to seasonal pulses, so that at times of planting, tending, and harvesting labor may be in high demand. Still it is routinely underused, so much so that subsistence farmers characteristically underproduce. During the agricultural labor season, Kapauku Papuan men, studied by Leopold Posposil (1963), work on average a little more than 2 hours per day at agricultural tasks, women somewhat less than 2 hours. Robert Carneiro (1970) observed that Kuikuru men, tending swidden plots in the Amazon Basin, spend about 2 hours a day on agriculture and 90 minutes fishing. The remainder of the day is spent dancing, wrestling, or resting. With a little extra effort the Kuikuru could produce much more manioc than they do. Many households in societies such as these fail even to provide for themselves. Their needs are met by interhousehold distribution (Sahlins 1972). Thus farmers working only a few hours a day may be supporting even more than their own households. Colonial administrators, confronted with such underproduction, often concluded that the native people they oversaw were incurably lazy, an ignorant view unfortunately not entirely gone.

It was to account for this contradiction between the availability of labor and the failure to apply it beyond subsistence that Ester Boserup wrote her classic work *The Conditions of Agricultural Growth* (1965). The key she found was the marginal return to increasing labor, a subtle concept brought forth with the brilliance and resources of Western economic science but understood all along by the lazy natives. Simply put, while nonmechanized agricultural intensification causes the productivity of land to go up, it causes the productivity of labor to go down. Each increment of labor produces less output per unit than did the first unit of labor.

Boserup developed a typology of agricultural land use, with fallow period the basis for classification. As fallow periods decline, labor must increase to provide services and nutrients that formerly came from the environment. Boserup's typology extends from forest fallow cultivation (swidden or slash-and-burn, as discussed), through bush fallow (6 to 10 years between cropping), short fallow (1 to 2 years), annual cropping, and multicropping. Through the sequence, labor increases because of the growing needs to prepare, fertilize, and water land.

Boserup's general principle has been amply illustrated by studies of subsistence agriculture throughout the world (Clark and Haswell 1966; Wilkinson 1973). In northern Greece, for example, labor applied at an annual rate of about 200 hours per hectare is about 15 times more productive per hour than labor applied at 2000 hours per hectare. The latter farmer certainly will reap more per hectare in absolute terms but will harvest less per hour of work.

If increasing labor is so inefficient, why undertake it? Boserup suggests that the factor driving agricultural intensification is population growth. Although population has not always been an independent variable in human history, the well-known tendency of peasants to produce large families suggests the power of Boserup's insight. In immediately pre-Revolutionary Russia, economist A. V. Chayanov (1966) studied the intensity of labor in 25 Volokolamsk farm families. Plotting differences in intensity per household against the number of consumers per household, Chayanov found economies of scale: The larger the number of workers per household, the less effort each individual must put forth. As Marshall Sahlins (1972:91) pointed out, productive intensity is inversely related to productive capacity. This is important in two respects. First, part of the incentive for peasants to have children is seemingly to reduce the per capita labor burden. Second, even under the infamous conditions in which they lived, Russian peasants underproduced. Those who were able to produce the most actually underproduced the most. Leisure was valued more highly than the marginal return to extra labor.

Farmers naturally don't meet the vicissitudes of life with an obdurate inflexibility of labor. They labor and produce as they must and typically use social institutions to buffer risk and spread surpluses. To develop and support social institutions means to increase complexity and the intensity of labor. Two studies of farming and risk in the prehistoric arid Southwest, the first by Timothy Kohler and Carla Van West (1996) and the other by Michelle Hegmon (1996), clarified the conditions under which farmers cooperate, pool resources, or face risks alone.

Among subsistence producers, noted British social anthropologist E. E. Evans-Pritchard (1940:85), "it is scarcity, not sufficiency, that makes people generous." Kohler and Van West pointed out that the value of marginal production grows faster than production itself when little is produced and, conversely, when the harvest is great. The marginal value of extra bushels of maize declines in a good harvest because it is difficult to store, degrades during storage, and, if it is to be exchanged, costs much to transport. Yet not to store or share (that is, not to overproduce) is risky in an arid environment where precipitation and growing season are both minimal and variable.

High mean production favors pooling of harvests. To the most productive households, the value of production shared in a good year is less than the value of production they expect to receive in a poor year. For underproducing households the logic is the same. High year-to-year environmental variability accentuates this tendency, for it is in such conditions that the greatest value is realized to storing food physically or socially. The latter is achieved by sharing food, so that a surplus is converted into a social obligation. So during years that are variable but yield large harvests on average, the mean value to a household of sharing each year can be shown graphically to exceed the average value over several years of not pooling.

When mean production is low the distribution of advantages is very different. Averaged over a series of years, the mean value of household self-reliance exceeds that of sharing. Moreover, not to share is to conserve expenditures on social overhead: the aspects of complexity that are necessary adjuncts of sharing, such as social and ritual institutions, specialists, and communal architecture. It is the productive households that defect when mean production is low, for the production they give away in a lean year is greater in value than the food they could hope to receive back in a year that is even worse. Our prehistoric households experiencing a series of poor years will be better off if they do such things as disperse and forage or go elsewhere.

Using one of the most precise paleoenvironmental data sets compiled anywhere, Van West estimated prehistoric agricultural productivity for an area of 1500 square kilometers in southwestern Colorado for A.D. 901–1300. In this sequence there were five periods of high production, each lasting from 24 to 50 years, and the same number of below-average periods, running from 10 to 50 years. Probably nowhere else is there a long-term data set of comparable precision; it provides unparalleled opportunities to understand intensification of production and social change.

The archaeological record does not easily yield knowledge of prehistoric motivations, but it provides firm examples of social cooperation. In southwestern Colorado, prehistoric farmers did much as the model predicted. In periods when pooling would have been advantageous they tended more often to build aggregated communities, public works, roads, and religious structures—all costly manifestations of complexity. They did less of these things when cooperation was disadvantageous and even went so far as to dissolve aggregated communities. The highest levels of social complexity occurred when the advantages of cooperation were highest (A.D. 1100–1129); the final period of community aggregation came in the mid-1200s, when cooperative behavior was also suitable; and the region was finally abandoned in the 1270s or 1280s, when the best strategy was defection. Thus these prehistoric people seem to have undertaken the higher costs of extra production, sharing, and increased social complexity when it was to their advantage to do so, and conversely.

Whereas for foragers sharing may be instantaneous, for farmers production entails continuous, prolonged cooperation and corresponding social institutions. Michelle Hegmon (1996) simulated the consequences of different strategies of sharing among the Hopi of northeastern Arizona. Hopi agricultural yields are notoriously variable, even though they plant across a number of landforms to ameliorate this variability. Pooling harvests allows households to reduce further the risks of productive variation, and it is this that Hegmon chose to simulate.

In Hegmon's simulation, a household that does not meet its subsistence needs 3 years in a row fails to survive. Independent households in one run of the model survive or fail on their own; there is no sharing. Over 20 years nearly half (46 percent) of simulated independent households would die out, an appallingly high rate with an unacceptable social cost. Unrestricted pooling (everyone shares) is demonstrably better: 72 percent of households survive for 20 years, although this is a rate on which any society would want to improve. With unrestricted pooling, one household that does poorly can imperil an entire group.

Restricted sharing is a strategy intermediate between independence and pooling. Food is shared within a select group, who in actuality

75

would be close kin. Hegmon simulated sharing among groups of five households, with the rule that a household first meets its own needs, then offers to share any surplus. After the needs of all five households are met, food left over is put into storage. Over simulated runs of years of plenty and insufficiency, similarity and heterogeneity, 92 percent of households still survived. Restricted sharing clearly is the best guarantor of survival, and it is quite similar to what the Hopi actually practiced. It limits the damage that a consistently underproducing household can inflict. Sharing within limits promotes the most advantageous distribution of yields. The costs of the social institutions that facilitate sharing ensure the survival of as many households as possible.

Simulated Hopi households do ill or poorly only because of computer-generated probabilities. They do not contain lazy or unskilled people or poor pieces of land, nor the converse of these. In the real world, exchanges are not symmetrical. Some groups consistently produce and share more than others. When this happens for a long time something like a patron–client relationship develops: Those who regularly contribute more receive elevated prestige or status. Unequal sharing is the stuff of which inequality is made and has always been a watershed development in social evolution. Hunter-gatherers generally display an egalitarian ethos and work hard to keep everyone on the same level. Once pervasive inequality is legitimized, positive feedback amplifies it and shifts the evolution of society onto a new course. The pace of social change speeds up. Social maneuvering generates further change, and a society grows more rapidly in differentiation, complexity, and costliness. Innocent steps taken to survive, such as overproducing and sharing, generate unforeseeable social changes that may take several generations to run their course.

People today are socialized to consider complex cultural organizations the normal, natural, and even desirable state of human affairs. As discussed earlier, we have come to believe that the development of complexity was solely a function of surplus food, leisure time, and human creativity. This view was developed before it was understood that complexity costs, and for a long time it prevented us from understanding the deleterious consequences of complexity, including collapse (Tainter 1988). The validity of this outlook is testable, for if it is so, ancient societies should have rushed to adopt complex characteristics at the fastest possible rate. Chief among these would have been state-organized societies. If complexity is sought rather than undertaken when it is advantageous or necessary, then there should have been no other consideration in ancient societies than to become states as quickly as possible.

Carol Raish (1992) studied how much time elapsed in 16 ancient societies between the establishment of village farming communities and the formation of states. The periods ranged from 400 years in what is now Mexico to 6200 years in central and southeastern Europe. The interval increased monotonically from the shortest case to the longest, but Raish found that this observation masks interesting variety. It turns out that there are significant differences between societies that possessed domesticated livestock (which are particularly those in Asia and Europe but also include Egypt and Peru) and those without (mostly New World societies but excluding Peru). The societies with livestock display an average of 3978 years between the establishment of village farming and the development of states. For societies without livestock the figure is only 2693 years. The difference is statistically significant. Having or lacking domesticated livestock clearly influenced how rapidly ancient societies became states.

Domestic animals increase the labor of those who must care for them, but they provide offsetting benefits. They provide a subsistence buffer so that when other resources fail one can sell one's stock or live off milk, blood, or meat. Livestock allow people to convert and store products such as grasses that humans otherwise could not directly consume. They allow intensification of field agriculture by providing environmental services in the form of pulling plows and providing manure. In times of localized scarcity they allow food or goods to be transported more easily.

States also provide services, including internal order, external defense, varying degrees of economic security, and overarching organization. This last feature allowed for such things as economies of scale in irrigation agriculture in Mesopotamia and north China, the establishment of universal languages (such as Latin, Manchurian, and Quechua), and the development of secure long-distance travel routes. But states also impose very high costs (Tainter 1988), which are borne primarily by increases in human labor. For this reason, as Robert

Carneiro (1970) pointed out, independent people never willingly accept domination by states. To the peasants who must support states, it is not always clear that the benefits they enjoy match the costs they must bear.

Village farmers with livestock thus appear to have used a rational strategy. Although caring for livestock always increases the complexity and cost of farming, it seems to have increased costs to a lesser degree than supporting a political hierarchy would have. With livestock to buffer the risks of farming, the need to develop more complex organization was deferred, sometimes for thousands of years. This is not to say that livestock raising and political complexity were alternative adaptations. Rather, the former buffered risk well enough that the positive feedbacks that can facilitate political hierarchy, such as inequality and dependency, took longer to emerge. Ancient people never rushed to develop complexity where it could be avoided. They developed it where necessary or advantageous.

In early states and empires, the capacity to organize labor and intensify production gave rise to both the ancient monuments that we marvel at today and the portable objects that make up the collections of famous museums. Yet in these societies information about productive capabilities always seems to have been less developed than the capabilities themselves. Rulers often seem not to have known the limits of the land and peasants to produce; if they knew they did not fully understand, or if they knew and understood, they believed that compelling peasants to greater labor would compensate for the declining productivity of land. The result was systems that produced wondrous accomplishments for a few generations, then underwent periods of prolonged economic stagnation, conquest by another state, or collapse. The Third Dynasty of Ur (ca. 2100 to 2000 B.C.) is a particularly telling example.

In southern Mesopotamia, intensive irrigation initially produces agricultural yields that support growing prosperity, security, and stability. The Third Dynasty of Ur pursued such a course. It expanded the irrigation system and encouraged growth of population and settlement. To capture this enhanced energy flow it established a vast bureaucracy to collect taxes and tribute. For a few years all was well— at least for the rulers. Unfortunately, this strategy plants the seed of its own demise. After a few years of overirrigating, saline groundwater rises and destroys the basis of agricultural productivity. The political system loses its resource base and is destabilized. Large irrigation systems that require central management are useless once the state lacks the resources to maintain and direct that which it created.

A few centuries earlier, in the Early Dynastic period (ca. 2900 to 2300 B.C.), crop yields per hectare averaged about 2030 liters. Under the Third Dynasty of Ur this declined to 1134 liters. At the same time, farmers had to plant their fields at an average rate of 55.1 liters per hectare, more than twice the previous rate. Badly salinized lands go out of production almost indefinitely, so the pressure intensifies to get the most from the remaining fields. As yields declined and costs rose, farmers had to support an elaborate state structure. It was a system that took a high toll in labor. The Third Dynasty of Ur was following a strategy of costly intensification that yielded clearly diminishing returns.

The Third Dynasty of Ur hung on through five kings and then collapsed. However unfortunate for would-be kings, the consequences were catastrophic for the larger population. By 1700 B.C. yields were down to about 718 liters per hectare. Of the fields still in production, more than one-fourth yielded on average only about 370 liters per hectare. The labor demands of farming a hectare of land were inelastic, so for equal efforts cultivators took in harvests about one-third the size of those a millennium earlier. Soon southern Babylonia was extensively abandoned. By a millennium or so after the Third Dynasty of Ur the number of settlements was down 40 percent, and the settled area contracted by 77 percent. Population densities did not rebound to those of Ur III until the first few centuries A.D., a hiatus of perhaps 2500 years (Robert McC. Adams 1978, 1981; Yoffee 1988).

Notwithstanding this experience, intensifying the production of resources or developing new ones sometimes yields surprising results. In regard to the material basis of society, the law of diminishing returns sometimes is upended by the law of unintended consequences. In one of the most significant works of economic history, Richard Wilkinson (1973) described the consequences of population growth and deforestation in England in the fourteenth through eighteenth centuries. A population growing through most of this era progressively cut its forests and intensified agriculture. As forests were cut to provide agri-

79

cultural land and fuel for ever more people, the supply of wood no longer sufficed. Coal came to be used in its place, although with reluctance. Coal was polluting, and it was costlier to obtain and distribute than wood. Coal was not available everywhere, so entirely new distribution systems had to be devised. Digging a fuel from the ground costs more than cutting a standing tree, and coal overall costs more per unit of heating value than wood. Many of those forced to rely on coal experienced a decline in their financial well-being.

As coal gained importance in the economy, the most accessible deposits were depleted. Mines had to be sunk ever deeper, until groundwater limited further penetration. This vexatious problem stimulated greatly the development of the steam engine, which in time was perfected enough to pump water from mines effectively. Thereafter the coal-based economy could not be turned back.

The serendipitous part of this transformation was that, with the development of an economy based on coal, a distribution system (canals and railways), and the steam engine, several of the most important technical elements of the Industrial Revolution were in place. It must be one of history's greatest ironies that industrialism, that great generator of economic well-being, arose in part from steps taken to alleviate resource depletion, supposedly a generator of poverty and collapse. The sense of irony is enhanced by the realization that both the wealth and the complexity of an industrial economy are made possible by the subsidy of a fossil fuel that people preferred not to use.

It is always interesting and sometimes illuminating to toy with conjectural history—to ask, "What if things had happened differently?" An exercise in such conjecture based on the case of industrialism leads to a dilemma in what constitutes sustainability. The dilemma is simply put: If deforestation started a process of adaptation that contributed significantly to the development of industrialism, what might have happened had resource conservation been in place? What if Renaissance Britons had practiced sustainable forestry and so didn't need to use coal, perfect the steam engine, or build canals and railways? No doubt industrial production processes eventually would have been developed. Yet without the *early* need to develop systems to produce, distribute, and use coal, how would the pattern of industrialism have differed in time, place, or form? Perhaps *fin-de-siècle* western Europe would have been like pre-Revolutionary Russia, a land of wood-dependent, underproducing peasants in which intensification, complexity, and industrialism were imposed from the top. We can never know, and carried too far the speculation is futile. Yet it is also highly suggestive. The lessons of Wilkinson's study are clear: Scarcity spurs innovation, and poverty leads to what we commonly call progress. Conversely, to live sustainably and preclude the incentive to innovate may be to forego potentially desirable futures. Surely few western Europeans today would prefer that their ancestors had practiced sustainable forestry if that would mean that their economic status would be no higher than that of the average Russian.

This case is of the highest significance to our efforts to determine what it means to be sustainable. Many economists argue that for creative human societies, resources do not matter. The economic rewards to those who are ingenious enough to alleviate scarcity guarantee that new resources will always be produced (e.g., Barnett and Morse 1963; Gordon 1981; Sato and Suzawa 1983). Thus resources can be harmlessly dropped from sustainability equations; the only variable worth considering is the extent to which governments distort the free-market mechanisms that, unfettered, will guarantee resources. This, in condensed form, is the view of technological optimists. There are many points in such reasoning on which to be uncertain, and the argument as a whole has polarized our society and political system. Yet the historical experience of Britain in practicing unsustainable forestry does not prove the technological optimists wrong. Resource depletion in this case proved not to be an unsustainable practice for British society. A new resource came along, with the technology to produce and use it, just as the technological optimists would predict.

It will seem odd, in a book on sustainability, to point out that seemingly unsustainable practices sometimes improve peoples' well-being. Yet this is part of the dilemma of sustainability. Intelligent and well-meaning people can review indicators of sustainability and reach quite opposite conclusions about our future. Moreover, they could all be correct. We return to this dilemma in our final chapter.

RESOURCES, INTENSIFICATION, AND SUSTAINABILITY

We see in these examples peoples who lived by very different means but who uniformly avoided complexity and unproductive labor. Hunter-gatherers such as the Kalahari San exploit a highly productive resource to the fullest, eating their way through it and then moving on. When foragers gather resources, as among the Cree, they do so with a clear conception of returns to expenditure. Resources that yield lower returns, and habitats that contain such resources, were not used by Australian Aborigines until more productive habitats were full. And when subsidies were provided by the Australian government, the lowest-return resources were dropped. Agriculturalists rely on similar economic logic. They may use ecosystem services to subsidize their own efforts, as among the Hanunóo, provided that their population density is low enough to allow for long fallow. Where shorter fallow is necessary, labor must become more intensive to substitute for services formerly provided by the environment. Productivity per unit of land may rise, but productivity per unit of labor declines.

Social organization often is a component of producing resources. We have discussed how agriculturalists use sharing and other cooperative behaviors to buffer risks and increase the likelihood of survival. Cooperative social institutions, however, are another form of intensification. They carry costs that necessitate increases in production. Although they may increase the likelihood of survival, they do so at a cost of diminishing returns to labor.

Coercive institutions exacerbate this condition. They have their origin in cooperative institutions, but once established they propel social evolution in new directions. Coercive institutions such as states bring economies of scale to such things as irrigation, defense, and other public investments. They do so at such high costs in labor and environmental damage, however, that catastrophes of unprecedented scale sometimes occur, as with the Third Dynasty of Ur. States seem to grow either by conquest or because their neighbors must also become states to avoid being absorbed. It is hard to point to an example of people willingly accepting the diminishing returns to labor that accompany state organization. State organization seems to have been facilitated when effective means of buffering risk, such as range expansion, use of domesticated animals, or cooperative sharing, were exhausted.

These examples of resource production concern historical cases and contemporary subsistence producers, but they have implications of a general nature. After initial high-return phases, resource production moves from less to greater intensification and from more to less return per unit of effort. Thus, whatever resources societies consume today probably are the most productive and efficient resources they will ever consume. As the most accessible supplies of these resources are exhausted, continued consumption at the same levels necessitates an increase in expenditures and a decline in productivity. Under our present way of life, our experiences in consuming resources will be the same as that of the Australian Aborigines or Mesopotamian farmers. If we must become more complex to be sustainable, we must expect diminishing returns to our efforts. Thus the paths before us are either to intensify what we are doing, or hope to do what eighteenth-century Britons did: stumble upon a new energy subsidy.

PRODUCING KNOWLEDGE

If monitoring and predicting are aspects of being sustainable, then the production of knowledge stands equal in our discussion to the production of resources. Information can substitute for resources or allow them to be used more efficiently. Fuel injection is an example. It gives each cylinder of an engine just the right mix of gasoline and air, monitoring and compensating continuously for variations in speed, load, engine temperature, altitude, and the like. The one-time cost of developing the technology is repaid handsomely each day by many millions of efficient engines, each revolving thousands of times per minute and saving volumes of fuel that are minuscule in isolation but great in the aggregate.

Still, information has costs, which are often hidden or indirect. The cost in our example is in assembly, diagnosis, and repair of a device that depends more heavily on information and precise conditions than did its predecessor, the carburetor. Automotive mechanics have now become information analysts, supported by complex diagnostic equipment that itself must be developed, manufactured, and occasionally repaired. There are still mechanics who can diagnose an engine's ills by the sound as one drives into the garage, but they are a declining species. Still, it is usually advantageous to bear such costs. Peasants of ancient Mesopotamia no doubt would have wanted their rulers to have precise knowledge of the salinization of their fields, of their labor and tax burdens, and of their returns to planting. Considering the disaster that followed the Third Dynasty of Ur, such knowledge would amply have repaid whatever it cost.

Terms such as *information age* and *information economy* reflect the fact that our economy and society are responding, without much central direction, to the imperatives to produce and consume more information and fewer resources to maintain the industrial standard of living. Semiconductors are now everywhere, substituting information for energy, materials, and labor. The development of today's information-processing technology is an underappreciated serendipity of our age, for it came at just the right time. Had the costs of energy and labor risen as they did in recent decades without commensurate developments in information technology, we would all now be like the U.S. Postal Service: an antiquated behemoth struggling to control its metabolic costs (oil and labor) by a belated and awkward retrofit of information controls. Anyone who works in an office to which computers have been introduced knows how quickly work and vital processes come to depend on them. When the computers malfunction, most other activities cease as well. A grocery store can no longer sell you so much as an apple unless its computer is working. It is the same throughout our economy and society, which could no longer function without computers. Under such a condition, maintaining the flow of information becomes as critical to sustainability as maintaining the flow of resources. If the Third Dynasty of Ur could collapse for a failure of information, surely the risks of such a failure today are immeasurably greater.

As natural resources become less accessible or abundant, or stressed in the case of biological systems, our response typically is to develop and apply knowledge. This is seen clearly in the case of coal development, as discussed earlier. When water levels prevented economical mining, research to improve the steam engine became imperative. The research was successfully done and the problem solved. There is a certain inspirational value to such incidents. Optimistic writers such as Paul Hawkens (1983) have argued that our current need to develop information carries great potential to generate economic growth. We hope he is right. Yet the information needed to improve physical systems can hardly be more complex than the systems themselves. With apologies to chemists and physicists, biological and social systems present challenges of a very different order. A biological or a social system incorporates both the complexity of the physical subsystems of which it is composed and its own great complexity.

The law of requisite variety applies also to information. Former Forest Service Chief Jack Ward Thomas was fond of noting that ecosystems are complex. To manage them successfully we need obviously to understand them, and that presumes knowledge that is rich and diverse to a degree commensurate to the systems themselves. Much of this book concerns the biological knowledge that we need to manage ecosystems sustainably. No environmental scientist would suppose that we have enough such knowledge. The challenge is to produce it at a cost that is politically and socially acceptable. This is a serious challenge indeed, for the development of knowledge is itself subject to the law of diminishing returns. This fact is another central element of understanding sustainability.

Those of us whose role in society is to produce information believe deeply in the value of what we do and argue endlessly for it. On the face of it, our dependency on information should make such arguments unnecessary. Yet each year we fight new battles with legislators, administrators, deans, and philanthropists to get the funds we need. Each year the fight seems to become more arduous, and we must devise ever more clever justifications in an endlessly escalating contest between the forcefulness of our message and society's resistance. That a society that needs information to sustain itself should be reluctant to invest in it is a telling fact. It suggests that there is only so much information we can afford and that, like groceries, fuel, and motor homes, we must limit the amount of it that we enjoy. Information competes with other things that cost money. Unfortunately, in deciding to limit information it is likely that we also limit our opportunities to become sustainable.

Christopher Toumey (1996) pointed out that for some centuries science has been a part of popular cosmology, justifying and reinforcing aspects of daily life that range from religion to commerce. Anything so entrenched in our symbolic repertoire is hard to analyze objectively. A historical perspective will help, for it allows us to see how the costs and benefits of information have changed.

We first consider education. Most of us believe it is a good thing and, more importantly, a useful thing. As our society increases in complexity, becomes more dependent on information, and becomes more competitive, workers need higher levels of education. In 1924 S. G. Strumilin assessed the productivity of education in the nascent Soviet Union. The first 2 years of education, Strumilin found, raise a worker's skills an average of 14.5 percent per year. The productivity of education declines by adding a third year, for this raises the worker's skills only an additional 8 percent. Four to six years of education raise skills only a further 4 to 5 percent per year (fig. 2.2). Clearly in this case there were diminishing returns to additional education.

We would all like to think that a dynamic democracy might do better on such measures than the workers' paradise. This might prove so were it not for the fact that there are underlying factors that are unrelated to political regime. We now understand that the greatest quantities of learning occur in infancy, at no extraordinary cost to society. The quality of early learning is crucial also, for it determines the ease with which the child takes to later education. Overall the learning imparted at home to young children brings the broadest and most general benefits to both the child and the society. It is our most productive investment in education.

The most comprehensive study of the role of information in our society was published in 1962 by Fritz Machlup. It is a study that is regrettably little known today but still enormously informative. Machlup showed that in 1957–1958, home education of preschool children cost the United States $886,400,000 per year for each age class from newborn through 5. This cost is mainly in foregone parental income. In elementary and secondary education the costs grew to $2,564,538,462 per year per age class for years 6 through 18. For those who progressed to higher education (33.5 percent of the eligible population in 1960), a 4-year course of study cost the nation $3,189,250,000

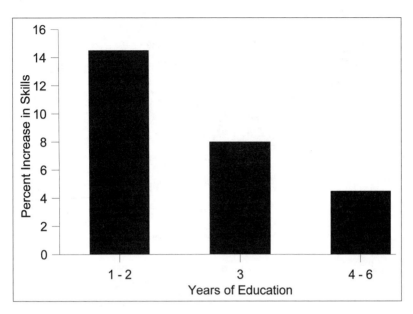

FIGURE 2.2

Productivity of educating workers in the early Soviet Union.

Data from Tul'chinskii (1967:51–52).

per grade per year. Thus, the monetary cost of education between pre-school, when the most general, broadly useful education takes place, and college, when the learning is most specialized, increased in the late 1950s by about 360 percent in actual terms. On a per capita basis it grew by 1075 percent. Yet from 1900 to 1960 the productivity of this investment for producing specialized expertise showed disturbing trends. Taking ratios of students who earn higher degrees to those who earn lower ones (Ph.D./M.A., Ph.D./B.A., M.A./B.A.) and dividing by expenditures, the productivity of educational investment is seen generally to decline throughout (fig. 2.3). Higher levels of educational investment yield a declining marginal return.

Research complements education in producing knowledge and is subject to much the same forces. Most readers of this book will have some familiarity with the history of science, especially its explosive growth in the past two centuries. Early societies bore minimal costs to

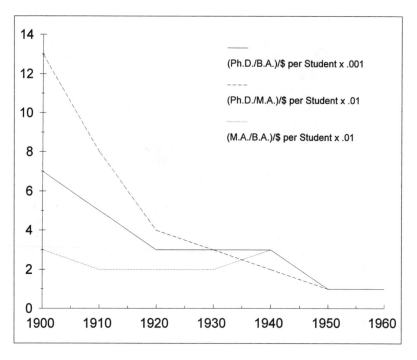

FIGURE 2.3

Productivity of educational investment for producing specialized expertise in the United States, 1900–1960.

Data from Machlup (1962:79, 91, 104–105).

support knowledge producers. Indeed in some cases, as Derek de Solla Price (1963:107) cogently noted, "society almost dared [scientists] to exist." Generally, from the time of the ancient world through the European Middle Ages, the cost to support knowledge producers amounted to little more than the support of individual naturalists or mathematicians and their students or the support of religious specialists who also developed secular knowledge. Certainly these costs were minuscule compared, say, to the cost of supporting the medieval or Renaissance church or aristocracy. Indeed, the cost of supporting knowledge producers tended to come from the wealth of precisely such institutions. From the nineteenth century came the image of the gentleman-scholar, a person with the wealth and inclination to engage in self-supported, often reclusive inquiry. Surely the finest example was Charles Darwin himself. Scholars with similar inclinations but lacking an independent income might pursue inquiries with the support of other institutions. Thus in the eighteenth and nineteenth centuries the income and free time of a parsonage often allowed its holder to conduct inquiries into local antiquities or natural history. William Stukely, first secretary of the Society of Antiquaries of London, entered the clergy in 1730. The best clergyman-scholar was of course Gregor Mendel, whose cost to society would have been the same had he been allergic to peas.

The image of the gentleman-scholar persists to this day and powerfully influences the public perception of science. Mad scientist films always seem to involve such a person. Yet the species itself is nearly extinct. The few remaining specimens make genuine contributions mainly in fields that require such things as peering many hours through a telescope or at images made through one. Even here the role of such scholars has receded as the parts of the universe closest to earth have become better known. Gentleman-scholars and other lone wolf scientists make themselves obsolete as they diminish the stock of scientific questions that are simple and inexpensive to resolve. This was expressed best by Garvin McCain and Erwin Segal when they noted that science is not likely to be advanced much further by peering through a homemade microscope or flying a kite in a thunderstorm (1973:158).

In every scientific field, early research concerns the apples easiest to pluck: the most easily resolved questions and the establishment of basic knowledge. As general knowledge is established early in the history of a discipline, that which remains axiomatically becomes more specialized. Specialized questions become more costly and difficult to resolve. Research organization moves from isolated scientists who do all aspects of a project to teams of scientists, technicians, and support staff who need specialized equipment, costly institutions, administrators, and accountants. Pick any recent issue of *Science* and count the number of authors per paper. As research becomes more specialized so do its results, so that findings are of interest to ever narrower branches of a field and of benefit to ever smaller segments of society. Although specialized science can yield astonishing results, it is hard to say that it

89

is equal in value to the general results of early, inexpensive science. Moon landings, the space shuttle, and close-in images of Jupiter's moons are surely remarkable, but they are hardly of greater net benefit to society than the theory of gravity or the principles of geometry on which they rely. The versatility of automobiles, railroads, and ships is essential to an industrial society, but every unit of benefit comes in part from the first watercraft, the first wheel, or the perfection of the steam engine. And although the field of genetic engineering is now starting to fulfill its promise, each decoded, spliced, and altered gene owes its manipulation in part to the cost-free work of Gregor Mendel. This is why the most famous scientists in each field are always founders or early practitioners. In many cases they were not more brilliant than those practicing today. Their advantages were that they practiced at a time when the threshold costs to becoming famous were lower than the are today (because problems were less costly to resolve), and their research, being of a general nature, was of the most widespread value (Tainter 1988:112).

Thus fields of scientific research follow a characteristic developmental pattern, from general to specialized, from wealthy dilettantes and lone wolf scholars to large teams with staff and supporting institutions, from knowledge that is generalized and widely useful to research that is specialized and narrowly useful, and from low to high societal costs. Considering why economic growth slows in developed countries, Walter Rostow (1980) produced the curve shown in fig. 2.4, which he considered the characteristic way in which a field of learning develops.

Scientific research yields diminishing returns as a field resolves simple, inexpensive questions and moves to more complex, expensive ones. As this happens to the scientific enterprise as a whole, it must consume an ever larger share of a nation's wealth just to maintain a constant production of new knowledge. Nicholas Rescher (1978, 1980), who has written effectively on this phenomenon, inspected the average costs of a research project at the National Institutes of Health from 1966 to 1971. During this period constant-dollar expenditures rose 13 percent without any increase in the apparent merit of the research (Rescher 1978:80, 1980:85, 92). As long ago as 1960 such developments prompted Dael Wolfle to pen an editorial for *Science* titled "How Much Research for a Dollar?" In the United States, investments in research

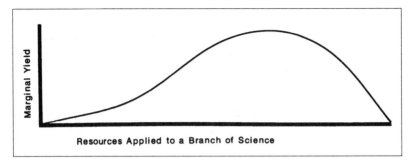

FIGURE 2.4

Marginal yield in a branch of science.

After Rostow (1980:171).

and development grew from 0.1 percent of gross national product (GNP) in 1920 to 2.6 percent in 1960 (Rescher 1978:67). When Derek de Solla Price published *Little Science, Big Science* in 1963, he pointed out that science was growing faster than either the population or the economy, and of all scientists who had ever lived, 80 to 90 percent were still alive at the time of his writing.

As complex, specialized research consumes ever more resources of funds and personnel, diminishing returns to the investment begin to appear. Once again the remarkable data sets compiled by Fritz Machlup (1962) prove their worth. Patents are one measure of the productivity of research, particularly in our society, which places a premium on practical knowledge. For example, Machlup found that between 1900 and 1954, as the numbers of scientists, engineers, and technicians rose in the United States, their productivity in acquiring patents dropped sharply. Between 1900 and 1950 the ratio of growth in patenting to growth in numbers of scientists and engineers declined by nearly 90 percent (Machlup 1962:172). In one particularly telling period, 1941 to 1960, the filing of patent applications fell nearly 50 percent per 100 scientists and engineers and 83 percent per million dollars of expenditure (fig. 2.5). The trend has continued. In the United States, research and development expenditures per scientist and engineer rose at an annual rate of 0.47 percent between 1964 and 1978 while patenting declined at an annual rate of 2.83 percent (Evenson 1984:107–108).

91

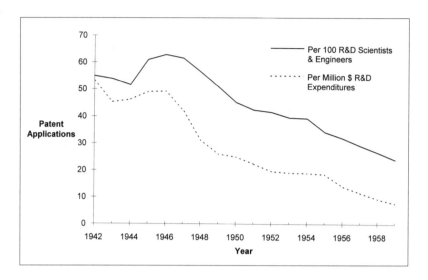

FIGURE 2.5

Patent applications and research inputs in the United States, 1942–1958.
Data from Machlup (1962:173).

Some economists attribute such patterns to a decline in the value of patents (which convey less protection in an age of reverse engineering) and to shifts in research investments. For example, research in ecosystem management rarely yields patentable inventions. Machlup thought that the decline in patenting was attributable to the rise of military research, which also does not usually yield patentable inventions. This last point is demonstrably not the case. Machlup's own data show that the pattern of declining productivity began well before the research and development efforts beginning during World War II. Moreover, similar patterns are evident in many other countries, most of which do not invest as heavily as the United States does in military research. Robert Evenson (1984) surveyed 50 countries and found that in nearly all, inventions per scientist and engineer have declined. This includes Japan, renowned for its productivity, inventiveness, and skill at picking profitable lines of research and development. Jacob Schmookler (1966:28–29) demolished the notion that military spending accounts for the decline in patenting by showing that, *excluding* government-financed projects, the number of industrial research personnel out-stripped the growth of patents by 24 times.

The other potential causes, including declining value of patenting and decline in the share of patentable inventions, are more subtle. They may account for some of the decline in patenting but surely not all of it. In a study published in 1945, when patenting had greater value and little funding went to environmental research, Hornell Hart examined the pattern of patenting in a number of technologies ranging from airplanes to weaving machinery. The characteristic pattern he found is a logistic curve in which, for any specific field, patenting first increases and then declines. It is thus hard to escape the conclusion that the allocation of resources to research and development eventually yields a declining marginal return and that patent statistics reflect this.

Medical research and practice are a good example of diminishing returns to increasing investments in a problem-solving field. The costs of our investments in medical research and application are easy to compile. The results are more difficult to specify because the benefits of medicine are defined primarily in the negative—that is, by the absence or cessation of disease. It is hard to know the productivity of a field in which higher and higher expenditures produce ever more of nothingness—the absence of illness. Fortunately, there is a clear way to express the benefits of medicine: life expectancy.

In 1930 the United States spent 3.3 percent of its GNP to produce an average life expectancy of 59.7 years. By 1982 medical expenditures had grown to 10.5 percent of GNP. This is a telling figure in itself: When one field of social investment takes an increasing share of a nation's wealth, the share available to all other sectors axiomatically declines. Looking at this as a problem of sustainability research, spending ever more money on the most immediate affronts to our sustainability as individuals means less to spend on challenges to our sustainability as a society. The growth in medicine's share of national wealth produced in 1982 a life expectancy of 74.5 years. This is a worthy figure, yet it represents a 57 percent decline in the productivity of our medical investments over a period of 53 years (fig. 2.6). If our investment in medicine in 1982 had been as productive as it was in 1930, average life expectancy would have risen to 190 years. As it is, each extra year of national life expectancy is bought at a cost of lessening our prospects for long-term

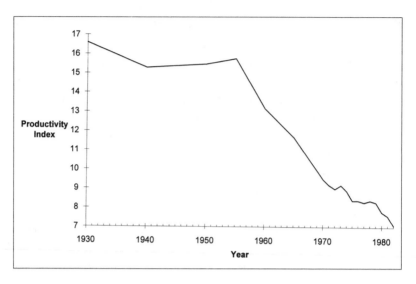

FIGURE 2.6

Productivity of the U.S. health care system, 1930–1982. Data from Worthington (1975:5) and U.S. Bureau of the Census (1983:73, 102). Productivity index = (Life expectancy)/(National health expenditures as percentage of GNP).

sustainability because there is less to invest in education, infrastructure, or other kinds of research. Yet clearly it would be impossible politically to design our public investments otherwise.

The declining productivity of medicine comes in part from the fact that the inexpensive diseases were conquered first, so that those remaining are more difficult and costly to treat. The basic research that led to the discovery of penicillin, for example, cost no more than $20,000. Declining productivity comes also from the fact that as life expectancy rises, further improvements are harder to come by. As each increasingly expensive disease is understood and conquered, the increment to life expectancy becomes ever smaller. The returns to medical investment diminish inexorably and will continue to do so.

Rescher (1980:94, 97) likens such trends to military competition:

Once all of the findings at a given state-of-the-art level of investigative technology have been realized, one must move to a more

expensive level. . . . In natural science we are involved in a technological arms race: with every "victory over nature" the difficulty
of achieving the breakthroughs which lie ahead is increased.

The similarity of science to the military is greater than even Rescher
may have realized. In each there is continuous pressure to understand
a wily adversary about which one can never know enough and which
becomes ever more grudging in each secret it yields. Every eventuality
must be considered, even those least likely, for often the improbable
strategy will be the one that is deployed. In the case of social and biological research the subject is adaptable, so that much knowledge is
impermanent and must continuously be brought up to date. In both
science and defense, increasing expenditures bring diminishing
returns, yet the practitioners of each feel uneasy if any line of inquiry
stays unexplored or any potential developments unexamined. Both justify themselves on moral grounds—that they are good for society—yet
both are potentially bottomless pits into which money may endlessly
be poured. For this reason the decision of how much to spend on science or armament can rarely be made on moral grounds but must be
made on political and economic criteria.

The fact that knowledge investments reach diminishing returns is a
little secret that many scientists and educators would be uncomfort-
able sharing with administrators or trustees or explaining to an appropriating committee. Yet the topic must be honestly addressed; our sustainability may depend on doing so. If there are economic limits to our
ability to understand complex systems or to conduct the research that
produces health, energy, semiconductors, or other dimensions of well-
being, then an early comprehension of the problem is much to be
desired. We shall return to this point.

SUMMARY AND IMPLICATIONS FOR SUSTAINABILITY

Human societies are problem-solving systems that, to survive, must
differentiate and become complex to a degree sufficient for the problems they face. The capacity of human societies rapidly to change in
complexity is one source of their great adaptability as well as of his-

tory's greatest cataclysms (as we shall see in chapter 3). This is the dilemma of complexity: It is necessary for human welfare, yet too much of it can be as detrimental to sustainability as too little. When confronted with a crisis for which more complexity is the only short-term solution, no society would choose extinction instead. Increases in complexity usually are small and incremental, so each instance of growth seems not excessively costly. Complexity works its worst damage in unforeseeable, cumulative effects. Thus the effects of growing complexity may be indiscernible in a lifetime. They may run their course over periods ranging from generations to centuries, which suggests that part of what it means to be sustainable will be found within the historical sciences.

Rational human populations (including scientists) must act with enough efficiency that they do not become extinct. In fact there is much evidence to suggest that people are consistently averse to investments of labor and complexity that are not productive. Thus human societies were at first small and had egalitarian relations, labor that was undifferentiated except by age and sex, and high mobility. Stresses such as those of growing populations, climatic perturbations, or hostile neighbors have led to such developments as social hierarchies, high levels of differentiation and specialization, and constrained territories. Each is a dimension of growing complexity and costliness. Often, as we will discuss in chapter 3, increasing complexity of organization yields no net benefit, for it is undertaken just to maintain the status quo. It is a little like medicine in this regard, where success is defined as the absence of disease.

Rational human populations first use sources of food, energy, raw materials, and knowledge that meet a society's needs at reasonable costs. When these no longer suffice it is necessary to shift to resources that are costlier to locate, extract, process, and distribute while potentially yielding no higher return. Thus extensive land use systems, such as hunting and gathering, have given way of necessity to agricultural systems that are increasingly intensive. All the while, human nutrition has remained fairly stable at about 2000 to 2500 calories per person per day. Under such constraints the emergence of complex societies is possible under only one of two conditions: an increase in human labor or the deployment of a new energy subsidy. The constellation of complex features that we call civilization would be impossible otherwise.

Most troubling is the fact that knowledge production is also subject to diminishing returns. This is the factor left out of the calculations of the technological optimists (Tainter 1988:211–212), who assume that, alone among human endeavors, creativity can always produce constant or increasing returns. Certainly it has done so in many cases, of which we have briefly discussed the Industrial Revolution. Yet our discussion of the development of science suggests that such happy outcomes may not always emerge. Field after field of intellectual endeavor seems to reach diminishing returns to investment (Hart 1945). This is the great uncertainty of the sustainability debate: whether creativity, given incentives, can always overcome resource limitations, or whether problem solving reaches such levels of complexity and costliness that societies cannot afford it. The evidence of history is not clear on this question. There are examples of societies that found their problems too expensive to solve and instead collapsed, and there are examples in which quite the opposite happened. We discuss both in the next chapter.

We began this chapter with a discussion of monitoring and predicting. Where sustainability is at issue, what one monitors and predicts matters a great deal. It is an important thing to get right. Whatever recommendations emerge from monitoring the wrong variables, or failing to monitor the right ones, clearly may be catastrophic. Scientists and managers today who concern themselves with sustainability tend to assume that the things that are important to monitor are stocks and services of the natural environment and the rates at which they change. Our discussion suggests a new consideration. Monitoring environmental conditions can take us only so far if our approaches to problem solving become so complex that we cannot afford to take corrective actions. As we have alluded in this chapter and will discuss in the next, problem-solving systems that experience prolonged periods of diminishing returns must either find new subsidies or face termination. Sustainability requires that our institutions of problem solving themselves be sustainable. To ensure a sustainable future we need to understand not only our resources and natural environment but also whether there are limits to our ability to understand and care for them.

In summary, sustainability is not free. It does not emerge automatically from recycling or from conserving biodiversity. Sustainability will colloquially mean maintaining a way of life that people desire. One must work at being sustainable and be willing to pay to achieve one's sustainability goals. Sustainability or collapse follows from the success or failure of problem solving. As complexity and costliness grow, problem solving seems regularly to reach diminishing returns—in producing resources or knowledge, or in organization. In societies dependent mainly on solar energy and human labor, complexity in problem solving meant that people worked harder just to maintain the same way of life. The costs of complexity may be offset by energy subsidies, which with industrialism produced great increases in general well-being that previously would have been unimaginable. Yet as we now know, these subsidies too are not free. With energy subsidies, the costs are dispersed through the air or the atmosphere or shifted onto the future. Energy subsidies appear free only when their full costs are not counted in. Sustainability will always cost, and achieving it will always be a cost–benefit calculation.

3

Complexity and Social
Sustainability: Experience

There was scarcely a solvent government during the twentieth century. In an age when each politician blames an opponent for this universal condition and journalists amplify the deception, it is worth pointing out that it has always been so. As problem-solving institutions that inherently attract challenges, governments have an inevitable tendency toward greater complexity. As complexity rises so do costs, inexorably squeezing all services and functions, and the system that makes them possible. We have discussed one example in the Third Dynasty of Ur.

Although contemporary experience offers examples of unsustainable governments, as in the former Soviet Union, we noted in chapter 2 that none of us has experience of a way of life or level of complexity being unsustainable. For examples of complete unsustainability we must search back in history, and fortunately there are several cases to discuss. The need to examine sustainability through historical cases actually is a great asset, for such examples have two significant advantages over experiences of the moment. The first is that the factors that influence sustainability or its absence develop over very long periods: generations or centuries. It is not possible to understand a momentary condition unless one comprehends how the moment came to be and its place in a long-term trend. Contemporary studies by themselves can never clarify whether a long-term trend is sustainable. Only historical research and monitoring can do that. The second advantage to historical studies is that one can see the outcome. It is not necessary to predict whether a system is sustainable. We know whether it was, and what remains is to determine why. It is for such reasons that we suggested in chapter 2 that sustainability is a topic requiring historical research.

We discuss in this chapter four historical cases of long-term problem solving. Each was of a different nature, and as a group they produced divergent outcomes. Accordingly, each conveys a different lesson for sustainability, which we discuss here and in chapter 7. Two of the systems proved to be unsustainable in the long term: the Roman Empire and the Abbasid Caliphate. The Roman Empire is our paradigm of an unsustainable problem-solving system. Because it is our best example of how a society comes to be unsustainable while striving consciously for the opposite, we describe it in some detail. It is a case that shows how trying to sustain one's society can actually bring on just the opposite. In the Abbasid Caliphate we have an example of how complexity and problem solving demand the production of resources and can drive productive systems to collapse.

We also discuss two systems that proved to be sustainable for long periods but used very different strategies to do so. The first is the recovery of the Byzantine Empire from calamities in the seventh century A.D. that nearly brought the empire to an end. The second is the spiral of competition and conflict that has characterized the states of Europe since the fall of the Roman Empire. The Byzantines and the Europeans both managed to solve their problems of sustainability, but did so at different ends of the energy consumption spectrum. The Byzantines found a solution to sustainability that was, for their world, stable and self-perpetuating. The European solution, in which we still participate, was to scramble continuously to find new ways to sustain complexity. The Byzantines and the Europeans can be viewed as surrogate examples in the contemporary debate over whether our future should consist of simplicity and less consumption or complexity and increasing consumption. Regrettably the European case is the only one we discuss for which we do not know the outcome. In a sense this book is part of the process of trying to comprehend that outcome. The retrospective task must fall to future historians.

In each of these examples, the sustainability goal was to perpetuate, or even expand, a political entity within a specific territory. Sustainability goals today are surely more diverse, but the value of these cases lies in illustrating the constraints that govern problem solving and how problem-solving institutions evolve in the long term to either sustainability or collapse.

COLLAPSE OF THE WESTERN ROMAN EMPIRE

The Roman Empire is paradoxically one of history's great successes and one of its outstanding failures. In the long run it proved not to be a sustainable system for solving the problems it faced, even though the Romans tried very hard to make it so. At the start, though, the Romans found a means of expansion that was fiscally self-perpetuating. As more peoples were conquered they came increasingly to provide the economic base for further expansion. In 167 B.C. the Romans captured the Macedonian treasury and eliminated taxation of themselves. When Pergamon was annexed in 130 B.C. the state budget was doubled, from 25 million to 50 million silver denarii (coinage denominations are discussed below). Pompey raised it to 85 million denarii after he conquered Syria in 63 B.C. Julius Caesar relieved the Gauls of so much gold that its value in Rome fell 36 percent (Lévy 1967:62–65).

Augustus (27 B.C.–A.D. 14) added more territory to the empire than any commander before or after, but he also capped its size (fig. 3.1). Later conquests were small and strategic. Augustus aimed to establish a small, stable army and government. His first task was to disband much of the enormous army he found himself commanding (Stevenson 1934a:221). Between 30 and 14 B.C. he settled 155,000 veterans at a cost of 215 million denarii* (Duncan-Jones 1990:125). To provide for mil-

---

*The monetary system of the early empire was as follows:
1 aureus (gold) = 25 denarii (silver)
1 denarius = 4 sestertii (brass)
1 sestertius = 2 dupondii (brass)
1 dupondius = 2 asses (copper)
1 as = 4 quadrantes (copper)

itary retirement he established a 5 percent tax on Roman inheritances. Yet out of his personal fortune (from the conquest of Egypt) he donated 45 million denarii to the cost of settling his veterans (Stevenson 1934b:195). After 13 B.C. a cash bonus of 3000 denarii was paid to each retiring soldier (the equivalent of more than 13 years' pay) rather than the earlier practice of finding them land.

Despite his modest administration, Augustus may have greatly increased the cost of Roman government. In the middle of his reign Rome's income was about 125 million denarii per year (Frank 1940:53). The major imperial costs included pay, rations, and fodder for the army; the civil service; public works; the postal service; education; and the public dole. Later costs came to include the workers in imperial arms factories and uniforms for the civil service (Jones 1974:35). The army became a professional, long-service force, entitled to regular pay and retirement. A civil service was created, and state employees began to receive salaries. Large sums were spent to provide Rome with wheat, water, police, and public buildings (Stevenson 1934b:191). The Roman dole was also a major expense. Julius Caesar found 300,000 beneficiaries, nearly one citizen in three. He reduced this to 150,000, but by 5 B.C. the figure had risen again to 320,000. From 2 B.C. to the reign of Claudius (A.D. 41–54) the figure was about 200,000 (Stevenson 1934b:203).

Occasional but significant expenses were donatives to the army on an emperor's accession or anniversaries and largesse to the people of Rome. When Marcus Aurelius came to the throne in A.D. 161, for example, his accession donatives were so costly that the silver content of the denarius was temporarily reduced from 83 to 78 percent (Walker 1978:125).

The Roman economy was based on agriculture, which in the late empire provided at least 90 percent of the state's revenue. Trade and industry were insignificant, mainly because of the high cost of land transport. Under the Edict on Prices, issued by Diocletian in A.D. 301, transport by road was 28 to 56 times more costly than by sea. A wagonload of wheat doubled in value in 480 kilometers, a camel load in 600 kilometers. It was less expensive to ship grain to the other end of the Mediterranean than to carry it 120 kilometers by land. Land transport was so expensive that it was often impossible to relieve inland

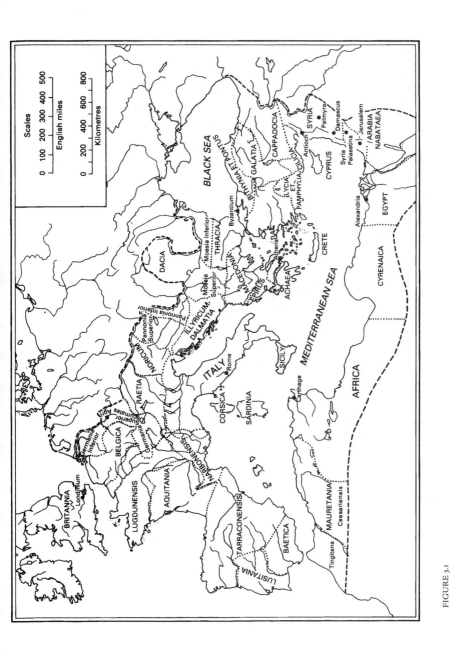

FIGURE 3.1

The Roman Empire at the time of Hadrian, ca. A.D. 117.

famines. Egypt's value came not just from its agricultural productivity but also from its access to water transport.

With transport costs so high, the only goods that could profitably be shipped long distances were those of small size and high relative value—that is, luxury goods. Most peasants could not afford such goods. Thus industry was undeveloped, and most needs were met by local artisans. In the late empire industry could provide only 5 percent of the government's revenues (Cumont 1936:629; Duncan-Jones 1974:1, 368; Hammond 1946:70–71; Jones 1964:841–844, 1974:30, 37–39, 83, 138).

The government financed by agriculturally based taxes barely sufficed for the ordinary administration of the empire. Reserves were difficult to accumulate. When fiscally prudent emperors did accumulate a surplus, as was done by Tiberius (14–37) and Antoninus Pius (138–161), it was quickly spent by their successors. Many new emperors found the treasury empty. Augustus often complained of fiscal shortages (Gibbon 1776–88:140; Hammond 1946:75) and relieved the state budget out of his personal funds (Frank 1940:7–9, 15).

Taxes in the early empire were levied at fixed rates that in many areas had long been set. Taxes generally were raised in the more wealthy and populous Mediterranean lands and spent on the army on the Rhine and the Danube and in Britain (Jones 1974:127). The government operated on a cash basis and rarely borrowed. Its budget was at best minimally planned (Jones 1974:189). When extraordinary expenses arose, often the supply of precious metal was insufficient to produce the necessary coins. Facing war with Parthia and rebuilding Rome after the Great Fire, Nero (54–68) began in A.D. 64 a policy that later emperors found irresistible (fig. 3.2). He debased the silver denarius, changing the alloy from 98 to 93 percent silver. He also reduced somewhat the size of the denarius and the gold aureus. The denarius was the coin most worth debasing, for it was used to pay the troops, and army pay was the single biggest item in the budget (Jones 1974:194).

The period from Augustus to Antoninus Pius (138–161) was the height of Roman power and prosperity. Yet even this early there were indications of trouble. During Tiberius's reign (14–37) there was concern that the level of taxation was too high (Finley 1973:90). Already by the time of Augustus there were problems finding recruits for the army (Luttwak 1976:12). Perhaps in response, Nerva (96–98) established

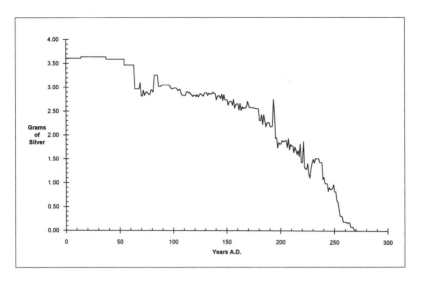

FIGURE 3.2

Debasement of the denarius to A.D. 269. From 238 to 269 the analyses are of nominal denarii, calculated as half the value of the antoninianus.

Data from Cope (1969, 1974, and analyses on file in the British Museum), King (1982), Le Gentilhomme (1962), Tyler (1975), and Walker (1976, 1977, 1978); see also Besly and Bland (1983:26–27).

a system to care for Italian orphans. He also provided land to poor citizens at a cost of 15 million denarii (Longden 1936:192, 210). Even so, by A.D. 100 few Italians were serving in the army (T. Cornell, personal communication, 1991), and during the second century there continued to be problems recruiting (Jones 1974:128).

Antoninus Pius (138–161) had left a sound treasury at his death and a surplus totaling 675 million denarii (Jones 1974:189). It is the last time for quite a while that we hear of a Roman emperor leaving his successor a surplus. Within a few months the Parthians invaded the empire. The Roman victory brought only short-lived rejoicing, for the returning troops carried a plague that devastated the empire for 15 years. In some of the affected areas as much as one-fourth to one-third of the population died. As late as A.D. 189, 2000 per day were dying in Rome. This plague, and later recurrences, initiated a population decline in

105

Mediterranean lands that was to last for 500 years (Boak 1955:19; Mazzarino 1966:152; McNeill 1976:116; Russell 1958:36–37). As plague ravaged the Roman army, Germanic tribes in 167 breached the northern defenses and penetrated as far as northern Italy. The situation was so urgent that Marcus Aurelius (161–180), in raising two new legions, accepted even slaves and gladiators.

The first economic consequence was to deplete Antoninus Pius's surplus. A couple of years of war consumed the careful savings of 20 years. Even so there were not sufficient funds for the campaigns. Marcus financed the empire's efforts by auctioning gold vessels and imperial art treasures. He was also forced to lower the silver content of the denarius (fig. 3.2), a strategy that thereafter had to be used quite frequently (Walker 1977:58–59, 1978:125–126).

Throughout the empire at this time there are indications of the strain. Fewer inscriptions were carved in the second century, apparently because ordinary townspeople could no longer afford them (Hammond 1946:75). Public construction declined in Italy in the second century, although it did not reach its nadir until the third. Recordkeeping seems to have been relaxed, as Egyptian records (where documents have survived best) show a precipitous decline at this time in the registration of such mundane matters as tax and land records (fig. 3.3) (Duncan-Jones 1990:68–69).

The denarius continued to fall under Marcus Aurelius's son and successor, Commodus (177–192). The debasements were certainly inflationary. A slave who in Commodus's reign cost 500 denarii sold for 2500 a few decades later (Mazzarino 1966:153). In Ephesus the price of bread doubled from the early second to the early third centuries (Duncan-Jones 1990:147). Under the principle that bad money drives out good, the older, more valuable coins would have been withdrawn or melted down. People would naturally prefer to pay obligations in newer coins of less intrinsic worth.

The death of Commodus in late 192 marked the start of a period of civil wars. Septimius Severus (193–211) ultimately emerged on top, but to secure his position he courted the military. He increased the size of the army (on paper about 350,000 men) and raised pay. His successors continued this policy, and by 238 army pay had risen more than four times. Severus also augmented the Roman dole, adding oil to the list of

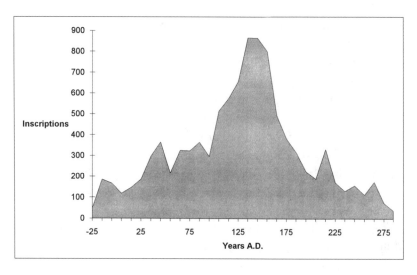

Figure 3.3

Volume of surviving documentation from Roman Egypt, 30 B.C. to
A.D. 282.

Data from Bureth (1964), grouped by midpoints of 10-year intervals.

free commodities. The most profound of Severus's actions went so far
beyond anything his predecessors had done that it has come to be
known as the Great Debasement. At the end of Commodus's reign
(192) the denarius still stood at 73 percent silver. Between late 194 and
198 Severus reduced this to about 56 percent silver (fig. 3.2) (Walker
1978:51, 59, 61). Even this was insufficient to meet the government's
obligations. Caracalla in 212 extended Roman citizenship to all free
inhabitants of the empire. This increased vastly the pool of those liable
to the Roman inheritance tax, which he incidentally doubled (Jones
1974:172). In 215 he introduced a new silver coin, known today as the
antoninianus. It was declared to be worth 2 denarii but weighed only
1.5 denarii (Bolin 1958:49; Jones 1964:16; Walker 1978:62-64). It proved so
unpopular that soon it ceased to be minted.

The half century from A.D. 235 to 284 was a time of unparalleled crisis during which the Roman Empire nearly came to an end. There were
foreign and civil wars, which at times followed one upon another

107

almost without interruption. Many provinces were devastated. The army and the bureaucracy grew in size, and taxes were increased. Financial crises led to the final debasement of the currency and to ruinous inflation. By prodigious effort and sacrifice the empire survived the crisis but at great cost. It emerged at the turn of the fourth century A.D. as a very different organization.

It was a period of violent political instability. In this 50-year period there were at least 26 legitimate emperors; they ruled an average of 30 months. Histories of the period are so inadequate that the number of usurpers may never be known. Estimates run as high as 50, or an average of 1 per year.

From the mid-240s until 272 there were continuous Germanic incursions, some reaching deep into Italy itself. In 247 the celebrations for Rome's 1000th anniversary had to be postponed because Emperor Philip (244–249) was fighting in the Balkans. In 251 the emperors Trajan Decius (249–251) and Herennius Etruscus (251) were killed in battle with the Goths, along with much of their army. Gallienus (253–268) campaigned yearly from 254 to 259 along the Rhine and the Danube, crushing a massive incursion of the Alemanni in 259 at the very outskirts of Milan. His father, Valerian (253–260), was less fortunate. In 253 the Persians captured and sacked Antioch. Valerian went east, where he spent his reign campaigning. While he was occupied with the Persians, Goths attacked the undefended cities of Asia Minor. Finally in 260 Valerian was captured by the Persian King Shapur and taken into captivity. He never returned.

It was the low point of Rome's fortunes, and for a time the empire disintegrated. A Rhine general named Postumus (260–269) seized Gaul and Britain and for a time held Spain and even Milan. To the east, Odenathus of Palmyra was nominally an ally, but he and his son, Vabalathus (271–272), exercised independent authority over the eastern parts of the empire.

The Roman Empire shrank to Italy, the Balkans, and Africa (fig. 3.1). Even in this reduced empire there was much to do. In 267 the Goths returned to western Asia Minor and the Aegean. They destroyed the great Temple of Diana at Ephesus, then captured and sacked Athens. In 270 an invasion of Germanic peoples again burst into Italy. Rome itself was defenseless, having long outgrown its 500-year-old walls.

Emperor Aurelian (270–275) drove the invaders from Italy and took care to see that the threat to Rome did not recur. He ordered that new walls be built around the city, the walls that are still seen today.

The walls of Rome signify a new historical phase: a Roman Empire on the defensive, whose borders were no longer secure and whose interior was unprotected. Within a few decades cities everywhere built walls, at great effort and expense, and often hastily (fig. 3.4). Walls were built around important places such as Nicaea and the other cities of Asia Minor devastated by the Goths and even around small inland towns in Britain (Frere and St. Joseph 1983). At the time of the Gothic sack of 267, a small area of Athens was hastily walled. As at other cities, the wall was built in part by tearing down public buildings and reusing the masonry (Williams 1985:20).

With cities sacked, lands ravaged, and the empire's resources marshaled for defense, the population could not recover from the plague of 165 to 180. The catastrophes of 235 to 284 fell on a declining population, which suffered further when the plague returned from 250 to 270. In a province so essential as Gaul the rural population declined, being killed or captured by barbarians, starving, or deserting their fields to join bands of outlaws. Town populations fell also, sometimes to the size of the Celtic villages that preceded the empire (Boak 1955:19, 26, 38–39, 55–56, 113; MacMullen 1976:18, 183; Rostovtzeff 1926:424). Vienne, for example, shrank from 200 to 20 hectares, Lyons from 160 to 20, and Autun from 200 to 10 (Randsborg 1991:91). Paris was restricted to the Île de la Cité (Williams 1985:20).

Government expenses rose in the areas still under imperial control. The size and payroll of the army grew, as did campaigning costs. The government's fiscal obligations may have doubled (MacMullen 1976:102–104, 107–108). The silver currency plummeted in value. In 238 the antoninianus was brought back and soon replaced the denarius as the standard silver coin (fig. 3.5). In A.D. 253 the antoninianus still had about 35 percent silver. By 260 it held 15 percent. The nadir was reached in 269, when the antoninianus contained only 1.5 percent silver (Cope 1969) (fig. 3.2). "The Empire," wrote Harold Mattingly (1960:186) of this period, "had, in all but words, declared itself bankrupt and thrown the burden of its insolvency on its citizens." By the time of Diocletian (284–305) the state was so unable to rely on money that it collected taxes in the form

109

FIGURE 3.4

Late Roman wall at Conimbriga, Portugal, enclosing an earlier wall.

Photograph by J. Tainter.

FIGURE 3.5

Fiscal distress in the Roman Empire of the third century A.D. *Left,* Denarius of Maximinus (235–238), minted 235. *Right,* Antoninianus of Herennia Etruscilla (249–251), struck ca. A.D. 251 over a denarius of Maximinus. Maximinus' ear, flattened and inverted, can be seen to the right of the empress's ear, and under her neck are the letters XIM from Maximinus' name. The antoninianus supposedly was worth 2 denarii, yet the mint was taking denarii minted 16 years earlier and reissuing them as antoniniani. Maximum diameter of larger coin: 22 millimeters.

Photograph by J. Tainter.

of supplies usable directly by the government and the military, or in bullion, to avoid having to accept its own worthless coins. Money changers in the East refused to give small change for imperial coins.

The crisis called for drastic changes in the government and in the order of society. In his brief reign, Aurelian (270–275) pushed out the barbarians, reattached the Orient, Egypt, and Gaul, and reformed the currency. He issued larger antoniniani, with silver content raised to 4 percent. Equally important, Aurelian showed future emperors how to make the most of an economy in which money had no value. When the walls of Rome had to be built he conscripted craft guilds to do it. When the empire needed to bring vacant lands back into production, he ordered that they be farmed under the direction of local governments (MacMullen 1976:205–206).

By the late 270s the worst was over, although probably no one at the time knew it. The empire and its peoples had suffered terribly. Agriculture, rural population, cities, literacy, and trade had all declined. Frontier lands were abandoned, and many cities had been pillaged. With money having no value, the army made compulsory purchases at outdated prices, then resorted to outright requisitioning and ultimately plundering (Williams 1985:19). There were predictable increases in banditry. Rebellious bands formed in Gaul. They were suppressed by Maximian in 286 but reappeared in the mid-fourth century and remained to the end of Roman rule (Boak 1955:27, 38–40; Gibbon 1776–88:242–243; Rostovtzeff 1926:424, 437).

Diocletian (284–305) and Constantine (306–337) initiated sweeping political and economic changes that transformed the empire. They designed a government that was larger, more complex, and more highly organized. They commanded larger and more powerful armies. The government taxed its citizens more heavily, conscripted their labor, regulated their lives, and dictated their occupations. It was an omnipresent, coercive organization that subdued individual interests and amassed all resources toward one overarching goal: sustaining the empire.

The most pressing need was a larger military. In A.D. 235 the army was about 300,000 to 350,000. It stood at 400,000 when Diocletian took office (285). Diocletian and Constantine raised the level to between 500,000 and 600,000. The army may have been as high as 650,000 by the end of the fourth century (Boak 1955:91–94; Finley 1968:157; Jones 1964:57, 1974:129; Luttwak 1976:177; Williams 1985:97). A second transformation was in the administration of the empire. Diocletian subdivided provinces into many smaller ones and separated civil from military authority in each. This made it more difficult for provincial governors to rebel. He increased the size of the imperial administration, which now moved with an emperor as he rushed to trouble spots. The bureaucracy was perhaps doubled (MacMullen 1976; Williams 1985).

Money had become so worthless that the mints issued coins in bags of 1000 (Williams 1985:117). Responding to the problem, Diocletian "turned the entire empire into a regimented logistic base" (Luttwak

1976:177). He implemented Rome's first budget. Each year calculations were made of anticipated expenses and a tax rate established to provide the revenue. Just to establish the tax system was an immense affair, entailing a complete census of people and land across the empire. Even Italy was taxed again. Diocletian's system united two elements of taxability: land and people. These were categorized into abstract units in which goods, services, and produce became interchangeable by a single standard. The tax rate was established from a master list of the empire's resources, broken down province by province, city by city, field by field, household by household. Never before had the state so thoroughly penetrated its citizens' lives. The process caused Egypt to rebel in 297. Even so, peasants knew in advance what they would owe, and the empire's soldiers, having adequate supplies, no longer stole from them. Yet Diocletian's successors revised the rates upward. Taxes apparently doubled between 324 and 364 (Williams 1985:118–125; Jones 1974:82). Villages were held liable for the taxes on their members, and one village could even be held liable for another. Compulsory services existed along with taxes, and the two were sometimes interchangeable. Each citizen, each guild, and each locality was expected to produce whatever the state needed. This included military conscription: Cities, great estates, and groups of small farmers were expected to produce men just as they produced taxes.

Because the state could not pay inflated market prices, it could not rely on private enterprise to fulfill its needs. Voluntary associations of businessmen and artisans were transformed into obligatory organizations, subservient to the state. The process began with trades directly supplying the cities, or the armies, or other strategic elements. From there it spread to the whole of industry. Ship owners supplying grain to Rome found their occupation compulsory and hereditary, as did the guilds of millers and bakers. Later to be controlled were beef and mutton merchants, wool merchants, dyers, weavers, and transport merchants. By the fourth century heavy industries such as brickworks, mining, and quarrying were under state control. Ultimately even industries peripheral to the state's needs came to be regulated: glass, pottery, furniture, and ornament making. Where the government's needs were particularly acute, as in arms and uniforms, it established its own factories (Williams 1985:133–136).

Inflation continued unabated, and Diocletian made several attempts at monetary reform. He introduced a large, silver-covered copper coin of about 10 grams and 4 percent silver, known today as the follis (Meyer 1987). He also reintroduced a coin of nearly pure silver but may not have had sufficient bullion to issue enough of them. Under continuing inflationary pressures the value of the follis was halved in 301 (though without reducing its weight), and subsequent emperors reduced both its size (Fig. 3.6) and silver content (Van Meter 1991:47).

Diocletian was a master political strategist but a naive economist. As masses of these coins were produced, prices rose higher and higher. Probably no Roman official understood that the amount of money in circulation affected prices. In the second century a modius of wheat (about 9 liters) had sold during normal harvests for about one-half denarius. In Diocletian's Edict on Prices (301) the price was set at 100 denarii (Jones 1964:27). In 335 a modius of wheat sold in Egypt for more than 6000 denarii and in 338 for more than 10,000 (fig. 3.7). In Egypt an

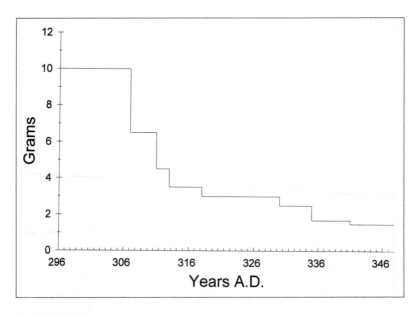

FIGURE 3.6

Reductions in the weight of the Roman follis, A.D. 296 to 348.

Data from Van Meter (1991:47).

artaba of wheat (about 40 liters) had risen 27 times over the second-century level (Duncan-Jones 1990:115), and from 250 to 293 the cost of a camel or donkey rose 60 times (Williams 1985:79). In 301 a pound of pork was set at 12 denarii; by 419 it cost 90 denarii (Mattingly 1960:222–223). In the 150 years before Diocletian the value of gold had risen 45 times and the value of silver 86 times (Jones 1974:201). Gold went from 50,000 denarii to the Roman pound in 301 to 504,000 in 450 (fig. 3.8). Finally in the fifth century the output of copper coins was reduced. With less currency available for everyday transactions, the rise in prices finally slowed.

In an attempt to control inflation Diocletian issued his Edict on Prices in 301. In general the edict set prices too low, favored creditors, and, much to the government's advantage, depressed the cost of transport (Jones 1964:27, 61; Lévy 1967:94; MacMullen 1976:122; Rostovtzeff 1926:463). Naturally it could not be enforced and, though never repealed, was allowed quietly to lapse. It is a remarkable document that reveals how deeply Diocletian's government monitored and regu-

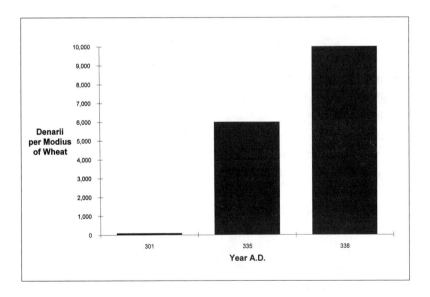

FIGURE 3.7

Roman inflation: denarii per modius of wheat in the early fourth century A.D.

Data from Jones (1964:119). A modius was about 9 liters.

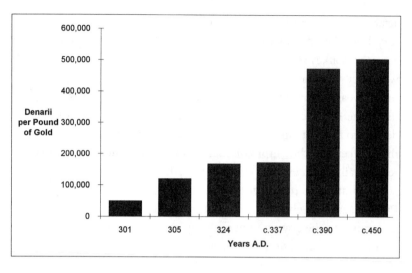

FIGURE 3.8

Roman inflation: denarii to a pound of gold, 301 to ca. A.D. 450. The
Roman pound consisted of 12 ounces, about 330 grams.

Data from Hodgett (1972:38).

lated everyday life. The edict lists about 1000 items in 32 sections,
including wages for a variety of workers and prices for meats, cereals,
game, seafood, grades of wine and beer, textiles, skins, and furs
(Williams 1985:128–131).

The changes of Diocletian and Constantine made the empire more
secure and efficient, but at substantial cost. By the fifth century the civil
service employed more than 30,000, yet it was by far the lesser cost.
Not only had the military doubled in size, but it included more cavalry,
which was particularly expensive. Fodder for a horse cost as much as
the rations of a soldier. A papyrus dating to A.D. 77 shows that it cost
677 denarii, more than twice a legionary's pay, to buy a single horse for
the army—and that for a pack animal, not a cavalry horse (Ferrill
1986:80; Jones 1964:51–52, 1974:129).

Rome in the fourth century had 300,000 people receiving public distributions (Mazzarino 1966:169). Public spectacles took place on 175
days a year, three times the rate under Augustus (Isaac 1971:158). As
Constantinople was built, 80,000 were placed on its dole. The cities of

116

the later empire were built and rebuilt from above, wherever there was military activity or the need for an imperial residence (Williams 1985:137).

These costs had to be borne by a depleted population. After the plagues of the second and third centuries the empire never experienced conditions favorable to recovery (McNeill 1976:116; Russell 1958:140). In 315 Constantine ordered assistance for poor and orphaned children, a policy first established by Nerva but perhaps allowed to lapse during the crisis. Yet the economic system created by the late emperors did not encourage population growth. In consequence there were shortages of labor for agriculture, industry, the military, and the civil service. Agricultural labor was so scarce that landowners bribed itinerants to enlist in the army in place of their own laborers. Height requirements for army recruits were lowered. By the late fourth century in the West even slaves were sometimes enlisted (Boak 1955:42, 97–98, 113–114; Jones 1964:149, 158–159, 1041–1043, 1974:87; MacMullen 1976:182–183).

In the late empire much cultivated land was abandoned. This problem first appeared in the late second century, perhaps because of the plague, and was a subject of imperial legislation from before Diocletian's time to that of Justinian (527–565). Aurelian had held city councils responsible for the taxes on deserted lands. In some eastern provinces under Valens (364–378), one-third to one-half of arable lands were deserted. The government addressed the problem by tying farmers to their plots. Yet as more and more farmers attached themselves to large estates this policy effectively established a system of serfdom.

The tax system of the late empire seems to have been to blame, for the rates were so high that peasant proprietors could accumulate no reserves. Whatever crops were brought in had to be sold for taxes, even if it meant starvation for the farmer and his family. Farmers who couldn't pay their taxes were jailed, sold their children into slavery, or abandoned their homes and fields. The state always had a backup on taxes owed to it, extending obligations to widows or orphans, even to dowries. It is no wonder that the peasant population failed to recover. Under such circumstances it became unprofitable to cultivate marginal land because too often it would not yield enough for taxes and a surplus. Faced with taxes, a small farmer might abandon his land to work for a wealthy neighbor, who would be glad to have the extra labor.

As the economic basis of the empire deteriorated, it became more and more likely that an insurmountable calamity would someday occur. It began in 376, when the Huns forced the Visigoths to petition for entry into the empire. Emperor Valens (364–378) gave them deserted land in Thrace, expecting to use them for the army. Rapacious Roman officials treated the starving Visigoths so badly, however, that they soon began to raid the countryside. Valens marched against them, but in 378 he was defeated and killed at the Battle of Hadrianople. As many as two-thirds of the soldiers of the eastern field army perished with him or were captured. It was the worst Roman defeat since the Battle of Cannae in 216 B.C. and had far more serious consequences. There was no choice but to accept the Visigoths into the empire as allies, not as subjects, and to pay them a subsidy. Within a few years this had disastrous consequences.

By the end of the fourth century the armies of the Eastern and Western Empires still totaled at least 600,000, of which about two-thirds were stationary frontier troops (Williams 1985:213). The western army may only have been about 200,000 (Ferrill 1986:22). The field army came increasingly to consist of Germanic allies. The Roman infantry itself abandoned its heavy armor and no longer trained in the old formations and maneuvers. It became suitable only for undisciplined, open-field fighting and came increasingly to resemble its Germanic allies and foes. When Attila and the Huns were defeated at the Battle of the Catalaunian Plains in 451, the Roman infantry still maneuvered well, but most of the fighting was done by Rome's Germanic allies (Ferrill 1986:47, 128–129, 148–149, 153; Luttwak 1976:175, 177).

The economic trends set in motion by Diocletian and Constantine culminated in the West in the fifth century. Peasants continued to flock to large landowners, and by A.D. 400 most of Gaul and Italy were owned by less than a dozen senatorial families (Williams 1985:214). In A.D. 395 one-eighth of the land in Campania was out of cultivation (Isaac 1971:6). In Africa, deserted lands ranged from 10 to 15 percent in some provinces, up to 33 to 50 percent in others (Boak 1955:45–46; Jones 1964:1039). Delinquent taxes were remitted occasionally in the early and middle fourth century but so frequently after 395 that it appears there was a general agricultural disintegration in the West (Boak 1955:52).

Civil services were breaking down in the fifth century: road repair, the postal service, and the supply of public animals (Dill 1899:237–240). By the mid-fifth century the frontier troops simply disappeared for lack of pay (Ferrill 1986:153). The population of Rome itself declined from about 1,000,000 in A.D. 367 to about 400,000 in 452 (Hodges and Whitehouse 1983:50–51). From the mid-fourth century on few monumental buildings other than churches were built in Gaul and almost none in Britain. Decaying masonry buildings were repaired or demolished and replaced by timber structures (Hodges and Whitehouse 1983:82). In Gaul there were widespread revolts: in 417, from 435 to 437, and in 442 (Jones 1964:812).

In November 401 the Visigoth leader Alaric led his people into Italy, right up to the walls of Milan. For the next several years the Roman general Stilicho was compelled to address this threat. There was a profound shift in policy: Hereafter the priority changed first to defending Italy and ultimately to defending little more than the emperor and his court. When Alaric sacked Rome in 410, the Emperor Honorius (394–423) remained safe in Ravenna.

To defend Italy Stilicho withdrew troops from the Rhine and Britain. The repercussion was not long in coming. On the last day of 406 an alliance of Vandals, Suevi, and Alans crossed the frozen Rhine, and virtually overnight a major part of Gaul was lost. In time some of these people moved on to Spain or north Africa, while others remained in Gaul as nominal Roman allies. Rome eventually was able to reestablish a measure of control over Gaul and Spain. In 429, however, the Vandals crossed to north Africa and took Carthage in 439. Rome's north African food supply was gone forever.

The Western Empire was by this time in a downward spiral. Lost or devastated provinces meant lower government income and less military strength. Lower military strength in turn meant that more areas would be lost or ravaged. By 448 Rome had lost most of Spain (Barker 1924:413–414). After the fall of Majorian in 461, Italy and Gaul had little connection. The empire shrank to Italy and adjacent lands. The most important ruler in the West was no longer the Roman emperor but the Vandal King Gaiseric (Ferrill 1986:154; Wickham 1981:20).

Whereas some civilians of the Western Empire resisted the barbarians, many more were simply inert, and some actually joined the invaders (Jones 1964:1059–1061). In 378 Balkan miners went over en masse to the Visigoths (Lévy 1967:99). Often the invaders were welcomed as saviors from Roman oppression (Frank 1940:303–304; Weber 1976:7). In mid-fifth-century Gaul a deputation of property owners and municipal authorities invited the Burgundians to occupy some of their lands (Isaac 1971:127). Perhaps the most notorious event of all came in 406, when Roman subjects in Pannonia actually moved west to take part in the great invasion of Gaul. "By the fifth century," concluded Robert Adams (1983:47), "men were ready to abandon civilization itself in order to escape the fearful load of taxes."

In the 20 years after the death of Valentinian III (455), the Roman army proper dwindled to nothing. The government came to rely almost exclusively on troops from Germanic tribes. When finally these could not be paid, they demanded one-third of the land in Italy. This being refused, they revolted, elected Odoacer as their king, and deposed the last emperor of the West, Romulus Augustulus, in 476. The Roman Senate informed the Eastern Emperor Zeno that an emperor in the West was no longer needed (Jones 1964:244). In northern Gaul a small territory remained under Roman administration until annexed by the Franks in 486. Its leader, Syagrius, was called the Roman King of Soissons—that is, just another petty chieftain.

Outside Italy the only cultural traditions to survive the Roman collapse were those supportable directly by land ownership. Taxation declined because there was no longer much on which to spend money. The Roman army, that great consumer of wealth, was gone, and Germanic warriors provided their own equipment and sustained themselves from their estates. Political hierarchies simplified, and except for the church, most aspects of Roman administration disappeared. Towns were sparsely populated. They took on a country-like atmosphere, and fields were cultivated within the walls. There is little evidence that towns even existed in most parts of western Europe outside Italy after A.D. 600, although such places still had names and the Roman roads still converged on them. The only substantial town buildings were churches, and many "towns" consisted of little more than monasteries (Hodges and Whitehouse 1983; Pfister 1913; Randsborg 1991; Vinogradoff 1913; Wickham 1984; Tainter 1999).

Understanding Roman Unsustainability

The Roman expansion was at first highly successful as the subject peoples paid for further conquests. It was a policy with a high return but one that inevitably could not be maintained. Once any empire has looted its conquests, it must undertake thereafter to administer, garrison, and defend the provinces. Typically this cost must be paid out of yearly income. For the conqueror, costs rise and benefits decline. For a one-time infusion of wealth from each new province, Rome had to undertake administrative and military responsibilities that lasted centuries and had to be supported by the empire's yearly agricultural output. The empire had to maintain a far-flung administrative and military structure on the basis of an agricultural system of varying yields, which could produce little surplus per capita, and in the face of increasingly hostile and capable neighbors.

From Augustus on, the empire regularly faced fiscal insufficiency. The empire could not accumulate significant reserves, so when emergencies arose the only alternatives were to raise taxes directly or to raise them indirectly by debasement and inflation. As the empire grew more complex and controlling, sustaining it came to cost more and more. The extra expenditures, however, yielded no extra return on investment. The empire was not expanded, no major treasure was acquired, no new wealth was created. The higher costs from the third century on were incurred merely to maintain the status quo. It was a textbook example of diminishing returns to sociopolitical complexity (Tainter 1988:148–152).

With the economic system established in the late empire, there was more than simply a problem of diminishing returns to the cost of being the Roman Empire. During the fourth and fifth centuries the empire sustained itself by consuming its capital resources: producing lands and peasant population. Whereas early emperors had taxed the future to pay for the present (through debasement and inflation), the later emperors paid for the present by undermining the future's ability to pay taxes. The social consequences were predictable. The empire at least partly lost legitimacy. The cost of empire had risen dramatically, but the protection the state could offer proved increasingly ineffectual. To many, there were no remaining benefits to the empire, as barbarians, imperial armies, and tax collectors crossed and ravaged their lands.

Self-sufficient villas sprang up across western Europe, to which poor farmers fled for protection. The Germanic kingdoms (once they had taken the lands they wanted) held out the hope of security, a simpler fiscal structure, and local services. Once the local Romans became subjects rather than victims, the German rulers generally treated them fairly. Roman landowners offered their services to the new kings. Peasants found rent paying more attractive than tax paying and chose feudal social relations as the lesser evil (Wickham 1984:16–18).

There was nothing mysterious or romantic about the fall of Rome. It was a mundane economic process driven by besieged leaders who could not foresee the future and who for the most part could not have acted otherwise. As historian Moses Finley (1968:61) wrote, "This is neither a dramatic nor a romantic way to look at one of the great cataclysms of history. One could not make a film out of it."

THE EARLY BYZANTINE RECOVERY

The debacle in western Europe during the fifth century meant the end of the Western Roman state but not the end of the imperial concept. The Eastern Roman Empire persisted under its own emperors, changing greatly and coming to an end only when the Turks took Constantinople in 1453. For much of its history the Eastern Roman Empire lost territory, so that by the end the state consisted only of the city itself. Yet during the tenth and early eleventh centuries Byzantium was on the offensive and doubled the territory under its control. A lesson in sustainability is to be found in the steps that made this expansion possible.

Although its citizens and rulers called themselves Romans until the end, the Byzantine Empire (an eighteenth-century term) was Roman only in ancestry. Latin dropped from official use by the seventh century. Urban residents spoke Greek: Byzantium was, in fact if not in name, a medieval Greek empire. Yet the Roman legacy exerted lingering influences, among which was an unwillingness to cede claims to territory once subject to Rome. Even when such territory could no longer be held or recaptured, the Eastern Romans insisted on the status of the leading state of medieval Christendom.

The Eastern Empire survived the fifth century, but it did not do so easily. The East survived because it was economically stronger than the West and strategically less vulnerable. The provinces of the Eastern Empire included the older, more populous, and more developed of the Mediterranean lands. These provinces were better able to bear the costs of administration and defense. In the later empire the West had a budget only one-third that of the East and an army merely half as large. Yet the northern frontier it had to defend was more than twice as long. The East's mortal enemy was Sassanian Persia, but this was a sophisticated state with which long peace treaties could be made. The tribal peoples of northern Europe could be kept in place only by constant demonstrations of strength and, when these failed, always tested the frontier.

The most urgent need of the eastern emperors was to develop the economic base on which military security depended. The first task was to reestablish a sound currency, particularly in the base metal denominations on which daily life depended. This was undertaken by Emperor Anastasius (491–518), who in 498 adopted an innovation of the barbarians. The Vandals in north Africa and the Ostrogoths in Italy had begun to issue copper coins in various denominations, each stamped with its value and fixed in its relation to gold and silver. (Some coins of Ostrogothic Italy carried the now-ironic inscription "Invincible Rome.") The existing Roman copper currency had been degraded to tiny (0.56 gram) coins (called nummi) so lacking in value that they had to be exchanged in sealed purses containing thousands. In Italy in 490, 12,000 of these traded against a gold solidus (a coin about the diameter of a U.S. quarter dollar). The new denominations were calculated in multiples of the tiny coppers but anchored in a large coin called a follis (worth 40 nummi). They gained quick public approval and spurred the economy by making small transactions possible again. For today's readers accustomed to making transactions in currency it may be difficult to imagine an economy in which purchases must be made by bags of coins (or bags of notes, as in 1930s Germany) and in which merchants cannot make change. That was the problem solved by Anastasius's reform (Harl 1996).

The Roman army in the fourth and fifth centuries was divided between static frontier troops, who were little more than a trip wire, and the mobile field army, which did the real fighting. By the fifth century the field army consisted largely of Germanic mercenaries and allies, who often proved unreliable and even treacherous. Their generals meddled in Roman politics and during the fifth century regularly made and deposed emperors. As part of his financial reforms of 498, Anastasius gave the troops cash to buy rations, uniforms, and arms rather than issuing them. The allowances evidently were generous, so much so that the army attracted large numbers of native volunteers. Barbarian mercenaries and their generals continued to be employed but became much less important (Treadgold 1995). Within a few decades these economic and military reforms had produced such results that Justinian (527–565) could both increase the size of the follis (one of the few occasions when the intrinsic value of Roman coins was increased) and, after defeating Persia, attempt to recover the western provinces.

An army sent to north Africa in 532 conquered the Kingdom of the Vandals within a year. The Vandal king, the Vandal treasury, and the surviving Vandals were sent to Constantinople. As imperial taxes were reimposed, though, the prosperity of north Africa declined. Scarcely pausing, the Byzantine general, Belisarius, was sent in 535 to reconquer Italy. He had taken Rome and Ravenna, captured the Ostrogothic king (also sent to Constantinople), and conquered all of Italy south of the Po when he was recalled in 540 to fight the Persians again.

In 541, just when the job in Italy seemed essentially done, bubonic plague swept over the empire. It was a disease never before seen in the Mediterranean, and it took 4 years to run its course. Like any disease introduced to an immunologically naive population, the effects were devastating. Just as in the fourteenth century, the plague of the sixth century killed from one-fourth to one-third of the population.

For the Byzantines the only good news was that the plague affected the Persians too. The enormous loss of life among taxpayers caused terrible financial problems. A reserve of 29 million solidi amassed by Anastasius and Justin (518–527) was soon spent on Justinian's wars. Army pay fell into arrears, and troops either mutinied or handed conquests, such as the city of Rome, back to the enemy. The Ostrogoths recovered and retook most of Italy. Twice the field army of Byzantine Italy had to be rebuilt. The Moors took much of Byzantine Africa.

The continuous wars forced Justinian to make great expenditures on men, ships, and matériel. The population of the empire was so depleted by plague that more barbarian mercenaries had to be recruited, and they demanded pay in gold. By debasing the currency and slashing expenditures, the emperor was able to send another army to Italy in 552, and even to back a rebellion against the Visigothic king of Spain. Italy was reconquered by 554, but the last remnants of the Ostrogoths held out until 561. Yet in 558 the plague returned, and again military pay fell into arrears. As a result Justinian managed to conquer only about the southern fifth of Spain.

At his death in 565 Justinian left a greatly enlarged empire (fig. 3.9), but the new conquests proved hard to hold. Within 4 years the Visigoths attacked in Spain and the Moors in Africa. The Lombards invaded Italy and took most of the interior by 572. War resumed with Persia. Slavs and Avars (a coalition of tribes related to the Huns) crossed the Danube. The Byzantines again defeated the Persians, but the Slavs raided all the way to Greece. To pay for these wars, the alloy of the gold solidus had to be debased by adding silver, and the weight of Justinian's follis was progressively reduced (fig. 3.10).

The wars also took a toll on the Persians, and in 590 rebels overthrew the Persian king. The Byzantine Emperor Maurice Tiberius (582–602) put the king's son on the Persian throne but had to attend to problems in the Balkans. Byzantine troops defeated the Avars and Slavs and by 599 practically cleared them from the Balkans. But the empire's resources were sorely tried by these campaigns. Even after the victories over Persia and in the Balkans, there was no money to send troops to Italy. Maurice ordered his troops to accept arms and issues in kind in lieu of paid allowances, but when they mutinied he was forced to pay them anyway. As was the case in all ancient societies, the high cost of transport was the reason for the expense of Byzantine wars. Much money could be saved by not having to ship supplies. In 593 Maurice told his men campaigning north of the Danube to live off the land but had to retreat when they threatened to mutiny again. In 602 he ordered the troops to winter north of the Danube. This time they not only mutinied but also marched on Constantinople, where they killed Mau-

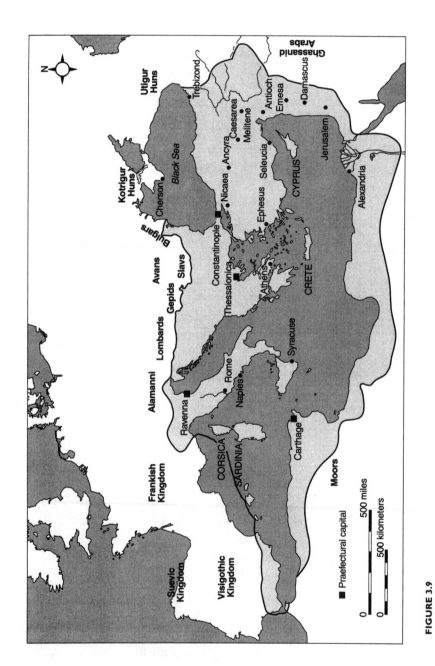

FIGURE 3.9

The Byzantine Empire in A.D. 565.

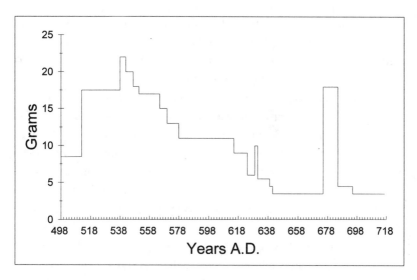

FIGURE 3.10

Weight of the Byzantine follis, A.D. 498–717.

Data from Harl (1996:197).

rice. Persian King Khosrau II, grasping at the pretext, vowed to avenge his benefactor and began to snatch Byzantine provinces. Thus began a crisis that lasted for more than a century and nearly brought the empire to an end.

The empire was so disorganized by its internal troubles that there was a general military breakdown in the Balkans and Asia. The Slavs and Avars again overran the Balkans. The Persians spread through Asia Minor. North Africa and Egypt successfully rebelled and placed Heraclius (610–641) on the throne. The empire he took over lay in ruins and was financially exhausted. The Persians reached the Bosporus in 615. In 619 they began the conquest of Egypt, the empire's richest province. Constantinople was besieged from 618 to 626.

The existing fiscal structure could not allow for recovery. In 615 church treasures were melted down to meet government expenses, from which silver coins were issued with the inscription "God save the Romans." Heraclius cut the pay of troops and officials by half in 616. This time the army accepted the reduction. Bronze was needed for

127

arms and armor, so Heraclius followed his predecessors in progressively lowering the weight of the follis (fig. 3.10). Many times the mint simply took larger coins minted in the sixth century, chiseled them into fragments, and restruck each piece as a follis (fig. 3.11).

The inflation caused by these expedients destroyed purchasing power and wreaked havoc on exchange rates. The folles of Heraclius were hastily and sloppily made but circulated at nominally equal value with the heavier coins of the sixth century. Sometimes coins worth one-half or even one-quarter follis were restruck as folles (even with the original value sometimes still visible), so that the mint made profits of two to four times merely by taking coins in taxes and reissuing them as pay. By the mid-600s the bronze currency was reduced to such chaos that folles had to be exchanged against gold by weight.

Heraclius's economic measures bought time for his military strategy to work. When the Persians took Jerusalem and carried off its

FIGURE 3.11

Reduction of the Byzantine follis. *Left,* Follis of Justinian, minted A.D. 538–539. *Right,* Follis of Constans II, minted A.D. 655–656. Maximum diameter of larger coin: 40 millimeters.

Photograph by J. Tainter.

relics, the war took on a religious tone and was conducted with corresponding fervor. In a series of campaigns beginning in 622, Heraclius counterattacked with increasing success. In 626 the siege of Constantinople was broken, and the following year the emperor began to advance into Persian territory. In 627 Heraclius destroyed the Persian army and in 628 occupied the Persian king's favorite residence. The Persians had no choice but to agree to peace. The Byzantines secured the return of all their territory. The war had lasted 26 years and resulted in no more than restoration of the previous status quo.

Later Byzantines remembered Heraclius as one of their greatest emperors. He is also one of history's tragic figures. The empire was exhausted by the struggle, and his great victory did not last. Newly Islamized Arab forces broke into imperial territory in 634 and defeated the Byzantine army 2 years later. Syria and Palestine, which had taken 18 years to recover, were lost again. Egypt was surrendered in 641. The most prosperous provinces were permanently gone, and soon the empire was reduced to Anatolia, Sicily, and parts of Greece and Italy (fig. 3.12). The Persians fared even worse. The Arabs conquered their empire completely.

Under Constans II (641–668) and his successors the strategic situation continued to deteriorate. The Arabs built their first fleet in 641 and with it took the capital of Cyprus. They ravaged Rhodes in 654 and in their first encounter defeated the Byzantine fleet the next year. After a civil war the Arabs returned to raiding Asia Minor in 663 and continued to do so nearly every year for two centuries (fig. 3.13). Constantinople itself was besieged each year from 674 to 678 (fig. 3.14). Finally the Arab fleet was driven off when the Byzantines deployed the incendiary compound called Greek fire. A new enemy, the Bulgars, broke into the empire from the north. In 697 the Arabs seized Carthage and the next year drove out the Byzantine force that had retaken it. From 717 to 718 an Arab force besieged Constantinople continuously for more than a year. The Byzantines used Greek fire again and in the summer of 718 ambushed reinforcements sent by the caliph. The Arabs were forced to withdraw and were never again able to mount such a challenge.

In the century before the victory of 718, the political and economic life of the eastern Mediterranean had been changed irretrievably. The huge empire that the Romans had put together was mostly gone.

129

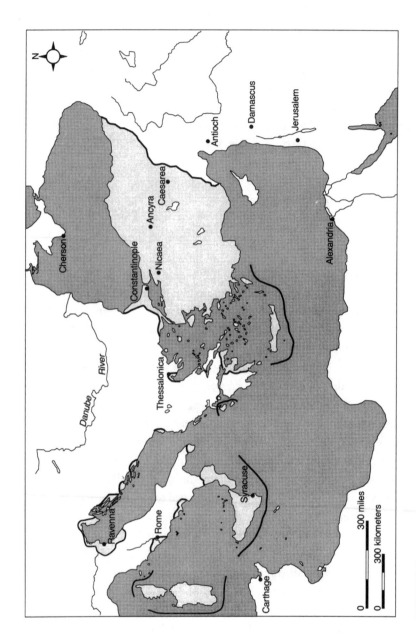

FIGURE 3.12

The Byzantine Empire in A.D. 700.

FIGURE 3.13

Ninth-century gate of Byzantine Ankara, incorporating architectural spoils from the destroyed Roman city.

Photograph by J. Tainter.

FIGURE 3.14

Land walls of Constantinople ca. A.D. 1200.

Digital reconstruction by A. Tayfun Öner (www.byzantium1200.org). Reproduced by permission.

Around 659 Constans cut military pay in half again. Debasements and inflation had ruined monetary standards and the fiscal and economic institutions that depended on them. There were no longer standard weights to copper coins. When the official value stamped on coins came to have no meaning in taxes or trade, monetary exchange was undermined. With army pay by the 660s cut to one-fourth its level of 615, the government no longer pumped coins into the economy. By 700 most people within or formerly within the empire had ceased to use bronze coins. In most Mediterranean lands the economy thereafter ceased to have a monetary basis. The economy developed into its medieval form, organized around self-sufficient manors. The magnitude of this transformation cannot be overstated. As historian Kenneth Harl (1996:205) observed, "urban life languished and population declined to its lowest level since the eve of the Iron Age."

One can scarcely imagine the difficulty of the transformation needed to save what was left of Byzantium. A way of life to which the peoples of the eastern Mediterranean had been accustomed for more than a millennium had to be given up. The emperors of the late third and early fourth centuries had responded to a crisis of similar magnitude by increasing the size of the army, the complexity of administration, and the regimentation of the population. The cost was covered by levels of taxation so high that lands were abandoned and peasants could not establish families large enough to replenish the population. Such a burden naturally did not endear the empire to its citizens. When the Arabs captured Egypt, always the most heavily exploited province, the Egyptians seem not to have been sorry to see the Byzantines go. Constans II and his successors could hardly impose more of the same exploitation on the depleted population of the shrunken empire. Instead they adopted a strategy that is truly rare in the history of complex societies: systematic simplification.

Arab civil war from 659 to 663 caused the caliph in Syria to purchase a truce. The respite allowed Constans II to direct fundamental transformations and to attend to matters in the western provinces. Even at one-fourth the previous rate, the government could not pay its troops. The attempts of Maurice Tiberius to give arms and rations in lieu of pay had led to mutinies and his death. Constans's solution was to devise a way for the army to support itself. He lacked ready cash, but the imperial family possessed vast estates, perhaps one-fifth of the land in the empire. There would also have been much land abandoned from the Persian attacks. The available lands were divided among the troops. In Asia Minor and other parts of the empire, divisions of troops—called themes—were settled in new military zones that came to have the same name. The soldiers (and even sailors) were given inalienable grants of land on condition of hereditary military service. It was apparently at this time that Constans halved military pay, for he now expected the troops to provide for their own livelihood through farming. Correspondingly the Byzantine fiscal administration was greatly simplified.

The transformation ramified throughout Byzantine society (fig. 3.15), as any fundamental economic change must. Both central and provincial government were simplified, and the transaction costs of government were reduced. In the provinces, the civil administration was merged into the military. Cities across Anatolia contracted to fortified hilltops. Aristocratic life focused on the imperial court rather than on the cities that no longer existed. The army lost much of the engineering ability of its Roman predecessor. Barter and feudal social relations replaced the millennium-old monetary economy. There was little education beyond basic literacy and numeracy, and literature itself consisted of little more than lives of saints (Haldon 1990, 1999; Treadgold 1988, 1995, 1997). The period is sometimes called the Byzantine Dark Age.

The results were evident almost immediately. The system of themes rejuvenated Byzantium. Byzantine forces began to put up much stiffer resistance to the Arabs, as evident in the victories of 678 and 718. In 664 the soldiers of north Africa, finding Constans's offer to their liking, put down a revolt of their governor. A new class of peasant-soldiers was formed across the empire. The new farmer-soldiers had obligations to no landowners, only to the state. They became producers rather than consumers of the empire's wealth. They formed a new type of army in which military obligation, and the lands that went with it, were passed to the eldest son. A soldier's other offspring were additional peasant laborers who were naturally interested in cultivating any other available lands and who were potentially further recruits. From this new class of farmers came the force that sustained the empire.

FIGURE 3.15

Changes in Eastern Roman imperial iconography. *Left,* The emperor triumphant: gold solidus of Theodosius II (402–450), minted between A.D. 430 and 439. *Right,* The emperor as religious hermit: gold solidus of Constans II (641–668), minted between A.D. 652 and 654. Each approximately 20 mm. in diameter.

Photograph by J. Tainter.

The lands these soldiers were given apparently were well worth having. A law of the tenth century specified that cavalrymen had to maintain lands worth at least 4 pounds of gold. (The actual specification was 5 pounds, but this was allowed to slip in practice.) Marines of the fleet were to hold lands worth 2 to 4 pounds. This was much more than a minimal peasant plot. Five pounds of gold would buy 144 acres, far more than a single cavalryman could work himself. Indeed, the minimum labor force for such a holding was seven men, and it could support up to 30 peasant families while still leaving some surplus to sell. By the standards of local communities, a cavalryman was a man of substance, free to undertake his military duties while others worked his lands. By the early tenth century cavalrymen often had cash savings of 3 pounds of gold. Because such a sum would have taken up to 24 years to accumulate from military pay, clearly most of it came from the sale of farm produce. A military treatise of the late tenth century advised generals to drill their soldiers regularly, for otherwise they would sell their equipment and best war horses to invest in their farms. Even lowly infantrymen had enough land that they could campaign during planting or harvesting, leaving others to work their farms.

This system clearly would have had drawbacks. Soldiers could not have been trained as well as when they were a full-time, professional force. Theme commanders often rebelled, particularly those close to Constantinople. Themes were hard to assemble for an offensive operation and less effective far from home. Yet the Byzantine army consisting of peasant-soldiers became a more effective force early on. The empire immediately began to lose land at a much slower rate. The Arabs continued to raid Anatolia but were unable to hold any of it for long. Soldiers were always near at hand. Fighting as they were for their own lands, they had a much greater incentive to repel invaders. After the establishment of the themes the Arabs made progress in Anatolia only when the empire had internal troubles from 695 to 717. By 745 Constantine V was able to invade the Caliphate, the first successful invasion of Arab territory in a generation.

The reestablishment of a small, professional force in the 760s (Haldon 1999:211) gave Byzantine forces once again an offensive capability. Campaigns against the Bulgars and Slavs extended the empire in the Balkans. Greece was recaptured for the empire. Pay was increased after 840, yet gold became so plentiful that in 867 Michael III met an army payroll by melting down 20,000 pounds of ornaments from the throne room. When marines were added to the imperial fleet it became much more effective against Arab pirates. In the tenth century the Byzantines reestablished disciplined, heavy infantry and added heavy lancers to their force (Haldon 1999:217–218). Soon they reconquered parts of coastal Syria. Overall after 840 the size of the empire was nearly doubled. The process culminated when Basil II (963–1025) conquered the Bulgars and extended the empire's boundaries again to the Danube. In two centuries the Byzantines had gone from near disintegration to being the premier power in Europe and the Near East, an accomplishment won by decreasing the complexity and costliness of problem solving.

Ironically, this success revived the problems that had caused late Roman and early Byzantine fiscal problems in the first place. A professional army was expensive, and the middle Byzantine military became increasingly fiscalized and burdensome (Haldon 1999:225, 238). Thematic forces were allowed to deteriorate in favor of mercenaries. Peasant-soldiers again began to lose their lands to the wealthy. Emperors after Basil II discerned no worthy threat and allowed the army to deteriorate. Based on a miscalculation of where danger lay, the thematic soldiers in Armenia, some of the best, were relieved of military duties and subject to taxation instead. The disasters that followed set in motion the final, slow decline of Byzantium, until by the fifteenth century the state consisted only of Constantinople. In an odd way these reverses confirmed the wisdom of the policy set in place during the crisis. The seventh-century system of themes proved a sustainable way to run, and even expand, a small empire, and the neglect of the peasant-soldiers in the eleventh century contributed to misfortunes that, in the long run, the empire could not survive (Ostrogorsky 1969; Hodges and Whitehouse 1983; Harl 1996; Haldon 1990, 1999; Treadgold 1988, 1995, 1997).

COLLAPSE OF THE ABBASID CALIPHATE

The Arabs, like the Sassanians and Germans before them, broke into a Roman world that had been declining economically for some time. The center of Arab political power during the early contests with the Byzantines was Damascus. Like any political entity, the Caliphate had its own internal conflicts. Just as the Arabs advanced during periods of Byzantine internal conflict, so also the Byzantines took advantage of Arab disputes. The system of themes apparently was established during the lull afforded by civil war among the Arabs from 659 to 663. The first Islamic dynasty, the Umayyads, was overthrown in 750. Damascus was full of Umayyad supporters and too close to the Byzantine frontier, so in time the new dynasty, the Abbasids, built their capital at Baghdad.

Rulers of Mesopotamia, from the last few centuries B.C. into the Islamic period, repeated the strategy tried 2000 years earlier by the Third Dynasty of Ur. There was an irregular but generally sustained increase in the scale and integration of the agricultural regime. By the Sassanian and early Islamic periods city building, population density, and other manifestations of complexity had reached their highest levels. Population increased fivefold in the last few centuries B.C. and the first few centuries A.D. The number of urban sites grew 900 percent. These trends continued through the Sassanian period, when population densities came to exceed significantly those of the Third Dynasty of Ur. At its height the area of settlement was 61 percent greater than in Ur III.

The fullest development of urbanism and agriculture began in the early to mid-sixth century A.D. Under Khosrau I (A.D. 531–579) the Sassanian dynasty reached its height, but his policies were reminiscent of those of the later Roman Empire. The needs of the state took precedence over the ability to pay. Taxes were no longer remitted for crop failure. Because the tax was fixed whatever the yield, peasants were forced to cultivate intensively. State income rose sharply under Khosrau II (590–628). This level of income would have been needed for the perpetual wars with the Byzantines. Under the Abbasids taxation became abusive. Tax assessments increased in every category. Fifty percent of a harvest was owed under the Caliph Mahdi (775–785), with many supplemental payments. Sometimes taxes were demanded before a harvest, even before the next year's harvest.

The Abbasid caliphs had a global influence in the world of their day. They traded with China in one direction and, in the other, exchanged embassies with Charlemagne. The Caliphate had enormous sources of silver, which it used in external trade. Viking traders brought Abbasid silver coins in great quantities to Scandinavia; thousands have been found in Russia and the Baltic. Chinese silk has been unearthed from a ninth-century Swedish grave, and a large Buddha now resides in the State Historical Museum, Stockholm. The Carolingian renaissance may have been financed in part by Abbasid silver coins, which were laundered, before reaching Christian Francia, by being melted by the Vikings and passed through middlemen. As the Abbasid economy expanded, trade emporia grew around the Baltic and the North Sea. Historian Sture Bolin once likened the Viking traders operating the Dnieper route to the sixteenth-century Spanish treasure fleets bringing silver from the New World. So great was the prestige of Abbasid coins that Offa, king of Mercia from 757 to 796, issued near-perfect imitations of them. His engraver, copying what he thought to be a design, produced the only English coin ever to proclaim, "There is God no but Allah and Mohammed is his prophet" (Hodges and Whitehouse 1983).

Under the Abbasids there was unprecedented urban growth. Baghdad grew to five times the size of tenth-century Constantinople and thirteen times the size of the Sassanian capital, Ctesiphon. Yet the Islamic rulers built their imperial structure on an unstable base. It was a costly regime, made worse by frequent civil wars. The capital was moved often and each time built anew on a gigantic scale. The Caliph al-Mutasim (833–842) built a new capital at Samarra, 120 kilometers upstream from Baghdad. In 46 years, he and his successors built a city that stretched along the Tigris for 35 kilometers. It would have dwarfed imperial Rome. Workers, artisans, and nobles were brought from across the empire to create a city overnight. The scale of construction was monumental. The Jausag al-Khaqani, al-Mutasim's palace built between 836 and 842, enclosed 175 hectares. It was larger than Versailles. The Balkuwara palace, built 849 to 859, was enclosed within a rectangular area walled 1250 meters to a side. These complexes had gardens along the Tigris, enormous palaces, courtyards, and construction in precious marbles and gilded woods (Hodges and Whitehouse 1983:151–156).

Yet as the state undertook these constructions it did not always fulfill its irrigation responsibilities. As the irrigation system grew in size and complexity, maintenance that had once been within the capacity of local communities was no longer so. Communities came to depend on the imperial superstructure, which in turn became increasingly unstable. Peasants had no margins of reserve, and revolts were inevitable. Civil war and rebellion meant that the hierarchy could not manage the irrigation system. Mesopotamia experienced an unprecedented collapse. In the period from 788 to 915 revenues fell 55 percent, from 479,550,000 to 217,500,000 silver dirhems. The Sawad region, at the center of the empire, had supplied 50 percent of the government's revenues. This dropped within a few decades to 10 percent. Most of this loss occurred between 845 and 915. In many strategic and formerly prosperous areas there were revenue losses of 90 percent within a lifetime. The perimeter of state control drew inward, which diminished any chance to resolve the agricultural problems. By the early tenth century irrigation weirs were listed only in the vicinity of Baghdad.

In portions of Mesopotamia the occupied area had shrunk by 94 percent by the eleventh century. Population dropped to the lowest level in five millennia. Urban life in 10,000 square kilometers of the

Mesopotamian heartland was eliminated for centuries (R. McC. Adams 1978, 1981; Waines 1977; Yoffee 1988). Robert McC. Adams (1981:xvii) described eloquently the aftermath of this collapse.

Much of the central floodplain of the ancient Euphrates now lies beyond the frontiers of cultivation, a region of empty desolation. Tangled dunes, long disused canal levees, and the rubble-strewn mounds of former settlement contribute only low, featureless relief. Vegetation is sparse, and in many areas it is almost wholly absent. Rough, wind-eroded land surfaces and periodically flooded depressions form an irregular patchwork in all directions, discouraging any but the most committed traveler. To suggest the immediate impact of human life there is only a rare tent. . . . Yet at one time here lay the core, the heartland, the oldest urban, literate civilization in the world.

As the Abbasid Caliphate had a global influence, so its collapse had proportional consequences. The Abbasid capital had become the economic center of the world. Yet in the century between Harun al-Rashid (786–809) and the accession of al-Mutadid (892), the Abbasid caliphs went from fabulous wealth to bankruptcy. Around 820 the supply of Abbasid silver to the Viking trade declined and remained low for the rest of the ninth century. The economy of northwestern Europe responded almost immediately. At Dorestad, a trade center and Carolingian mint at the mouth of the Rhine, the influx of Carolingian silver coins stopped around 830, and the mint itself closed. The port of Hamwih (Saxon Southampton) declined about the same time, and correspondingly the West Saxons invaded Kent and Cornwall. The Carolingian Empire plunged into civil war in the 830s. Reacting fiercely to the diminishing opportunities for wealth from trade, Vikings greatly intensified their raids in the 830s and 840s, repeatedly attacking coastal Frisian, Flemish, East Anglian, and West Saxon sites. In the 860s pillaging turned to invasion when the Danes began to conquer the eastern English kingdoms. In 911 Charles the Simple was forced to cede much of northern France to the Norsemen. The western Mediterranean Moslems, losing access to Abbasid wealth, began their own series of attacks on Crete, Sicily, southern Italy, Provence, and southern Anato-

139

lia. St. Peter's Basilica was sacked in 846, and an Islamic emirate was established at Bari, on the Italian mainland, from 847 to 871 (Hodges and Whitehouse 1983:149, 156, 158, 163–168). When world centers collapse the systemic repercussions are widespread, and those on the periphery usually have no understanding of the sudden rise of poverty and violence in their world.

DEVELOPMENT OF MODERN EUROPE

History is substantially a chronicle of reactive processes. Complex political systems adapt on a day-to-day basis less to the natural environment than to the maneuverings of other political entities. The process of keeping up a competitive stance, or preserving territory or citizens, drives the consumption of resources regardless of cost. Resource conservation as an abstract principle is never considered in a competitive environment if the cost of protecting resources for the future is to lose autonomy, territory, or identity today. Where competition is maintained over long periods—generations or centuries—it has profound consequences for the development of complexity and for sustainability (Tainter 1992).

Arms races are the classic example of diminishing returns to complexity. Any competitive nation will quickly match an opponent's advances in armaments, personnel, logistics, or intelligence, so that investments in these areas typically yield no lasting advantage or security. In an arms race, each competitor strives for advantage over its rivals, and the rivals strive to counter them and develop advantages of their own. Usually no one can gain an overwhelming advantage that lasts very long. More and more money, resources, and personnel are spent on that most fleeting of products: military advantage. The costs of being a competitive state continuously rise, while the return on investment inexorably declines. All the while a state must search continuously for the resources to remain competitive and develop an organization to deploy those resources effectively. The unfolding of this process in Europe of the last millennium not only altered European societies but ultimately changed the entire world. We will outline the development of this process from the fifteenth through the nineteenth centuries.

Europe before 1815 was almost always at war somewhere; scarcely a decade went by without at least one battle. From the twelfth through the sixteenth centuries France was at war from 47 percent of all years in some centuries to 77 percent in others. For England the range was 48 to 82 percent; for Spain, 47 to 92 percent. Even in the most peaceful centuries these nations were involved in war, on average, nearly every other year. In the whole of the sixteenth century there was barely a decade when Europe was entirely at peace. The seventeenth century enjoyed only 4 years of total peace; the eighteenth century, 16 years (Parker 1988:1; Rasler and Thompson 1989:40).

In the fifteenth century, siege guns ended the advantage of stone castles (fig. 3.16) and necessitated changes in the strategies and technology of defense. From the early fifteenth century, fortification builders designed walls that could support defensive cannon. A short time later walls were built that could also withstand bombardment. By 1560 all the elements of the *trace italienne* had been developed. This was a fortification system of low, thick walls with angled bastions and extensive outworks (fig. 3.17). It was an effective but expensive method of fortification, taking much labor to construct. In 1553 the city of Siena found it so expensive to build such fortifications that no money was left for its army or fleet. Siena was annexed by Florence, against which, ironically, its fortifications had been built (Creveld 1989:101–103; Parker 1988:7, 9, 12).

Trace italienne fortifications, if one could afford them, were a worthy investment. To capture a place defended in this way could take months or years. Offensive tacticians responded with more complicated siege methods, and their costs rose as well. A besieging force of perhaps 50,000 had to be kept in place for weeks or months. Such a force needed 475 tons of food per day, to which must be added ammunition, powder, and building materials. From this time on, local lords could not afford to build and defend an effective fortress or to attack one. No longer were the resources for war to be found in the feudal countryside. They had now to be sought in capitalist towns (Creveld 1989:106–108; Parker 1988:13).

There were comparable developments in open-field warfare. In the fourteenth and fifteenth centuries massed archers and the pike phalanx made the armored knight obsolete. These in turn were gradually

FIGURE 3.16

Tarascon castle, Provence, France, begun in the thirteenth century.

Photograph by J. Tainter.

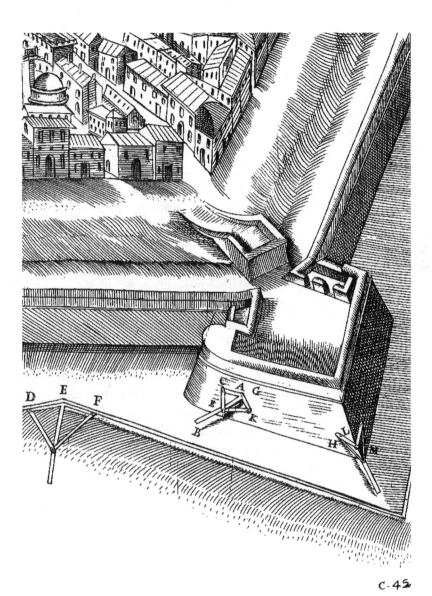

C-45

FIGURE 3.17

Geometry of a *trace italienne* bastion. From *Trattato del Radio Latino* by
Egnatio Danti and Latino Orsini (Rome, 1583).

Reproduced by permission of the Museum of the History of Science, Oxford.

superseded by the use of firearms. To make the most effective use of firearms took organization and drill. Infantry came to be drawn up in closely coordinated ranks. Those in the rear reloaded while the lead musketeers fired, and quick changes of position yielded an uninterrupted application of firepower (Creveld 1989:89–91; Kennedy 1987:21; Parker 1988:16–20). As commanders maneuvered for battlefield advantage, tactics were developed to increase the efficiency and effectiveness of firing. Training and battlefield coordination became more important. Uneducated soldiers had to be familiar with what were, at the time, history's most advanced weapons. Ranks had to open and close on signal. Victory came to depend on the right combination of infantry, cavalry, firearms, cannon, and reserves. Textbooks of military drill were published across the continent, and tactics became a gentleman's topic of study (Creveld 1989:92–94; Parker 1988:18–23).

War came to involve ever larger segments of society and became correspondingly more burdensome. Several European states saw the sizes of their armies increase tenfold between 1500 and 1700. Louis XIV's army stood at 273,000 in 1691. Five years later it was at 395,000, and nearly one-fourth of all adult Frenchmen were in the military. Between 1560 and 1659 Castile lost about 11 percent of its adult male population in the constant wars (Sundberg et al. 1994:13). Each day, a field army of 30,000 needed 100,000 pounds of flour and 1500 sheep or 150 cattle. This was more than was needed to feed all but the largest cities of Europe (Creveld 1989:112–113; Parker 1988:2, 45–46, 75).

Yet throughout these developments, and despite or because of them, land warfare became largely stalemated. There were few lasting breakthroughs. The new technologies, and the mercenaries to deploy them, could be bought by any power with enough money. No nation could gain a lasting technological advantage. When a nation such as Spain or France threatened to become dominant, alliances formed to counter its power (Kennedy 1987:21–22). Major wars of the time therefore were long and tended to be decided by cumulative small victories and the slow erosion of the enemy's economic base. Yet defeated nations quickly recovered and soon were ready to fight again. Land warfare had to be augmented by what amounted to global flanking operations. European wars were expanded into contests for power and influence overseas (Parker 1988:43, 80–82).

Europeans used the great wealth from the New World to sustain their ever more costly competition (Kennedy 1987:24, 27–28, 43, 46–47, 52; Tainter 1992:110, 124). In 1760 the Duke of Choiseul noted, "In the present state of Europe, it is colonies, trade and in consequence, seapower which must determine the balance of power upon the continent" (Parker 1988:82). The development of sea power and acquisition of colonies became aspects of strategy in stalemated European warfare. Because of this, European war ultimately changed the entire world. By 1914 the nations of Europe and their offshoots controlled fully 84 percent of the earth's surface (Parker 1988:5).

The naval powers of the time were England, the Netherlands, Sweden, Denmark/Norway, France, and Spain. From 1650 to 1680 the five northern powers increased their navies from 140,000 to 400,000 tons. By this time England had to import critical supplies for its fleet, including masts, tar, and pitch from Sweden, and tried to develop a tar industry in its North American colonies. In the 1630s the Dutch merchant fleet needed to build 300 to 400 new ships each year, about half of which were used in trade in the Baltic. Between the 1630s and 1650 the Dutch merchant fleet grew by 533 percent (Sundberg et al. 1994:38, 42). The naval strategy also involved problems of increasing complexity and cost. In 1511, for example, James IV of Scotland commissioned the building of the ship *Great Michael*. It took almost one-half of a year's income to build and 10 percent of his annual budget for sailors' wages. It was sold to France 3 years later and ended its days rotting in Brest harbor (Parker 1988:90).

As the armies continued to grow through the eighteenth and nineteenth centuries, new fields of specialization were needed. There was demand for skills such as surveying and cartography. It was necessary to have accurate clocks and statistical reporting. In the eighteenth century some armies carried their own printing presses. Organization became more complex. Staff and administration were separated. Armies no longer marched as a unit but could be split into smaller elements that traveled, under instructions, on their own. Battles came to last up to several months. In France, the *levée en masse* was begun in 1793. In 1812 Napoleon invaded Russia with an army of 600,000, including 1146 field guns, on a 400-kilometer front (fig. 3.18) (Creveld 1989:114, 117–122; Parker 1988:153).

In 1499, as he was embarking on a campaign in Italy, Louis XII asked what was needed to ensure success. He was told that three things alone were needed: money, money, and still more money (Sundberg et al. 1994:10). As all things military grew in size and complexity the main constraint came to be finance. In the decades before 1630 the cost of putting a soldier in the field increased by 500 percent. Nations spent ever more of their income on war, but it was never enough. In 1513, for example, England obligated 90 percent of its budget to military efforts. In 1657 the figure was 92 percent. In 1643, expenditures of the French government, mainly on war, came to twice the annual income (Kennedy 1987:58, 60, 63). England's wars in the 1540s cost about 10 times the crown's income (Kennedy 1987:60). In the mid-eighteenth century Frederick the Great also spent 90 percent of his receipts on war. He found it necessary to debase his currency and to extract both contributions and plunder from civilians.

FIGURE 3.18
Napoleon's cannon in the Kremlin, Moscow.

Photograph by J. Tainter.

146

Sweden used its vast forest resources to fund its wars. Copper, steel, and tar produced by the use of forest resources amounted to 90 percent of Sweden's exports. The currency earned in this way was the major basis for Sweden's war efforts. In 1701, 87 percent of England's tar came from Sweden. Between 1658 and 1814 England sent 20 fleets to the Baltic to secure the trade in masts and timber (Sundberg et al. 1994:28–29, 40, 42).

Sweden enjoyed a combination of low population, untapped forest reserves, and eager markets for its products. The major states, lacking such advantages, had rely on credit to finance their wars. Notwithstanding the flow of precious metals from its New World colonies, Spain's debts rose from 6 million ducats in 1556 to 180 million a century later, and bankruptcy often undermined Spanish military operations. The cost of war loans grew from about 18 percent interest in the 1520s to 49 percent in the 1550s. Both France and Spain often had to declare bankruptcy or force a lowering of the rate of interest. Governments coerced bankers into extending new loans by refusing to make payments on existing debts. From the sixteenth through the eighteenth centuries the Dutch, followed by the English, overcame fiscal constraints by gaining access to reliable short-term and long-term credit. They were careful to pay the interest on loans and so were granted more favorable terms than other nations. They used this advantage to defeat opponents, France and Spain, that were wealthier but were poor credit risks (Parker 1988:63–67; Rasler and Thompson 1989:91, 94, 96, 103).

It seems that the wars did not really augment net national wealth. Sweden's expenditures on wars, calculated as energy, exceeded the return by 240 percent (Sundberg et al. 1994:25). The wars raised permanently the cost of being a competitive state, and war-induced debt levels persisted long after the fighting ceased. Power always shifts, and victorious nations were never able to dominate for very long (Kennedy 1987; Rasler and Thompson 1989:106, 175–176). Many people of the time understood the futility of European wars, but a competitive spiral is not easy to break. In 1775 Frederick the Great eloquently summed up the state of affairs:

The ambitious should consider above all that armaments and military discipline being much the same throughout Europe, and alliances as a rule producing an equality of force between belligerent parties, all that princes can expect from the greatest advantages at present is to acquire, by accumulation of successes, either some small city on the frontier, or some territory which will not pay interest on the expenses of the war, and whose population does not even approach the number of citizens who perished in the campaigns.

Quoted in Parker (1988:149).

Consequences of European Wars

Because land warfare in Europe produced no lasting advantages, the expansion of competition to the global arena was a logical consequence. Competition expanded to include trading, capturing overseas territories, establishing colonies, attacking adversaries' colonies, and intercepting the wealth that flowed from them. Overseas resources became necessary to sustain European competition, and the balance of power in Europe came in part to rest on access to them. These were societies that, until the nineteenth century, were powered almost completely by solar energy. Sweden's support base in the seventeenth century consisted nearly entirely of renewable resources: 87 percent, nearly the same fraction as in the Roman Empire 1200 years earlier. Lacking colonies outside its region, almost half of Sweden's forest-based exports went to finance foreign wars (Sundberg et al. 1994:18, 20). For societies powered by solar energy and using that energy so heavily within the limits of their technology, the main way to increase wealth was to control more of the earth's surface where solar energy falls. It became necessary to secure the produce of foreign lands to subsidize European competition. New forms of energy, and nonlocal resources, were channeled into a very small part of the world. This concentration of global resources allowed European conflict to reach heights of complexity and costliness that could never have been sustained with only European resources (Tainter 1992:123–125).

European rulers of the time faced extraordinary pressures to extract wealth from their colonies. Yet even that wealth could not meet the cost of some campaigns. In 1552 the Hapsburg Emperor Charles V spent 2.5 million ducats on a campaign at Metz, an amount equal to 10 times his

American income. By the 1580s Phillip II was receiving 2 million ducats a year from American mines, but the ill-fated armada of 1588 cost five times that (Kennedy 1987:46–47). Even with this massive transfer of bullion from the New World, Spain's debt grew 3000 percent in the century after 1556, and bankruptcy caused Spanish military operations to fail. Clearly they would have failed much earlier (or not been undertaken) if Spain had not been able to draw on New World wealth.

Competition not only forced Europeans to search for foreign lands and resources but virtually guaranteed them success in doing so. Such conflict, particularly when it is stalemated in balances of power, selects for continual innovation in technology, organization, strategy, tactics, and logistics. Any power that does not match its competitors in these areas risks defeat and domination. A nation that survives this process will be so proficient at making war that, outside its group of peers, there may be no other military force that can withstand it. The inexorable pressure on European states to become ever better at making war meant that when they ventured outside Europe, they had a competitive advantage over other powers, whether organized as states or not (Kennedy 1987:16–30). Time and again over the past 500 years, comparatively tiny European forces have defeated much larger forces in the New World, Africa, and Asia. A state that survives such prolonged competition may find the rest of the world at its command (Tainter 1992:125).

European competition stimulated technological innovation, development of science, political transformation, and global expansion. For better or worse the repercussions of centuries of European war are a legacy in which we still participate and will do so for the foreseeable future. Unlike the Roman Empire, these centuries of war produced no collapse. The European system, though costly, proved sustainable over many centuries. The reasons, as we will discuss, lie in innovation and energy.

IMPLICATIONS FOR SUSTAINABILITY

There may be as many ways to think of sustainability as there are things one might want to sustain. It is a term akin to *ecology:* once it escapes to the public arena it takes on a variety of symbolic meanings. These range from recycling one's trash and using public transportation to doing whatever diplomats decide is sustainable forestry. In the rural American West, sustainability is considered to mean an end to making a livelihood from natural resources. From such misunderstandings, emotional connotations tinge public discourse at all levels from local to international. To borrow a phrase from Christopher Hitchens (1988:301), one simply unveils the term *sustainability* like a Medusa's head and turns all discussion to stone. Soon no doubt consumers will be able to choose from among a wide variety of "sustainably produced" goods, ranging from breakfast cereals to toothpaste. As suggested in chapter 1, when confronted with the degradation of a term never suitably defined to begin with, one should always ask, "Sustain what, for whom, for how long, and at what cost?"

In this chapter we have discussed attempts to sustain certain political structures: the Roman Empire, the Byzantine Empire, the Abbasid Caliphate, and the squabbling nations of Europe. Such cases are useful for understanding broader issues, for each attempt to be sustainable entails the same exercises: monitoring historic and current conditions, predicting future ones, and deploying problem-solving institutions to ensure that the future develops as one wishes. The constraints of complexity and costliness in problem solving apply regardless of the type of system in which problem solving is applied. We have reviewed several ancient societies that each responded to challenges of sustainability in quite different ways. The Romans and the Abbasids ultimately were unsuccessful in their attempts to maintain their states. The Byzantines were forced to admit that the age-old society of Greco-Roman antiquity, founded on a state-controlled, monetized economy and organized around cities, was no longer viable. In its place they developed a new system that made more effective use of internal resources. They might have kept more of their empire had they made the conversion earlier, but people will rarely acknowledge that an accustomed way of life is unsustainable except in the face of prolonged, devastating failure. The petty, bickering states of Europe are a case quite different and uniquely illuminating. Their continuous attempts to expand at a neighbor's expense, or prevent the neighbor from doing likewise, forced them to develop strategies of sustainability that have been surprisingly durable, though at great cost to many non-European peoples and the global environment.

Building on our discussion in chapter 2 and in previous work (Tainter 1988), figure 3.19 graphs the characteristic curve of diminishing returns to investment in complexity. As a society resolves problems by increasing complexity, it invests in greater production of resources, more processing of information, and greater sociopolitical control. Provided that its people are averse to complexity and unproductive labor, the benefit–cost ratio for these investments will at first increase favorably. Yet as the least costly resources and organizational solutions are progressively exhausted, solutions to further problems will have to come from developing more costly resources, increasing complexity of organization, processing more information, and working harder to defend oneself. Costs rise along with the complexity of problem solving.

On a few rare, precious occasions, societies encounter opportunities to invest in growing complexity that yields an increasing return. Sometimes this comes when a technological innovation allows us to tap a previously unused or underused energy subsidy, as in the development of the waterwheel or the steam engine. Often it comes from an energy

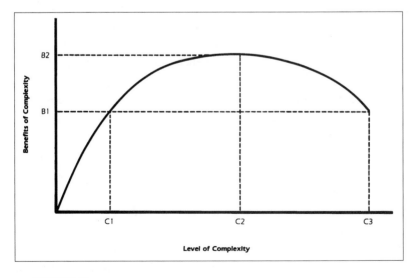

FIGURE 3.19

Diminishing returns to complexity.

subsidy that we have known about all along but have used minimally because its great potential was not realized or the initial cost to use it was too high. The development of agriculture was such a case; another was the reluctant shift to coal in post-Renaissance England. In the ancient world, where labor was usually abundant and technological development glacial, increasing returns commonly were realized from the few times when a state enjoyed a temporary military advantage over its neighbors and could expand at their expense. This amounts to an enhanced subsidy of solar energy.

Societies usually increase in complexity to solve the problems of their own existence and to maintain the status quo. Such endeavors are considered successful when the status quo is indeed maintained: The frontiers are defended, internal order has been restored, or people again have enough to eat. If the costs and complexity of problem solving grow merely to restore a system to its previous condition, axiomatically the marginal return on investment in complexity declines. This is seen clearly in recent efforts of international intervention. In Somalia the United States and other nations deployed a complex military machine, voracious in its consumption of oil, merely to provide Somalis with stability and a daily food allowance of 2500 calories or so per capita. Some years before, Somalis had all these things without the cost of an international effort. For all the expenditures, a stable government could not be imposed from without, and securing the delivery of food with helicopters yields nearly the lowest imaginable return on caloric expenditures. The result is that Somalia became a place that cost much more to sustain than ever before, yet its people were no better off than before its civil war. Much effort went to restore a status quo.

Nothing is forever, and nowhere is this more true than in the world of living systems, including human societies. Because the stock of challenges that a society may confront is, for practical purposes, infinite in variety and endless in number, there is no question of maintaining investment in complexity at a static benefit–cost ratio. It is always necessary to address some new problem, often by becoming more complex. Barring the development of new energy subsidies or more efficient use of existing ones, any problem-solving system ultimately comes to the point of diminishing returns shown in figure 3.19. At this point, although overall productivity can still rise, marginal productivity, within the constraints of technology and energy, can no longer do so.

Beyond point B1,C1 on the curve the decline in marginal productivity begins to deflect more sharply. A society at this point becomes increasingly vulnerable to collapse. In an ancient cultural system activated primarily by solar energy, the cost of increasing complexity was built on diminishing returns to labor. To pay taxes or meet other obligations peasants were forced into something they would otherwise avoid: unproductive work that yields them no clear advantage. Regardless of the type of economy, it is an unenviable situation. A contemporary problem-solving system that is part of a larger whole (such as an institution or a government agency) at this point will find increasing skepticism of its work, hard questions from appropriators, and a serious threat of extinction.

Two general processes make a society more liable to collapse at this point. First, a society in this condition finds its accumulated surpluses depleted and productive capacity sapped as it spends more and more to accomplish proportionately less and less. When major adversities arise, as inevitably they do, there is no longer surplus economic capacity with which to confront them. If the crisis is survived, there may be less ability to meet the next one. Collapse thus becomes a matter of mathematical inevitability. In time an insurmountable perturbation will come along. If the Roman Empire, for example, had not been toppled by Germanic tribes, it would have been later by Arabs or Mongols or Turks.

Second, declining marginal returns make complexity a less attractive strategy. The ruthless extraction of resources needed to sustain such a course makes separation or disintegration increasingly attractive. In the later Roman Empire peasant revolts in Gaul went on for generations. Perhaps more typically, peasants simply become apathetic to the well-being of the state. In both the later Roman and Byzantine empires the overtaxed peasantry offered little resistance to invaders.

Progressing along the curve in figure 3.19 there is a continuum of points and segments. In the segment from B1,C1 to B2,C2, the curve of marginal productivity deflects sharply to the right. To a society in this condition it will be clear that something is awry. Attempts to solve problems yield little net benefit, and almost anything that is tried appears inept. There will be ideological strife over what has gone wrong and what the solutions might be. Within any such society there will be people who *know* that whatever is wrong can be solved simply by reverting to the religion and lifestyle practiced earlier. Beyond B2,C2 there is a series of points, such as B1,C3, where the costs of complexity continue to rise but the benefits have actually declined to those available at lower levels of complexity and investment. The benefits of investing in complexity at B1,C3 are no higher than those at B1,C1, but the productivity of the latter point is far preferable. A society at B1,C3 is living on its capital and vulnerable to rapid collapse (Tainter 1988:119–123, 194–195).

The collapse of the Western Roman Empire is understandable in this framework. An empire built on the expectation of high returns to conquest and defense by client states found by the latter part of the second century A.D. that its enemies had grown stronger, the scene of fighting might be within the empire itself, and ordinary budgets would not suffice to keep enemies out. Still, most of the empire's inhabitants at this time would not have felt that they were seeing the start of a relentless slide. This did become clear to some in the third century, when the forces of Persians, Germanic tribes, and contending Romans crossed the empire. The first strategy to contain the crisis included debasement of the currency so severe that the monetary economy was ruined. The crises came so fast, so frequently, and from so many directions that there was no choice. The costs of containing the crises had to be paid whatever the true cost to the future.

Victories in the 290s gave Diocletian and his successors the opportunity to implement the second strategy, which was to increase the size and complexity of the problem-solving system (government and the army) and to organize the empire to produce the necessary resources. The strategy to accomplish this was to census every unit of potential production (whether person, plot of land, cart, or ship), establish the level of output required of it, then set the empire's agents to ensure that obligations were fulfilled. Nothing was allowed to interfere. If peasants abandoned their fields, they were to be brought back or the lands assigned to other cultivators. If an occupation was considered essential then it was made hereditary, regardless of the aspirations of those commanded to fill it. It was a system of extraordinary detail for its time. Outputs were what the empire needed, so outputs were ordered, arranged for, and monitored. When the outputs failed, laws were proclaimed that sought to restore them. The well-being of the peasants who produced the outputs was not of concern. As a consequence the productive system declined as lands were abandoned, peasant populations stagnated, and power shifted to large landowners. The peasants had little reason for loyalty or incentive to resist invasions. As the support system weakened, its outputs declined until collapse was inevitable. The lesson of the Romans is that a state or other institution can be destroyed by trying to sustain itself. Collapse can emerge from normal system functioning, including reasonable attempts at self-preservation.

The story of the collapse of the Abbasid Caliphate sounds similar in many details. Yet the costs in human and environmental terms are much clearer. In the Roman Empire we see indirectly—in abandoned lands, imperial legislation, and shrunken cities—the costs of managing for outputs to the detriment of the productive system. For the Abbasids the consequences are clearly detailed in settlements abandoned, taxes lost, and lands destroyed for centuries. Historians of the Roman Empire have sometimes asked why the Romans didn't intensify production to generate the resources necessary for survival. In the Abbasid Caliphate we see a society that did intensify production, with devastating consequences. The Abbasids, like the Third Dynasty of Ur, attempted to force greater production through greater delivery of water. In the Mesopotamian alluvium this leads in the short-term to great increases in production, both absolutely and per hectare. Governments always assume that incomes are permanent and build superstructures and levels of expenditure to use incoming funds fully. Like the Romans, the Abbasids managed for outputs rather than for the productive system. Finding the outputs necessary and to their liking, they built a structure that could not be maintained without them. This in turn reinforced the emphasis on production, so that outputs came to be demanded even before they had been produced. Only with difficulty can one imagine a government so remote from the means of production as to requisition the fruits of agricultural labor before the seed was even sown.

In their previous book, Allen and Hoekstra (1992) discussed the relations of context, control, and information flow between levels of a hierarchy. In most living systems, information flows upward, to higher levels, and controls operate downward. Of necessity the higher levels change more slowly than the lower ones, for to change they must respond not to primary forces but to information about these forces. The higher levels stabilize the system as a whole, buffering the endless shifts and microadaptations that arise as lower levels adjust to a constantly changing world. From these relations living systems assume their character.

Human social systems are somewhat different, which is why the generalizations of the preceding paragraph were qualified. Ecological systems often share matter or energy with other like systems, but they cannot share information to the same degree that human systems do. The matter and energy, taken out of context, cannot generate the same signals to higher levels. In some sense these signals must be generated internally by each ecological system (although the arrival of an invading species is one form of information and represents a counterexample). Human societies routinely exchange matter, energy, *and information* with other societies. Information flows more richly in human systems than in ecological systems, such as invaded plant communities. In human exchanges, information about matter and energy can be shared, information largely devoid of matter and energy can be shared, information can be concealed or made misleading, matter and energy can be given new meanings in the system that imports them, and information can be given new meanings. The exchange and transformation of information produce a dynamic character to human systems that other living systems do not possess. In part for this reason, human systems can never be understood fully by the same relations of hierarchy through which other living systems are understood.

The levels of a human system do not participate equally in the exchange of information with other systems. The information flows primarily to the upper levels, and it is often mediated by specialists (merchants, travelers, scholars, ambassadors, or intelligence agents). Moreover, the information flowing to elite levels tends only minimally to concern the foreign systems as wholes. It concerns mainly other elites. The higher levels in a human social system routinely receive information from and about the higher levels of other societies. This is the information to which elites usually respond, and it is the information that conveys the most importance and personal meaning to them. We are reminded of something told to one of us (Tainter) by a Russian colleague. As the Soviet Union crumpled about him and the citizens of the Union and its security forces battled in the streets, the information that mattered personally to Mikhail Gorbachev was not the demands of lower levels but his reputation in the minds of Ronald Reagan, Margaret Thatcher, and George Bush. It was this information about which he cared and that influenced his political decisions.

Elites respond to information about other societies, and particularly about the elites of other societies. This information routinely concerns behaviors to emulate, particularly consumptive behaviors, and the other society's competitive stance. When empires compete, such as the Romans and the Sassanians or the Byzantines and the Abbasids, the elites must display outward trappings of power to match their competitors. In the Near East the Romans and Byzantines often fared well on the battlefield but poorly at competitive display. The everyday currencies they could produce, in copper or gray, debased silver, compared poorly to the pure, brilliant silver of Sassanian or Islamic dirhems. Long after Roman and Byzantine victories had faded from memory there were good silver coins to suggest that just over the border was a wealthier empire that gave its citizens sound currency.

As we discussed in the case of European warfare, elites must also respond to actual threats posed by their neighbors. Or at least they must do so if they want to remain elites. This imposes a need continuously to demand resources from lower levels regardless of ability to pay. The competition between the Byzantines and Arabs would have forced the caliphs to demand higher outputs even if there was no other stimulus to do so.

In a world in which the political environment changes continuously, higher system levels in a society actually change *more rapidly* than lower levels. The ability to change rapidly in complexity, as we have seen, is one of the characteristics that make human societies so successful. The most rapid changes in complexity are imposed from above. In human societies it is the lower levels that often prove conservative and enduring, buffering the rapid transformations demanded from higher up. Peasant societies are noted for being unenthusiastic about change. For example, there are villages in the Near East where people still speak Aramaic, an ancient language spoken widely at the time of the Romans.

One of the tensions in every complex society is the contradiction between demands from the top for intensification and rapid change and the inability to intensify or resistance to change from the bottom. Responding as they do to horizontal signals from other systems, elites are always taken by surprise by rebellions or separatist movements from below. So it was with the Romans and the Abbasids, so it was in prerevolutionary Cuba, and so it is today with militias and tax protestors in the United States.

Having a continuous need to amass resources and responding more readily to horizontal than to vertical information, complex social systems manifest an inherent instability. Vertical feedback loops may be so poorly developed that the needs of the moment drive rulers to deplete long-term productive capability. In some cases, such as the Romans, there may be no choice. This tendency drove both the Romans and the Abbasids to live off their capital resources: producing lands and peasant population. Both confronted their problems by driving the return on investment in complexity into the realm of negative returns: the area on figure 3.19 from B2,C2 to B1,C3. For the Abbasids, overirrigating brought on salinized fields and declining yields, while demands for outputs drove the peasantry into ruin. As the elite levels became unstable, irrigation systems that depended on high-level management could not be maintained. The productive system, which had for so long been ignored, entered a final decline, taking with it all that depended on its outputs. In many areas the effects of this calamity have yet to be reversed.

It takes crises of unprecedented proportions to convince the rulers of an empire that they can no longer compete at a former level. Such a crisis must be significant enough to overcome the disparity between information conveyed horizontally and that conveyed vertically. The signals from below must be strong enough to overcome tremendous inertia; rulers of the ancient world had been accustomed for millennia to ordering outputs and having them delivered. The Byzantines finally perceived such signals during the crises of the seventh century, in which they lost half their empire and stood to lose it all. The population of the empire had not recovered from the losses of the sixth-century plague when the Persian invasion of the early 600s destroyed urban life in Asia Minor. Both the Persians and the Arabs took back into slavery as many of Anatolia's remaining inhabitants as they could catch and transport. Suddenly the rural economic basis of the Classical world was gone, as well as the urban basis for its culture and administration. Outputs dwindled, and the government could not meet its obligations. The resources for government and war could not be had regardless of how many commands came from above.

The signals from below broke through, and Constans in the early 660s ordered one of history's most remarkable transformations. By distributing his soldiers across the empire Constans created a force that was more devoted to defense, more effective, and largely self-supporting. Thereafter the empire stopped losing land so rapidly and in a while was able to take the offensive. The empire now relied on the native population rather than mercenaries. Historians have lavishly praised this development. "Byzantium was to emerge from the crisis," wrote George Ostrogorsky (1969:86), "in an essentially different form, able . . . to draw on new and vigorous sources of strength."

In ecological terms the Byzantine government shifted most of its employees down the political food chain, from consumers to producers of the empire's wealth. No longer did peasants have to produce enough to support both themselves and a recently ineffectual army. The army became a subsystem of the peasantry and integral to it. Soldiers were members of local communities. The land that they defended was their own. The people whom they defended were their kin and neighbors. Accordingly, they fought better than before, and the government secured a better return on its investment in them. No longer was the Byzantine government concerned exclusively with outputs. It now fostered the well-being of the system that produced resources. Because the state through its soldiers both produced and consumed outputs, it had clear incentives for conservation. Emperors legislated against attempts by the powerful to absorb the lands of small holders. By managing effectively for the productive system, the empire quickly revived. In centuries to come it became prosperous and amassed great wealth, in which the army shared. Soon there were advances on several fronts against formidable foes. As Islamic power waned the Byzantines retook lands in Syria and Armenia that they had lost centuries before.

Sustainability in the case of warring Europe was a richly complex matter. Here was a feudal and emerging capitalistic society with the potential of the Byzantine recovery for durability and of the Abbasids for disaster. War is ordinarily such a net consumer of wealth that it is a wonder modern Europe ever came to be. War consumes wealth not only, as is commonly thought, through physical destruction but also more insidiously through the costs of preparing for and conducting it. Nations can destroy themselves through competition even if they win. Nothing in all of history is equal to the competitive spiral experienced in Europe during the last millennium. Here was a system in which complexity and costs were driven ever upward, to the highest levels ever seen. If ever there was a system vulnerable to collapse, the political environment of Europe seems to be it. Of course it has yet to collapse, and paradoxically Europe and its former colonies, and some other nations that have adopted its technology, have produced the most widespread prosperity ever known. It has developed from a system whose sustainability seems miraculous, to one with a standard of living that is most desirable to sustain. How could today's well-being emerge from so many centuries of misery?

The answer is that the competitive need constantly to innovate and refine brought Europe to levels of technological prowess and organizational capability never before seen. Initially this gave Europeans a competitive edge over all other societies. Although they could not permanently defeat each other, they overwhelmed the rest of the world, a part at a time, rather easily. The wealth of the world—or as much of it as could be seized and transported—went to subsidize the nations of Europe and their wars. In the nineteenth and twentieth centuries these wars came to be subsidized further by the use of fossil fuels. Thus from the fifteenth century to the present day, Europe found energy subsidies to support levels of complexity and costliness that would have been unattainable based on the solar energy falling on this small appendage of Asia.

The history of Europe conveys a number of sustainability lessons. In competitive systems, being sustainable for the long term is always secondary to surviving today. Competition drives consumption of resources—human and other—regardless of the consequences for the future. In a competitive system or in any other, the key to sustaining growing levels of complexity and costliness is to find energy subsidies. This is true regardless of what drives increasing complexity.

European history is enigmatic. Europe developed political and economic systems that are the envy of the world but whose sustainability has been questioned. At the same time it gave us systems of problem solving that allow us to become the first people in history to comprehend what makes a society sustainable. Our urgent need is to produce that knowledge more rapidly than the growth of unsustainable complexity.

SOME CHARACTERISTICS OF SUSTAINABILITY

As we emphasized in chapters 2 and 3, societies wishing to be sustainable or to sustain their environments will monitor historic and current conditions, predict trends, and take steps to see that the future develops as it should. The Abbasids and the Third Dynasty of Ur both failed to monitor. Europeans did not predict the consequences of their competitive spiral, but they got lucky nonetheless. Having reached the level of complexity sustainable by their own resources, over the horizon they found another world to use, then new forms of energy. Had neither been there, Europe today would be a very different place. The Romans took steps to ensure the future of their empire, but these steps involved levels of complexity and costliness that made sustainability impossible. To be sustainable clearly is a delicate business. Furthermore, although none of the elements of sustainability (monitoring, predicting, and problem solving) can be left out, even undertaking them all does not guarantee success.

We do not yet fully know what it takes to be sustainable, but the discussions in this chapter and the last suggest that problem-solving abilities have much to do with it. Problem solving, like any activity that experiences increasing complexity, is subject to the law of diminishing returns. When increasing expenditures on problem solving fail to yield a commensurate return, a society's ability to address challenges has become impaired. Taken far enough, diminishing returns to problem

solving lead to unsustainability. Thus, sustainable institutions of problem solving are prerequisite to sustainable environmental management or a sustainable society. To borrow a colloquial expression, lacking sustainable problem-solving abilities may be the mother of all
problems.

In the case of environmental matters, sustainability requires that
institutions of research and management avoid becoming so complex
that research can yield only small increments to knowledge at great
cost or that management has become so hindered by procedural matters that little can be accomplished. This is a matter of balance.
Processes for public contribution to decision making exemplify the
dilemma. Given that public opinion is diverse and often self-serving
and that there are many opportunities to suggest, appeal, and litigate
decisions, a land management agency today achieves much less environmental manipulation per unit expenditure than was the case a generation ago. Public participation generates its own competitive spiral.
The more the public disputes decisions, the more agencies must invest
in justifying decisions, in training their employees in public relations,
and in litigation. As agencies get better in the public arena and in court,
those challenging decisions must themselves invest more resources in
their cause and devote more effort to securing those resources. It
becomes more complex and costly both to promulgate decisions and
to dispute them. Yet in a democracy, not to allow public participation
is unthinkable.

Complexity in decision making is part of the cost of democratic politics. Some people argue that the decisions made by this tortuous
process are better than they would be otherwise. This may often be so,
but to satisfy ourselves with better decisions is to miss an important
point. Are today's decisions *better enough* to warrant the extra complexity and cost? We do not know, but if they are not then environmental management is producing diminishing returns and is a less efficient problem-solving institution. Compounding the dilemma, there is
no right or wrong answer to issues such as public participation. There
are only subtle weighings of costs and benefits to taxpayers, to interest
groups, to the environment, and to the future. The only surety is that
too much complexity will guarantee that the problem-solving institution is not sustainable.

The solutions to complexity, as we have seen, have been to collapse (as did the Romans and the Abbasids), systematically simplify (the Byzantines), or find subsidies of technology and energy and develop the organization to use them (Europe). In environmental management there is a subsidy readily at hand: the ecological system itself and the solar energy that drives it. In a previous book two of us (Allen and Hoekstra 1992) suggested that in managing ecological systems, the best approach is to identify what is missing from natural regulatory processes and provide only that. Let the ecological system do the rest. Let the ecological system (that is, solar energy) subsidize the management effort rather than conversely. In conventional land management one makes adjustments for the sake of outputs. If the outputs are unsatisfactory, further adjustments are made to return them to a satisfactory level. Management becomes an endless cycle of making adjustments and harvesting outputs. Then something unexpected goes awry (such as a species in decline) and the management effort grows correspondingly (or even disproportionately) in complexity. The outputs may still come, but they do so at a higher cost. Managing for outputs, as the Romans and the Abbasids found, leads to higher complexity and declining marginal returns.

The approach we suggest is to let natural processes subsidize the management effort as much as possible and manage for the system that produces outputs rather than the outputs themselves. In a framework in which one manages for the productive system, outputs flow by themselves and are an automatic byproduct of the management effort. This is what we mean by the term *supply-side sustainability*. It is the secret discovered by the Byzantines when they shifted most of their government from being a consumer to a producer of the empire's wealth. Once Byzantine soldiers became farmers it was no longer as necessary to squeeze peasants for more outputs. No doubt taxes were still resented and difficult to collect, as they always are. The difference is that outputs were supplied and consumed by peasant cultivators, which are among the most durable of human institutions. They needed only good soil, moisture, and solar energy. The Byzantines provided what the system lacked (motivated cultivators), removed barriers to efficiency (a costly fiscal system and army), and let solar energy directly subsidize the empire. Sunlight energized all ancient states, but often with many intermediate transformations and transaction costs. A typical transformation sequence might run from solar energy to peasants, crops, coins, tax collectors, the government, and the army. Each transformation entailed net losses, just as in a trophic pyramid. The Byzantines shortened this sequence. In the system of themes, solar energy supported the peasant army directly without the losses and overhead that intermediate transformations entail. It was a supply-side enterprise.

There are many obstacles to a supply-side approach to environmental management. Not the least of these is the large initial investment in knowledge of the systems to be managed. Yet managing for the system rather than its outputs seems to be the only acceptable way to reduce the costs of complexity in environmental problem solving. For too long we have asked how to manage an environmental system to increase its outputs. The increasing complexity of this approach signals that we are asking the wrong thing. Sustainable problem solving in environmental management now requires that we learn how to ask new questions. These will be questions about the structure and function of the ecological system itself rather than about the myriad, ever-changing outputs we request of it. Throughout this chapter we have described the responses of various societies to crises. In supply-side sustainability one manages for the productive system in such a way that perturbations become crises less often. Crisis management is a worthy field of endeavor, but we would do well to make it obsolete.

II

A HIERARCHICAL APPROACH TO

ECOLOGICAL SUSTAINABILITY

4

The Criteria for Observation
and Modeling

I t would be convenient if the material world told us the criteria for
what to sustain and how to do it, but that cannot happen because
definitions and policies must of necessity come from the observer.
Even in hindsight, the material system does not tell us whether we
made the right decisions; it only gives us results, which are adequate or
not, again based on our decisions as managers or scientists. Given the
central and inescapable intrusion of human decisions, we emphasize
flexibility while being explicit. Being explicit provides a precision of
definition and purpose that allows flexibility because choices can be
made more deliberately. This approach is in contrast to the realist
agenda that uses a more rigid framework. Realism uses the whole
material system in its infinity of detail as the reality that policy and
management must address. Reality is thus asserted as the benchmark,
albeit left undefined. With undefined terms of reference, the realist is
less than fully explicit. Furthermore, because there is only one ultimate
reality for most realists, the referent is held inflexibly, leading to the
worst of all worlds, where the scientist is vague but stubborn about it.
By contrast, our approach is explicit and flexible, with clear advantages.

In our explicit approach, there are two separate components to
observing a system: the criteria that are used to adjudge significance of

degrees and types of discreteness, and the bounds on the scale of the observation. Criteria used to adjudge significance define what is in the foreground and therefore what is tacitly in the background. For example, with focus on the landscape criterion, populations become a lesser consideration of what sits on the landscape. Landscapes are ordered on spatial considerations, whereas populations assemble homogeneous groups of individuals, sometimes spatially but often not. Scale is a separate matter, the use of which may invite certain criteria to be used, but scale and criteria are separate components of observation. On the upscale side, the scale is bounded by the scope (extent). In the downscale direction, the scale is bounded by the fineness of distinctions (grain) (Allen, O'Neill, and Hoekstra 1984). The grain and extent of observation set the scale but do not determine what is recognized as being significant and therefore in the foreground.

A change in scale can occur because the observer changes the observation protocol, perhaps using binoculars to make the critical object appear bigger and so reveal more of its details. Alternatively, a change in scale may occur because the structure defined by the observational criteria changes size, perhaps in a process of maturation. For example, Allen's dog Raven arrived as a cuddly little puppy and is now an impressive adult rottweiler of more than 100 pounds. A change of the scale of an observational protocol has an effect only on the observation, and we presume there is no direct effect on the existence of the material system beyond observation. Meanwhile a change of size of the observed material entity may affect directly what appears in the observation, although not the scale-based decisions as to the observation protocol.

A change in the size of the material system may demand new observation protocols, but that is indirect and goes through the decisions of the observer that the old protocol has failed. A large organism is too large to be seen whole under a microscope, demanding a protocol for observation that does not use the techniques and hardware of microscopy.

Changes in the size and scaling of the entity under observation can have subtle or profound changes for science and management. They generally alter what is observed, but if such changes pass unnoticed by the observer, they may be responsible for failed experiments or predictions. There are underlying scalar properties in the deep subsystems, such as the constancy of the aggregation properties of water molecules, despite the water being a component of a larger organism. The size of an organism does not change the surface tension of water, although the significance of surface tension can change from crucial to trivial as organisms become larger. Despite a change in physical size of the entity under observation, the deep subsystem relationships between water molecules remain constant. The change in size of the whole changes the relationship between the whole and constant subsystem character- istics, and this in turn changes what happens and so what is observed. Surface tension of water can be a significant force for insects but not for a wading elephant. Very large animals are significantly constrained by gravity, such that elephants do not jump but fleas jump hundreds of their body lengths. A flea can fall through air more or less indefinitely with no harm, but a fall of more than inches is fatal for an elephant. It is not the falling that hurts but the stopping; more accurately, it is not the stopping that hurts, it is that all of you does not stop at once. The flea stops more at once than does an elephant, whose body keeps going down after its feet and then its belly have hit the ground. A standard question on the veterinary board examination is the six injuries sus- tained by cats falling from great heights. For a given size of animal, gravity has uniform consequences. The elastic properties of the mate- rial from which all animals are made are of the same order, and so larger animals under strong deceleration are dismembered more easily. Thus a change in size changes what happens when the observation is made.

If the observer keeps the scale observation protocol but changes the criteria for the type of system under observation, what emerges under observation can be radically different, even though the material system is exactly the same. Under some criteria for observation, the material observed in the foreground might remain stable. Even so, the same material happenings can mean that the system has become unstable, but under new observational criteria. For instance, a collection of species at a site may remain constant, although the system is neverthe- less losing critical amounts of mineral nutrients. The difference is not in the material situation but in the community versus the ecosystem criterion that is used for observation and assessment of stability.

Criteria emerge from decisions of the observer, and in that sense they are arbitrary but not capricious. For example, the decision to model an organism as opposed to something else is arbitrary. More than that, the definition of what constitutes an organism is all of the observer's choosing. Lewis Carroll's Humpty Dumpty said that his words mean what he says they mean, and the authors here agree with his being so explicit (fig. 4.1). However, once a definition is chosen, observations in the light of the formal model cannot capriciously change that definition and still be meaningful with regard to the model.

In the case of observing an organism, the model determines how discrete, how physiologically coherent, and how genetically identifiable a mass of biological material has to be to meet the criterion of being an organism. Changing the definition of what one means by an organism might change a single hydra with a bud on the side into two hydras. Recognizing more than one organism in the budding hydra might change what emergent properties are recognizable, but it does not affect the material hydra itself.

FIGURE 4.1

Taking responsibility for definitions. Humpty Dumpty not only had his words mean what he said they meant, he even claimed to pay more for words that had more complicated meanings. We approve of his taking responsibility for his definitions.

Photograph by T. Allen.

In the preface we indicated that this volume follows from *Toward a Unified Ecology*. There, Allen and Hoekstra attempted to unify ecology by looking at the consequences of scale applied to different ecological criteria. We made an effort to reject actively the conventional hierarchy of biology from cell to biosphere because only a very small part of the material flows in ecological systems follow the particular path of aggregation conventionally prescribed. For example, aggregation of organisms into populations captures some material flows, such as genetic material in a breeding system, but it misses much more. One would explain a different class of material flows if one scaled up from an organism to include its cohabitants of different species. Aggregation of heterogeneous collections of organisms is the model for community. Alternatively, integrate the organism into its physical setting and one studies nutrient cycles in a process-functional ecosystem. If that physical setting is a larger structure that is itself an organism, then the model could be epiphytism or parasitism, and so on. The separation of type of system from the effects of scale in *Toward a Unified Ecology* will be used again here.

In this chapter we use the same set of ecological criteria as in our last book, and we explore the scaling ramifications within each one. Within each criterion we contrast the effect of recognizing that criterion as opposed to another, with regard to sustainability. At each point we look at sustainability in three ways: how to continue to sustain, how we might fail to sustain, and how to restore sustainability once it is lost, all according to the criterion in that section of this chapter. The criteria, in order, are organism, landscape, population, and community. The criteria of biomes and the biosphere, and of the ecosystem, are described in chapters 5 and 6, respectively.

THE ORGANISM

In Allen and Hoekstra (1992), we identified three formal generalized attributes that make something an organism. First, organisms are usually structurally discrete; usually one can literally see them to be separate from one another. Second, organisms have a physiological coherence; there is usually a closure of processes inside the organism. Last,

171

an organism usually has a genetic identity. All of these characteristics are very general, but they are most clearly manifested in humans.

These three general attributes applied outside the mammals meet with various degrees of ambiguity and sometimes downright failure. For example, sea anemone relatives form colonies that are organically continuous, as one would hope for an organism, but consist of very autonomous individual polyps that appear as organisms themselves. Which scalar level is the organism? Plants are even more ambiguous as to what constitutes an organism. Sponges can be fed through a sieve and the parts spontaneously reconstitute the organism. Portuguese men-of-war are composed of cells from different genetic lines. For many units, the organism concept captures only part of the arrangement of biological material.

Often one or more of the general attributes for being an organism fail. In plants, genetic identity may not line up with physiological integrity or bounded form. Sometimes there is a special vocabulary to sort out the inconsistencies. The entity coming from a seed is called the genet to mark its genetic integrity, whereas the different physiologically autonomous and physically separate shoots joined at the root are called ramets. Some ramets are sufficiently independent to be called organisms in their own right (fig. 4.2). In some biological systems the physiological and bounded structure general attributes are so well met that complete lack of genetic integrity is ignored. Lichens are named to species based on discrete form and physiological coherence, but they lack genetic identity, being a symbiosis of at least two species from

FIGURE 4.2 (*opposite page*)

Ramets and genets. (A) Grasses in particular form physiologically separate branches all coming from a common root system. Although the whole comes from one seed and is therefore a genet, the separate branches are ramets, expressing the individuality of a physiologically separate entity, albeit part of a larger genetically defined entity. (B) Poplars are a clear example of organisms in their own right, though simultaneously arising from a common root system, as does this new clonal member derived from the root system of its larger neighbor.

Photographs by T. Allen.

A

B

different kingdoms (fig. 4.3). Thus various entities that are organismal may fail on one or more of the three general attributes listed earlier.

So if the organism as a concept is at best a fuzzy set, why is the organism so important and undeniably useful in explaining biological phenomena? Part of the answer is that, being such an anthropomorphic concept, the organism is an intellectually comfortable notion, even if it means ignoring the exceptions and half truths. The organism criterion invokes spectacularly tangible entities. This is very helpful when the investigation of sustainability is preliminary, and there is much uncertainty.

Sustaining the Umwelt

At some level, the sustainability of almost any ecological system depends on the healthy functioning of organisms within, so it is impor-

FIGURE 4.3

Lichens on a tombstone. Lichens are an amalgam of fungi and algae. They take on discrete bounded form, with the fungus providing a context in which the algae grow.

Photograph by J. Will-Wolf and S. Will-Wolf.

tant to understand how organisms read their ecological surroundings. In the heyday of behavioral biology in the 1930s, Jacob von Uexkull wrote a seminal piece (English translation 1957) that generalized the study of animal perception and animal responses to what they experience. He unified his ideas under the German name *Umwelt*, which translates awkwardly into English as "self-world." Each animal species lives in an experiential world, its Umwelt, defined by the type of energy or material that it uses to detect changes in its surroundings. The organism reacts to events in its self-world that may or may not be shared with other species, such as ourselves. In matters of sustainability, the world that the organism perceives is largely the reality that counts. It does not matter that there exist resources for sustenance or a context for successful breeding if the organism cannot recognize them. Even within our own species, an environment viewed as abundant by, for example, the Kalahari Bushmen might be seen as barren by strangers. If we humans cannot agree on what appears to be an environment conducive to human sustainability at a basic level, then it appears a nontrivial feat to identify sustainable environments for other species.

Part of the definition of a particular Umwelt comes from the rude mechanics of perception with different senses and the different material signal carriers that are processed. A simple contrast with ourselves is that bees cannot see red, but they experience a bouquet as a riot of ultraviolet color. As a result, flowers that may appear a homogeneous yellow to us have ultraviolet targets in their centers with ultraviolet lines on the petals that serve as bee landing lights, all unseen by the human eye. Accordingly, the world of a bee does not involve red fruits, for their role in pollination is completed by the time we primates serve as seed dispersers. Insects are not big enough to move large succulent fruits, so the less insects see fruits, the less mischief they can do to the process of seed dispersal.

Differences in the type of energy or matter used in perception are so basic that differences in self-worlds can be surprising. For instance, color-blind humans may not be aware of their deficiency. Coren and Ward (1989) reported a man who commented on the remarkable skill of cherry pickers, able to pick the fruit by shape alone; he did not realize the he himself was red–green color blind and that for the cherry pickers bright red stands out against green foliage. Color-blindness is a deficiency in the mechanical terms, an inability to distinguish between two wave bands. In a more positive sense, that lack of distinction can lead to a self-world where other distinctions can be made. For example, in World War II, color-blind young men were sought after and put at the front line as spotters. They had the ability to see past the deception of enemy camouflage. Mechanical differences in the perception of types of energy lead to very different experiences. The self-world of a wine connoisseur involves distinguishing between subtleties of flavor that are lost on most of us. Whereas the untrained palate focuses on the four tastes in the mouth of salt, sweet, sharp, and bitter, the connoisseur focuses on a different chemistry, the volatile organic flavors in the nose. If color-blind humans can be surprised by descriptions of the self-world of people with normal vision, and if occasional drinkers need convincing that wine connoisseurs are not making it all up, then it is clear that it is easy to overlook the differences between the human self-world and those of other animals, let alone plants (fig. 4.4).

Habits and Familiar Settings

Von Uexkull developed general principles for comparing the different ways in which animals, including humans, experience their surroundings. Beyond differences in mechanical senses, like bats and their sonic world, he noted that habits also define the individual's self-world. New experience is in the foreground, and what is familiar recedes into the background of a routine setting. Allen's rottweiler, Digby, barked defensively one day at a pizza box left some 30 yards away on a porch across the street from the window the dog regularly guarded. Only because of the dog did we notice the box at all. The differences between breeds of dogs is importantly a matter of alternative Umwelts, at least as much as within the human species. Rottweilers are visual trackers, whereas most dogs are scent trackers.

Again illustrating von Uexkull's point of Umwelts turning on familiarity, Allen's horse, Little Big Guy, once shied at a branch fallen down in the night next to a familiar trail. The rider gave the fallen limb no thought until it became important as the explanation for the horse's behavior. The horse, noticing the difference, seemed to be using a prudent strategy for an ungulate because the change might well indicate

Amorphophallus titanum
THE TITAN ARUM
BOTANY GARDENS AND GREENHOUSES · UNIVERSITY OF WISCONSIN · JUNE 2001

FIGURE 4.4

Reading Umwelts. Some plants depend on attracting pollinators with odors that we humans find repulsive. The titan arum (*Amorphophallus titanum*) is the largest inflorescence in the world and could well have the most vile odor. When this plant flowered in June 2001 in the University of Wisconsin's botany greenhouse, the stench on the night it opened was overwhelming, giving headaches to human observers. It attracts beetles at night for pollination and is called the "corpse flower," so strong is the smell. There are Umwelts that humans are not equipped to read properly.

Figure by Kandis Elliot.

an ambush. Looking at the same issue from a human predator's end, the effort is to leave no clues in a change of routine. Our colleague Curtis Flather is a consummate fisherman. As with all serious hunters, his strategy is to enter the self-world of the animal he seeks. Curt ice fishes by looking down the hole through the ice and has learned exactly how to move the bait so that it attracts attention without raising suspicion.

Particularly important in the present setting of sustainability is another of von Uexkull's principles, what he calls the animal's magical path. In that fanciful phrase, he means the critical paths through space and time that are the essence of the biology of the organism in question. He uses as a distinctive example an insect that infests human grain stores. Its magical journey is very short, from entering the grain in a juvenile form to emerging as an adult just millimeters away on the other side of the seed. The seed is very much the whole world of the developing larva. The world of migratory birds is much more like the size of the actual entire world. The vast size of their Umwelts has many implications for sustainability. For temperate birds that winter in the tropics, their Umwelt stretches half way round the earth. Sustainability for such species cannot be achieved by paying attention only to their summer homes. Thus sustaining a species entails paying attention to all major parts of its self-world.

Rare and Endangered Umwelts

The concept of Umwelt is crucial for many aspects of sustainability. Conscious efforts to sustain an organism must be focused on the Umwelt as much as on the material organism itself. Once the Umwelt is lost, conservation applied to the material organism probably is a last ditch effort, a gesture more than a real contribution. Once it comes to cleaning an oil-fouled seabird, then all other facets of sustainability are on the rocks. Rescue efforts of that sort are good publicity and contribute much to raising the consciousness of the general public, who do after all foot the bill for a significant part of efforts toward sustainability. However, in meaningful ecological terms, each recovery of an oil-fouled bird is a Pyrrhic victory. If the insult physically reaches the organism itself, then the Umwelt of the organism probably has been in trouble for some time.

Rescue attempts for species fighting their last battles for survival turn exactly on the issue of Umwelt. Keeping an animal alive in captivity is something that modern veterinary science can handle quite well, so the difficulty does not lie there. The problem is developing an effective breeding program (fig. 4.5). In captivity, the animals' Umwelts often are at odds with the conditions in which they are kept. The response is a refusal to breed, even at the mechanical level of showing no interest in sexual activity. The effort that must be put into coaxing many animal species to breed in captivity is itself a measure of how difficult it is to see what is crucial and what is incidental in an animal's environment. If we could read and implement the animal's Umwelt, then conservation and restoration would be easy.

Given the difficulty scientists experience in reading the world in the manner of other species, it is imperative that conservation efforts take action in the context of how endangered species themselves read and react to their world. At first glance, the best chance of a captive breeding program is to get all the remaining individuals of a disappearing species into captivity to form a breeding colony. On the face of it, leaving one or two pairs out in the wild cuts down the chances of success by the proportion of the potential breeding stock that they represent. In the case of the Mauritius kestrel, some of the dozen remaining birds were left in the wild (Temple 1978). The reason for leaving some birds in the wild is to instruct captive-bred birds in the Umwelt when they are released. Preserving an appropriate Umwelt is as important as preserving the genetic stock because the two exist together or not at all. If there is suitable habitat, perhaps on other islands, then capture and translocation can be effective. Umwelts work on particular types of place, but they can travel well between places of the same type. The rescue of rare remnants of species often is best achieved by a combination of actions, but almost all must be dictated in their particulars by the species' Umwelt.

For animals to be close to extinction, there is at least one and often more than one critical mismatch between the species and modern world in which it finds itself. At some level, the mismatch is between Umwelt and the changed environment. In the absence of sufficient primitive habitat, sometimes the Umwelt of the species is exactly the problem. Breeding captive birds but equipping them with the same old Umwelt only puts animals back in the same predicament that got the

species in trouble in the first place. In the case of the California condor, conservationists went so far as to use a puppet for a mother to rear chicks in captivity. Even so, there were significant behavioral problems with birds released from the captive breeding program as they landed on tin roofs and perched on high-voltage lines that electrocuted them. The habitat across the range was dangerous. Part of the problem was lead poisoning from food killed with lead shot. But in general the birds made poor decisions about avoiding human populations and artifacts (Snyder and Snyder 2000). Snyder et al. (1994) suggested that behavioral problems on the part of released birds is a general problem for rescue of endangered bird species.

The loss of an Umwelt can be almost as important as the loss of the species because it is a harbinger of actual extinction. Criticisms of research at the Ghombe chimpanzee reserve turn on Umwelt. Although the animals themselves survive, the chimpanzee Umwelt is being undermined. The presence of so many scientists studying the troops, despite efforts not to intrude, changes the world of the last savanna-dwelling members of that species. This breaks the last link in

FIGURE 4.5 (*opposite page*)

Crane restoration. (A) The International World Crane Foundation has a remarkable collection of rare cranes; it includes all the species in the world. These whooping cranes are as rare as any, displayed here to appear in a natural setting for visitors. (B) Many cranes are bred in the foundation's crane village, from which the public is excluded. Their breeding program follows a particular plan, wherein only some species are bred in a given year. The species that breed change over time as the plan prescribes. This might even involve sacrificing the eggs of rare species to have the adults raise the eggs of another equally 'or more rare species, the species that is part of the breeding plan for that year. The plan involves many aspects of maintaining the population, such that all species reproducing all the time has been identified as a suboptimal use of resources. Saving rare species involves counterintuitive decisions, for it is a complex management system wherein full-time maximization of reproduction is not the best plan.

Photographs by the International World Crane Foundation, reproduced for publication with permission.

the chain to an ecology that is homologous to the origins of our species. There are no other animals that are as closely related to us and live in the habitat that is thought to be the cradle of hominids.

When there is success in turning around a foundering breeding program of animals in captivity, there are very tangible signals to the scientist that important parts of the model are appropriate. Initial success in reading the Umwelt of an endangered species in the field is hard to detect, certainly not as easy as counting new babies in a zoo. The problem of developing understanding of the Umwelt of an endangered species in the wild probably is so complicated that it cannot be solved by having behavioral biologists take a reductionist approach to one facet at a time. And yet the problem of providing a physical setting that satisfies the Umwelt of an endangered species must be met; otherwise the species can decline for no reason that the conservationist understands.

The particulars of one species' decline are not often generalizable (Simberloff 1988). The reason is that such transfer invokes what has been called elsewhere (Weinberg 1975) a "middle number system." In so-called small number situations, such as planetary systems, one writes reliable equations for each part of the system, and a predictive solar system model emerges. In large number systems, such as those that invoke the gas laws, there are so many parts that the differences between them can be subsumed under some average or representative part. In physics, there must be more than Avogadro's number of parts, but once that condition is met the scientist can erect a mythical perfect gas particle that has no mass, friction, or volume. Although nothing like a perfect gas particle ever existed, an aggregate of perfect gas particles gives very good predictions of the emergent behavior of gases. Like small number systems, large number system models in general are predictive. In the third type of model, middle number systems, prediction is impossible. With too many parts to model each individually and too few to be able to rely on some representative number such as an average, it appears that middle number systems can have any one of their parts take over emergent system behavior. It is not just difficult to predict the behavior of a middle number system; it is impossible because the course of the outcome is importantly different every time. The particular path of decline of a species is middle number.

Any one of a very large number of small events can amplify and lead to extinction. After the fact one might be able to tell a convincing narrative of the decline, but that is of no value in preventing extinction before it happens in some new case. As the population becomes small, the influence of happenstance events can make all the difference. In the case of the heath hen (*Tympanuchus cupido cupido*), numbers were small but with habitat improvement had recovered to about 2000 (Bent 1932). Then a coincidence of fire killing birds and destroying habitat, a following harsh winter, predation pressure from goshawks, poultry disease, sex ratio imbalance, and inbreeding depression eradicated the species in 17 years despite human efforts to save it. The problem is that it is usually not a single missing factor in the Umwelt that is in principle the limiting factor. Any one of a large number of local factors are candidates for the cause of extinction, but if one could run the extinction of the heath hen through again, some other local factor would play the crucial role. This disconcerting state of our understanding does not mean that there is nothing to be done to avoid extinctions. However, remedial action must be found through a higher level of analysis than isolating critical factors. The key is the general condition that lowers thresholds, not the local factors that appear important only in hindsight. The general condition emerges with degraded habitats that put stress on the organism. The stress comes from the organism placed in a physical setting that is not consonant with its Umwelt so that the last of the species end up in the wrong place at the wrong time.

Stress and Unmet Umwelts

An Umwelt includes the ability to detect emergencies. This is a crucial part of an organism's existence, the recognition of the time for flight or fight. In plants it might be detection of insufficient water to continue the normal physiological status. At that point stomates on the leaves shut down, and water is conserved. The cost is that carbon dioxide is kept out in exactly the manner that water is kept in. Furthermore, temperature of the leaves may increase, with its own stress factors on enzymatic activity. However, that is the cost of doing business, and the plant will reopen its stomates when it has the water resources to sustain the transpiration stream. The price of closing stomates is a small loss of primary production, which is much better than structural damage because of wilting. A wilted plant is in great danger of permanently damaging the xylem, the plumbing system that serves the leaf.

It helps to distinguish disturbance from stress, noting the chronic nature of the latter and the critical cost that long-term accommodation entails. Wilting plants use temporary adjustments to deal with what amount to disturbance situations, but sometimes constant challenges lead to the chronic conditions. The cost of a permanent state of mobilization is telling. In primates, ascendancy to the social position of alpha appears to cause muscle gain and high testosterone levels. Conversely, for individuals low down in the social system, constant anxiety that comes with being assaulted regularly leads to chronically elevated levels of glucocorticoids, the hormones for flight and fight. This physiological imbalance causes hair loss, weight loss, and impaired immune function. Recent research in experimentally stressed monkey mothers indicates that in utero insults to babies have greater effect if the mother is emotionally stressed (Schneider and Moore 2000; Schneider et al. 1999; Anderson et al. 1986).

An inadequate physical setting for a fully functioning Umwelt will lead to stress. This in turn may lower resistance to disease and slow growth in conditions that would otherwise favor it. Thus the demise of a population could equally well be an endemic disease becoming epidemic. Perhaps the collapse occurs as a result of a harsh winter that normally would have been merely difficult but is disastrous when met after an inability to accumulate fat reserves properly in the previous summer. Whether it will be disease, a marginally difficult winter, or some other factor that pushes the species over the edge is not predictable. The reliable general cause is a dysfunctional Umwelt.

On the other hand, an environment that might be judged as stressful by human standards may well not be so for other species. A case in point is the burrowing owl, one of whose characteristic habitats is airports (Heintzelman 1984). Humans might find their Umwelt challenged living next to a runway with jumbo jets or military aircraft taking off day and night. However, from Florida to Utah, in commercial airports and Air Force bases, burrowing owls appear to find the new open vegetation that keeps out the interference of other animals very conducive to population growth. All this indicates that correctly reading the Umwelt of rare species would be a very helpful conservation tool. However, humans are wont to overestimate the insult to an organism's functional environment in some cases while equally underestimating the importance of environmental factors that are incompatible with the critical species' Umwelt.

Even as our species stresses the material world of other animals, we simultaneously insult ourselves. In 1978, Jack Vallentyne made a proposal that was quite radical for its day. In his account of the ecosystem approach (Vallentyne 1983), he noted that there is no longer any "somewhere else" to which to send modern societal waste materials. Thus poisoning of raptors with persistent toxic chemicals such as organochlorine compounds does not stop with thin eggshells for those species alone. It is only because the food chain of industrial humans is engineered to be short that our species does not suffer more from environmental insults. Indeed, at the periphery of modern industrial society, Native Americans have been forced to abandon traditional patterns of resource capture, exactly because traditional lifestyles put people at the top of long food chains. Bioaccumulation means that large fish in the Great Lakes Basin are subject to advisory notices.

Although chemical insults may not appear as an insult to the animal's Umwelt, they have the same effect as an attack on the Umwelt. Mother rats stressed by bright lights and restraints have lower testosterone levels that in turn lead to feminization of male offspring (Anderson et al. 1986). Feminization is one of the patterns that is emerging in wild animal populations that are insulted with persistent toxic chemicals (Minister of Supply and Canada Services 1991; Fox and Weseloh 1987). Thus the stress of chemical insult in the environment has the same effects as stress that comes from pressing organisms into a physical setting that is not compatible with the animal's normal, healthy Umwelt. Sperm counts in industrial nations are significantly down (Virtual Elimination Taskforce 1993; James 1980). Reproductive clinics in Madison, Wisconsin, have increasing difficulty in finding sperm donors that meet their standards for sperm concentrations.

The Human Umwelt and Sustainability of Other Species

Although species extinction rates are high almost across the board, some important classes of species appear to be in more trouble than

185

others. Most particularly, the large carnivores and the primates are very worrisome. Other groups, such as some insect classes, may be suffering more arithmetic species loss, but they have more species to lose and probably can bounce back fast enough, with evolution of new species within the group, that the entire order is not in danger. Not so our closest ecological equivalents or our closest biological relatives.

Cats, dogs, and primates have the misfortune to share many features of their Umwelts with our species. The cats and dogs overlap because of a general ecological equivalence. They suffer from being close ecological analogs of humans. By contrast, the primates suffer from being homologs. Many facets of their Umwelts that correspond to our own come from their being made the same as ourselves. Primate hands and human hands are homologous, not analogous. Carnivore stereoscopic vision is analogous to ours, coming from pressures of convergence and the need to tell distance in both lifestyles. The human hand has an analog in the gripping jaw and hasping teeth of dogs and cats. Both primates and big carnivores encounter our species head-on as we blunder into their worlds.

When humans were merely a minor tropical species, this sharing of resources with other macrofauna was of no consequence. However, 6 billion people is a very different matter. In first-order interactions, where there is direct competition for prey species, the cats and the dogs lose out to human capture of the resource. Many wild species of grazers are extinct as a result of human predation pressure. The extinction of whole macrofaunas on islands appears to involve humans, although not simply so. The carnivores that do survive must rely on prey that exist in smaller numbers of individuals than heretofore. Although bison survive in North America, they are not a functional food source for carnivores. Even for the people who live in their former range, bison is an exotic meat. Humans breed and feed cows; this is such a large operation that whole tracts of the earth have been changed to facilitate livestock rearing. Cows graze the landscape in a very different way from game species, leading to rangelands with greatly altered characteristics. Also supporting cattle are croplands that are made explicitly unavailable to all grazers except cattle by enclosure.

The problem for primates is that they are the closest homologs for humans. Accordingly, they make exquisite medical models and are being decimated in the cause of medical science and psychology. Primate juveniles also make appealing child surrogates, and there is a brutal market in primate infants of all sorts. Mothers are killed for their babies, and then only a very few infants survive to make it to market. Although homologs of ourselves, primate Umwelts fit in a human context only with difficulty. And yet there are poignant moments of exchange between them and us, as when Jane Goodall as a near bystander was reassured with a hand gently laid on her by a chimpanzee at the end of an impressive display of dominance directed at another chimpanzee (Goodall 1988).

The differences between human and nonhuman primate Umwelts are seen in their different capacity for language. In early research, language was mistaken as being necessarily spoken. In this primitive phase, not only were efforts made to get chimpanzees to speak, but in laughable work on large-brained sea mammals, a dolphin was trained to vocalize a few English words through its blowhole. A dolphin can vocalize in human phonemes through its blowhole about as well as humans can converse by blowing their noses. However, even when the mechanical problems are treated adequately, such as with the use of sign language or a computer interface, nonhuman use of language lacks richness and syntax.

Comparative studies of infant cognitive development in primates indicate that nonhuman primates first develop a solid understanding of material interrelationships. They progress fairly quickly to second-order interactions of a bat-and-ball type in which objects interact. However, nonhuman primate logical development in activities such as addition and subtraction occurs later and is always dominated by physical activity and material understanding. Early dexterity in nonhuman primates appears to interfere with the later protomathematical development. By contrast, human infants are slower to develop mechanical understanding, and they develop it in parallel with logical development. Mechanical manipulation and logical function in humans continue to stimulate each other in an open-ended development of technical dexterity with logic (Ahl 1993) (fig. 4.6). The process goes on through life and is essentially how empirical testing in science challenges old models and generates new models for further testing.

As to the preferences that the human Umwelt engenders, the model of savanna vegetation is helpful. Humans appear to enjoy sitting and

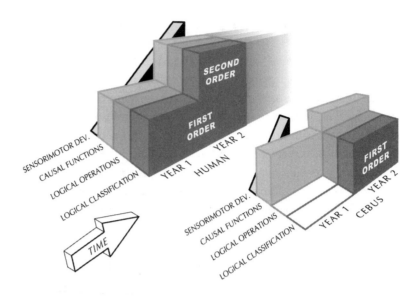

FIGURE 4.6

Comparing humans with nonhuman primates. Cebus monkeys show early development of physical mastery, but their ability to form sets remains very limited. Monkeys are tyrannized by the mechanical world. Meanwhile, human infants do not develop second-order mechanical skills, such as hitting one object with another, until they can also form sets and perform second-order protomathematics, such as exchanging items between groups. As a result, humans exist with a balance between conceptual and mechanical abilities, so that we challenge the material world with experiment.

Based on data from Ahl (1993).

watching from a position of safety, with a setting that covers the back and offers some shade. Law enforcement authorities in Canada commissioned a formal study from psychologists to guide them in finding lost children. Apparently one finds missing children regularly with something like a large boulder at their backs in a position that gives them a lookout. At the edge of the forest, one's back is covered and there is a view out into the open. In a statistical study of human preferences for tree shape, Orians (1998) found that his sample of people preferred the classic savanna tree shape, with branches close to the ground. Perhaps the tree shape is a model for quick escape from predators (fig. 4.7). With Heerwagen, Orians also found statistically significant differences in idealized landscapes in paintings depending on time

FIGURE 4.7

Savanna tree and a feline predator.

Photograph L. M. Puth.

of day and the sex of the painter (Orians and Heerwagen 1992; Heerwagen and Orians 1993). Evening scenes are close to images of refuge, and women artists skew toward refuge availability in their scenes, relative to their male counterparts. Creating successful contrived human environments depends on predicting human behavior, and the human Umwelt can be very helpful in guiding the design. Landscape architects use the form of savanna vegetation to create parks. For example, at the University of Wisconsin in Madison, there is behind the student Memorial Union a place known as the Terrace. It is a highly contrived lakeside savanna, full of hiding and watching places (Allen 1998b). As a test of what is in an idealized landscape, Orians (1998) reported that British classic garden designs opened up the landscape but maintained refuges, mimicking savanna vegetation.

Living Systems Theory

Compared with von Uexkull's (1957) Umwelt, a more medical general systems model of the organism appears in James Miller's living systems theory. Miller (1978) reduced the organism to about a score of generalized parts, each playing a role in the full function of the organism. Miller went on to use those generalized parts and their functions to make an analogy that identifies organism types of function in many other living systems. The organism analogy is applied to structures, such as cells that are parts of organisms. Above the level of the organism, Miller applied the organism analogy to elaborate human social units. At all these levels of organization, Miller applied the same 20 generalized parts and their general functions, as when a police force is identified with the function of antibodies in an organism. Miller identified subsystems that process information and subsystems that process energy and matter. Two subsystems, the reproducer and the boundary, process both information and matter/energy. Not only does the analogy offer insights into the generalized function of living systems at levels away from the literal organism, but it also facilitates a fresh look at organisms themselves in more general terms. This generalized view offers a fresh perspective on sustainability under the organism criterion.

In an organism that is not being sustained, the reproducer often is insulted. Disruption of the matter and energy set of subsystems would include lack of material resources—starvation—the consequences of which are obvious. Disruption of the information-processing subsystems is less tangible but just as devastating. The internal transducer, the channel/net, and the decider subsystems are particularly affected by chemical insults. Internal transducers pass information between different media in the system. Synapses connect nerves, and glands respond to nervous stimuli and other glands. These connection sites function with chemical signals and so are susceptible to chemical interference. The mode of action of poisons usually is to block receptor sites. Receptors are the stock in trade of internal transducer subsystems. Miller's scheme applies at many levels, and so the general class of transducers subsumes the notion of loss of a functioning Umwelt at the level of whole species. It is no surprise that blocking receptors is also important at the lower level of the cells. A cell becomes cancerous when it cannot pick up signals from other cells that the region is already fully populated and it is time for the cell to stop dividing. Mutagens strike at the heart of an organism's ability to transfer information.

Interference with information transfer not only affects real-time behavior of organisms but also interferes with signals in development.

Therefore chemical insults disorganize organismal structure through assaults on the information-processing systems. The default setting of birds and mammals is development into a female. Males are a deviation from that path. The feminization of male chicks that occurs in insulted bird populations is the failure of the male-inducing signals to be produced or received. Organochlorine rings in persistent toxic chemicals appear to be just the right size for insertion into critical positions to interfere with the systems that transfer information. Organochlorines are also chemically stable and are fat soluble, both factors that put them in exactly the wrong place in ecological systems (Virtual Elimination Taskforce 1993). Biological systems scavenge for uncommon yet valuable material, so fat is accumulated in food chains, along with its solutes. The consequences of bioaccumulation are disastrous for the generalized information-processing systems in organisms. It involves developmental pathways to produce monsters.

The channel/net and decider, in Miller's terminology, embody information connectors and switches. The channel/net system moves information significant distances inside the organism, and the decider deals with that information as it arrives. They are most susceptible to chemical disruption. The blood system as it transfers hormones is an example of the channel/net, but probably more heavily used is the nervous system, the other main channel and net subsystem. Nervous degeneration is one response to insult by toxic materials. The decider is a diffuse hierarchical system. At low levels it makes decisions as mundane as a reflex response in the face of a strong enough stimulus. The decision is whether to fire the motor arm of the reflex arc. It is no accident that one of the first lines of medical investigation is checking reflexes, with failure being an early sign of degeneration. At higher levels, the brain is the critical decider. In humans, loss of mental function is a common sign of a nonsustainable ecological situation (Virtual Elimination Taskforce 1993). Allergenic responses to pollution, such as high asthma rates in Windsor, Ontario (Bates and Sizto 1987), are again a disruption of the decider (fig. 4.8). In asthma it decides to constrict the bronchi to change air flow characteristics. Other insults produce autoimmune responses, a situation in which the decider is thrown completely off the mark and decides that part of the organism is alien and must be destroyed.

FIGURE 4.8

Exporting pollution. The incidence of asthma in Windsor, Ontario, is high. Windsor is just across the Detroit River from Detroit, seen here from Canada. Although there is light industry in Windsor, a significant proportion of the air pollution comes across the river, from the United States.

Photograph T. Allen.

Minimal Viable Systems

The organism often is used as an analogy for biological systems in general. Stafford Beer's (1979, 1981, 1985) version refers to minimal viable systems. This is a particularly appropriate general model for a discussion of sustainability because it concerns explicitly systems that are viable but cannot be sustained if any more parts are removed.

As with Miller's model, minimal viable systems apply explicitly to other levels. As a general systems theorist, Beer applied his model to business systems of all sizes, but reference to the organism was again explicit. Unlike von Uexkull but like Miller, Beer focused on having the requisite parts. Beer identified a small set of parts that are needed to keep a system autonomous while playing a role in a larger setting. He linked his minimal viable systems explicitly to higher and lower levels of organization. In his system diagrams, the same icons appear not just at the level of the system under consideration but also for the minimal viable systems above and below. Thus a given minimal system is sandwiched between lower-level minimal systems of which it is composed and upper-level systems to which it belongs.

Without using the term explicitly, Beer emphasized the importance of sustainability at all levels. In a minimal viable system such as a firm, the departments within it must be autonomous and self-contained if long-term survival is to be achieved in a competitive situation. For sustainable functioning of themselves and the firm they serve, departments must possess the critical general properties of the whole firm, although set in the context of the firm. Consider the marketing department as the part analogous to Miller's output transducer. Beer would emphasize that it must have all the components it needs to be a minimal viable system in its own right. For example, it must have its own output transducer through which it serves its firm. Beer's model is a minimalist version of Miller's scheme, where the highly specified parts of Miller are aggregated into more general functions. Beyond Miller's model, Beer insists on sustainability of functional units at all levels. In the tightly hierarchically nested structures that are organisms, Beer's general model reminds us that lack of sustainability can arise at any level and spread to bring other levels down.

Beer's model is reminiscent of Koestler's (1967) holons nested within holons, but Beer's are prescribed to be viable in themselves at all levels. Koestler spoke of his holons in a stable system as playing a dual role in which they assert constraints over their parts while playing a role as an integrated part of a larger system. Cancer is an example of a subsystem being overly assertive as to its autonomous action, at the expense of the integrative role. Normally cells in the organism are constrained subsystems, their capacity for reproduction being suppressed by the presence of the cells around them. The deficient input transducers of cancer cells propagate their pathology upward, making the whole inviable. This highlights one of the essential difficulties of predicting sustainability because the whole system may come crashing down because of the unexpected failure of some neglected subsystem.

Organisms as Fragile Systems

The organism is an example of a fragile system. Fragility relates to sustainability. Removal of the forest has left many landscapes fragile, but fragile does not mean unsustainable. In fact, the issue of ecological sustainability does not emerge until there is a large enough human presence to render the system fragile. In primeval forests, sustainability is moot in that the vegetation has embodied in it processes that recover from disturbance and maintain or regain the steady states of mature forest. Fragility is a property of systems that function far from equilibrium. Deforested systems are held far away from equilibrium forest cover by human activity. Properly organized human activity can allow such a system to be sustained, even if the human presence makes the system fragile. It is the high degree of organization that leads a system to be fragile, for it is the organization itself that is vulnerable while being necessary for sustainability. Thus all highly organized systems, from organisms to businesses, are fragile. Fragile systems persist and remain remarkably resistant to change if effectively organized around a sufficient and reliable energy source.

The essence of fragile systems can be captured in three characteristics associated with system collapse. Taken together these characteristics explain the distinctive patterns of dissolution of complex systems. Eventually the forces for organization in a fragile system will be overcome, as they will in any system. Disintegration occurs as organization becomes worn out or when some insurmountable external perturbation overwhelms all. The manner of the demise of fragile and complex systems is informative as to how they work over the period when they are fully functional. All three characteristics of the collapse of fragile systems show how fragility and complexity imply each other. The first character indicates rules that cannot be violated without inviting system collapse, and the other two are explicitly about the manner of collapse. Let us work through each character of fragile system collapse.

First, fragile systems have narrow tolerances on their parts. One of the characteristics of organization is that it provides a reliable setting for the parts of the organized system. The parts are adapted to that context, and in it they achieve high function. The enzymes in a warm-blooded animal are a case in point. The narrow tolerances of the parts set the limits of system resilience. Resilience is the aspect of stability reflected in a capacity to bounce back from a disturbance that radically changes the state of the system parts. Note that cold-blooded animals can return from lower body temperatures easier than can warm-blooded animals. As a counterbalance to low resilience of the parts, organized systems generally are more resistant to change, as when a warm-blooded animal increases its base metabolism in cold conditions to resist cooling. Fragile systems can manifest significant stability by virtue of being more resistant to change than their robust counterparts. Resistance holds system parts well inside their ranges of tolerance, even in the face of significant forces for disturbance or stress. An example of this in a human-dominated ecological system is crop plants of the Green Revolution. Primitive cultivars can withstand drought, poor nutrient soil, and pest infestation much better than their highly specialized modern counterparts. In the speech he gave on the occasion of his Nobel prize, Norman Borlaug explicitly stated that there is no Green Revolution without chemical fertilizers and pesticides, both of which hold the wonder varieties in highly favorable but narrow conditions. Narrow tolerances for system parts is a property of both fragile and complex systems.

The second general characteristic of fragile systems relates to the speed with which the system manifests instability. The term *fragile* conjures up images of something rigid and resistant to deformation but unable to deform in the face of stress so great that deformation cannot be resisted. It breaks instead of deforming. Glass is fragile, and the analogy to something like glass or porcelain is entirely appropriate in the use of the term in describing fragile ecological situations. The second characteristic of a certain rigidity of fragile ecological systems is also a characteristic of glass. When a fragile system does go unstable, it changes rapidly from a state of being whole to one that is fragmented. Fragile systems manifest a sudden lack of integrity.

The third characteristic of fragile systems is related to the second. It is that fragile systems give no warning that something is wrong until everything is lost. Thus not only is the process of disintegration rapid, but there is characteristically no warning of the coming catastrophe. There is an organized separation of the function of the whole from the functioning of the parts. When any system collapses, it is the normal functioning of the parts out of their normal context that exacerbates

195

the problem, and the separation of levels of function in fragile systems is responsible for the great and rapid change that occurs at system collapse. Fragile systems appear to be fully functional and behave perfectly normally until the critical, irreversible change occurs. The Green Revolution example is again useful. There is a cost to the capacity of wonder varieties to grow fast and deliver a huge return. In avoiding the normal costs of plant growth, wonder varieties encumber themselves with other overheads. Building natural botanical poisons and other unpalatable biochemicals is not free, for it takes energy to make them. Botanical poisons are part of the plants' budget. Some of the resources in the plant body invested in growth of a modern wonder variety have been diverted from generalized defense strategies that were part of the budget of the less productive primitive ancestors. Therefore Green Revolution plants stand naked, with no intrinsic protection from pests. They grow well, with little indication of the coming infestation, until suddenly the whole crop is destroyed by the pest outbreak.

One way to look at the fragility of human-dominated ecological systems is to pay attention to the way humans tinker with the frequency characteristics of the ecological system. Contrast the frequency characteristics of the direct human activity applied to the system with those of the desired outcome. Sustainability often is one of the goals of human action. Sustainability involves lengthening the low-frequency periodicity of the system or at least bringing intrinsic long-term periodicities to the fore in system functioning. By contrast, the actual human action is very high frequency, cutting a tree here, spending a day building a wall there. This is characteristic of highly organized systems in general; that is, high-level systems resist change and the influence of external forces by entraining high-frequency processes that preempt every turn of the external influences over the long term. The analogy of a high-wire act is appropriate; by constant minute adjustments, the tightrope walker stays aloft. It is the steadfast resistance to change achieved by fast internal activity that gives high-level systems a certain fragility. As long as the tightrope walker's pole can integrate the adjustments into the maintenance of stability, total system behavior is slow and remains within bounds. The special interaction of long- and short-term facets of high-level systems gives them the three sets of characteristics of fragile systems outlined earlier: narrow tolerances, rapid collapse if it happens, and no warning of impending collapse. Humans influence long-term aspects of ecological system behavior by vigilant and persistent local activity.

An organism is an example of a highly organized system and can illustrate the foregoing arguments about fragile systems. Note that an organism has all three characteristics of fragility just mentioned. First, the organism persists for a long time by constant, high-frequency activity. Biochemical activity generates high-level physiology that prevents the environment from directly influencing the inside of the organism. This gives the parts, the organs, a constant setting in which to work, and they have evolved to require these narrow tolerances. Second, system breakdown of an organism—death—occurs over a very short span of time, relative to the full life span. Third, death often occurs with little indication that catastrophic breakdown is about to happen. Heart attack victims can be fully functional until the coronary artery blocks. Though somewhat different in character, an automobile is also a system functioning at a high level of organization, and it too involves narrow tolerances and sudden breakdown, often with little warning. Deforestation and subsequent human compensation make ecological systems that are fragile—perhaps sustainable but certainly fragile.

THE LANDSCAPE

The distinctive character of the approach to ecology taken by Allen and Hoekstra in *Toward a Unified Ecology* is the separation of scale and type. One advantage of separating the scale of an entity from the class to which it belongs is the clarity that emerges in the distinction between types. The differences that arise in assigning a particular ecological system to one class of system as opposed to another become evident once the confounding issues of scale are excluded. Separating scale from the type of system puts the type of system into stark relief so that a particular characterization given in a particular type stands out clearly. The landscape is organized so that process is readily linked to pattern. A consequence of the ease of perception of landscapes is that landscape theory has had the opportunity to develop beyond what is intuitive. As a result, there is great difference between the lay perception of landscapes and the manner in which landscape ecologists study them.

Sustainability is more a matter of process or function than an event or structure. Although death might be seen as the ultimate nonsustainability of organisms, there is something unsatisfactory about an event, in this case death, marking a distinct end in a discussion of sustainability. Shakespeare likened the unfolding of a human life to a play, a structure on which a curtain comes down to end it all. Even so, for most questions of ecological sustainability, there is usually not one readily identifiable moment when the curtain comes down, when the system is finally not sustained. The essential dynamic nature of sustainability brings into question the moment of ultimate demise in a nonsustainable system. Both organisms and landscapes are tangible, but of the two criteria, only landscape has the enveloping, contextual quality that invites notions of sustainability as a process.

The distinction between organisms and landscapes becomes significant when one considers types of action that might enhance sustainability. Sustaining an organism in good health might require direct action from a veterinary surgeon, but it is less apparent that landscapes might also need explicit action to sustain them. If they do need remedial action to assist them, it is less clear what form landscape remediation might take. In most conventional usages of the concept of sustainability, there is no single thing to which one can point and say that is where the sustainability of the system resides. Accordingly there is generally no simple, single action that unequivocally enhances sustainability.

Another important difference between the application of the concept of sustainability to organisms as opposed to landscapes is the requirement for a discrete scaling in organisms but the opportunity for continuous scaling of landscapes. There is a continuous nesting of landscapes within larger landscapes. True, organisms can be found nested inside other organisms, as the bacterial flora in the cow's rumen illustrates. Nevertheless, not all smaller scaled systems allow the application of the organism criterion. For instance, the front leg of a cow is not a smaller organism in itself. By contrast, the back 40 acres on a farm is a perfectly good landscape that has been capriciously carved out by surveyors from a larger landscape, the 160-acre quarter section to which the back 40 belongs. For that matter, the back 39.5 acres also constitute a landscape by any standard. Thus parts of organisms generally do not meet the criterion for each being an organism in themselves, but almost all parts of any size of a landscape can be seen validly as landscapes in their own right.

In issues of sustainability, this continuity of the landscape criterion across scales presents a problem. Sustainability is relative to the spatiotemporal scale of the system, and it is the responsibility of the observer to be explicit as to the scalar aspects of what is to be sustained. Between the largest organism on the globe and the smallest there are organisms somewhere in the world, at the right stage of maturity, to fit just about every size. Despite this general scalar continuity of organisms, there is scalar discontinuity for organisms with regard to each particular local material system (Holling 1992). In a given place there are organisms of different sizes. For instance, in a barn, there are microorganisms at one scale, rodents at another, and domesticated ungulates at yet another; however, there may well be no organisms between those three general size classes. Accordingly, part of the responsibility for being explicit about the scale of the organism can be left to convention embodied in such terms as *microorganism*. On the other hand, because landscapes can be capriciously circumscribed and still be worthy units of scientific study, there are few general guidelines for defining the unit to be sustained according to the landscape criterion. This is an invitation to confusion that must be met with explicit statements of landscape scale for every study and every consideration of sustainability (Allen 1998b).

Historical Landscapes in Context

Of all ecological observations, those on the landscape are particularly robust. In some cases, the landscape phenomenon can be a surrogate for a population effect. Landscapes often can be the vehicle for an initial understanding of an ecological problem that belongs properly to another organizing criterion. One reason is the ease with which some aspects of scale can be read on landscapes. Contrast this with the many pitfalls in matters of scale that await students of other types of ecological system. When a situation looks unexpectedly, suddenly different, quite often nothing much has changed except the scale used by the observer. Scale changes may be imposed unwittingly by holding some measure constant in the face of some scalar change in the working of the material system. For example, counting individuals in population biology is a standard practice, but separate individuals have different scalar properties. The numbers of prey and predators as a measure of the respective states of their populations uses the same measurement device for differently scaled aspects of population function. The negative effect on a prey population of the capture of a prey individual corresponds to a positive benefit to predators; however, death of prey is instantaneous, but benefit to the predator from having eaten takes a while to manifest itself in increased predator numbers. In contrast to population and other types of ecology, landscape ecology offers ease with which scalar effects can be handled. Easy handling of landscapes means that hidden scaling effects are less troublesome. When using the landscape criterion, the ecologist is unlikely to be misled by unwittingly changing the scalar relationship between the observer and the material system.

Sustainability, as we have noted, is a historical process. Whereas it is difficult to address food webs or ecosystem cycles of the past, landscape patterns survive as a modern material presence in a way that brings ancient ecology to life. This is manifested in the subtle interpretations of glaciation, where snake-like ridges called eskers indicate rivers inside and eventually under the melting ice. The glacier is gone, and the river has dried up, but the pattern is still there to be interpreted. New ground-penetrating radar that can see through sand in the eastern Sahara from satellites surprised the scientific community by finding ancient, buried river beds (McCauley et al. 1982). Now the Egyptian desert is one of the driest places on Earth, some areas receiving no rain at all as far as modern records can tell. And yet the landscape holds the secret of a more verdant past whose signature is left under the sand. In the north of Mali, archaeological sites of the last millennia B.C. hold crocodile, hippopotamus, and fish bones in a landscape that today people can barely travel through safely, let alone inhabit (Petit-Maire et al. 1983). In a landscape orders of magnitudes more ancient, astronomers and geomorphologists infer a watery past for the surface of Mars from river beds that probably carried liquid water across the Martian landscape. Landscapes not only give insight into system sustainability but also preserve in fossilized pattern the death throes of ancient systems as they became unsustainable.

Ancient writers described landscapes better than they did almost any other type of ecological system apart from organisms. Fossil organisms capture the past in concrete terms, but much of the process of the living creature is lost. However, the process of living creatures is indeed captured in reports of the writing of Aristotle, where we read of catfish that nurture their young. In general catfish do not make very good parents, so the reports were at first dismissed. However, the landscape offered a tag on this unusual display of parenting, for in the very region mentioned in the ancient texts, a species of catfish was found that does indeed protect and rear its young (Hughes 1975).

Ancient descriptions capture the essential landscape ecology, whereas more esoteric ecological notions such as food webs and nutrient cycles are not reported because they are intangible and probably were not noticed. Sometimes the ancient chroniclers did not focus on reporting ecology, but landscapes are so easy to describe that a rich account comes down to modern times. For example, the traveler Pausanius in the second century A.D. reported groves of trees whenever he found them (Alcock 1993; Hughes 1975). Although Pausanius may well not have been conscious of it, the implication is that the process of deforestation to provide fuel for hearths and blacksmiths' forges and timber for ships was largely over by that time. Also there is evidence from treaties often mentioning timber as a critical commodity, again implying a shortage of forest on the landscape (Hughes 1975).

People routinely alter landscapes socially and economically, adding constraints that change ecological structure and function. Humans perceive landscapes in quite variable ways, seeing them as opportunities to achieve goals that may have no ecological basis. People mold landscapes to fulfill these goals, even where they do not consciously mean to do so. Sustaining a landscape often depends on understanding the behavior of people acting as the context for that landscape.

We discuss several historical examples that illustrate the social, political, and economic contexts of landscapes. Historical cases are preferable to those that are merely contemporary; what is important about a landscape is rarely limited to what is happening today. In these illustrations, the human context ranges from small groups of foragers, to political relations between contending states, to the international arena of a great empire. In each case the character of the landscape is shaped by the interaction of biophysical processes and human problem solving. It is in the framework of this relationship that landscape sustainability must be addressed. Our examples include the landscape preferences of prehistoric foragers and farmers in the middle Rio Grande Basin, hunter–gatherer manipulation of the California landscape, the constructed agricultural landscape of the Maya, and the role of domestic and international conflict in shaping the landscape of Roman-era Greece.

Landscapes of the Middle Rio Grande Basin. The rift valley in what is now New Mexico, through which flows the Rio Grande, has sustained human life for 12,000 years or more—as long a time as we know people to have been in this hemisphere. The central part of the basin near Albuquerque has two main north–south trending rivers: the Rio Grande itself and, to its west, the Rio Puerco. The latter joins the former about 90 kilometers south of Albuquerque. Between the two lies a broad, flat tableland that rises to a height of about 225 meters above the valley floor.

The native occupants of the basin subsisted on the land first by foraging and later by farming. Maize was known by the end of the second millennium B.C., but the economy was based primarily on foraging until the first few centuries A.D. After about A.D. 500, while foraged resources remained important, people relied on cultivated maize, beans, and squash. By about A.D. 700 prehistoric Southwesterners came to live in aggregated communities of apartment-like dwellings, made famous by ruins at Mesa Verde, Chaco Canyon, and other places. By about A.D. 1000 most of the rural parts of what is now the American Southwest sustained more people than live there today. The descendants of these people, the Pueblo Indians, continue to live in the Rio Grande Basin and other parts of the Southwest.

The cumulative data of decades of archaeological research show that the early foragers and the late farmers had very distinct landscape preferences (Tainter and Tainter 1996b). The foragers of what archaeologists call the Archaic period (ca. 5500 B.C.–A.D. 500) practiced a mixed subsistence economy based on hunting such animals as deer, antelope, and rabbits and gathering plant foods such as grass seeds, forbs, succulent seeds and fruits, early season greens, and pinyon pine nuts. The archaeological sites made by the activities of these people tend heavily to concentrate along the Rio Puerco, particularly along tributary drainages of its upper reaches (fig. 4.9).

The late prehistoric farmers of the Pueblo IV period (ca. A.D. 1300–1600) favored a very different landform. These people, living in pueblos of up to 3000 rooms, concentrated in and near the broad alluvial floodplain of the Rio Grande Valley (fig. 4.10). By 1300 or so the

FIGURE 4.9

The landscape of hunter–gatherers, ca. 5500 B.C. to A.D. 500, in central New Mexico. Hunter–gatherers preferred a diverse landscape that today is considered marginal for human use.

Source: Tainter and Tainter (1996b:24).

FIGURE 4.10

The landscape of Puebloan farmers, Rio Grande Valley, ca. A.D. 1300–1600. The agriculturalists preferred alluvial bottom land, much as people do today.

Source: Tainter and Tainter (1996b:28).

upper Rio Puerco basin was abandoned and seems to have been little used thereafter. This event was part of a broader regional process, for by the fourteenth century A.D. much of the upland areas of what are now the contiguous parts of Arizona, New Mexico, Colorado, and Utah were abruptly abandoned.

This sequence of landscape use is puzzling at first glance. To Euro-Americans, the most desirable landscapes in this area are the fertile river valleys. These are where Hispano- and Anglo-American colonists settled first. It is therefore surprising to learn that in 12,000 years of occupation, the Rio Grande Valley is where American Indians concentrated *last*. The early Native American use of the region appears actually to have focused on landscapes that, in a Euro-American view, are considered the most marginal. The resolution of this paradox conveys a lesson in the perception and use of landscapes.

The landscape preference of the post–A.D. 1300 farmers is easy to understand. The Rio Grande Valley is a broad alluvial plain that was regularly enriched by deposits of new silt. The water table near the river is high enough to be reached by cultigens' roots. It is where one would expect large-scale cultivators to settle.

The hunter–gatherer landscape preference is more subtle. It stems from spatial variation in the distribution of resources and from the energetics of transporting food. A forager's landscape is a mosaic of locations of foods and other resources, which become available at different places throughout the year. To move such foods to those who will consume them is much too costly. A trip of 100 kilometers or so will cost about one-third of the food being transported. This represents what those carrying the food need to eat. For this reason hunter–gatherers usually move consumers to resources rather than the reverse. They must be, and typically are, highly flexible and mobile, roaming throughout their landscape as the seasons progress. A yearly seasonal round for foragers in the middle Rio Grande Basin might have started with a move to lower elevations for early season greens, movement throughout mid-elevations from late spring through early fall to gather seed-bearing plants and succulent fruits, then to higher elevations in the fall to hunt ungulates and gather pinyon nuts.

The need to be mobile constrains and shapes hunter–gatherer societies. Large groups cannot coalesce for very long. The landscape atomizes society, which for much of the year consists of the smallest feasible social group: the family. Yet in some landscapes there is a way to minimize or even escape the mobility constraint. These are landscapes that offer a high degree of topographic diversity. In a landscape with significant variation in altitude, resources will be separated vertically, but the horizontal separation between vegetation zones typically is much less than in a homogeneous plain. Hunter–gatherers in a diverse landscape can forage for many resources from one or a few locations merely by moving up and down slope.

The upper Rio Puerco drainage system is characterized by many tributaries originating in higher elevations to the east and west. These tributaries have sculpted a dissected landscape of mesas, canyons, valleys, rock shelters, sand dunes, ponds, and outlying rock formations. It is a landscape much more diverse than the floodplain of the nearby Rio Grande Valley. This diversity attracted early foragers to settle preferentially in an area that Euro-Americans consider useful for little other than cattle raising and inferior even for that.

The broader value of this example lies in clarifying aspects of sustaining landscapes. Desirable landscape characteristics are fundamentally a matter of social definition. Landscapes sustain people in particular social and economic configurations and are valued for their ability to do so. Perceptions of landscapes are inescapably as variable as human society itself. No specification of landscape sustainability can be offered that does not arise specifically from a human social and economic condition. A human system is nearly always the context that delimits a desirable landscape configuration.

The Landscape of Native California. On May 31, 1793, Don José Joaquín de Arrillaga, captain of cavalry, interim governor, and inspector comandante of upper and lower California, issued the following proclamation.

With attention to the widespread damage which results to the public from the burning of the fields, customary up to now among both Christian and Gentile Indians in this country, whose childishness has been unduly tolerated, and as a consequence of various complaints that I have had of such abuse, I see myself required to have the foresight to prohibit for the future (availing myself, if necessary, of the rigors of the law) all kinds of burning, not only in the vicinity of the towns, but even at the most remote distances, which might cause some detriment, whether it be by Christian Indians or by Gentiles who have had some relationship or communication with our establishments and missions. Therefore I order and command all comandantes of the presidios in my charge to do their duty and watch with the greatest earnestness to take whatever measures they may consider requisite and necessary to uproot this very harmful practice of setting fire to pasture lands.

Quoted in Timbrook, Johnson, and Earle (1982:171)

With these words the full power and majesty of the Spanish Crown was deployed to end an ancient practice by which the landscape of California, and its native inhabitants, had coevolved.

Native California has always been an anthropological enigma (fig. 4.11), with some of the highest population densities in native North America. Whereas hunter–gatherers in various parts of the world have lived in average densities of about one person per 2.6 square kilometers, parts of California supported densities up to 10 times greater. Native Californians often lived in sedentary communities of up to 1000 people, whereas most foragers have had to move seasonally as resources sprout and are consumed here and there and are rarely able to aggregate many people for long. California overall has about 5 percent of the land of North America north of Mexico but held at least 18 percent of the native population (Kroeber 1925:885). This remarkable density and way of life was accomplished by regimes of manipulating vegetation that gently molded the California landscape into a mosaic of patches capable of supporting not only extraordinary numbers of people but also densities of flora and fauna that have always awed Europeans and Americans.

Hunter–gatherers are generally characterized by egalitarian social relations. That is to say, decisions are made by consensus (one scholar described egalitarian life as a continuous encounter session), and lead-

FIGURE 4.11

Native California hunters.

Source: James C. Cowles, *The Natural History of Man* (London: H. Ballière, 1845), after a painting by Louis Choris, 1816.

ership is achieved, ad hoc, and temporary. When a circumstance such as conflict or a hunt requires leadership, a capable person is temporarily allowed to direct the affair. At other times foragers work actively at maintaining egalitarian relations. A foraging way of life, as noted, calls for flexibility. People must be free to move—to aggregate at places where resources are abundant and to disperse when that is necessary. The social landscape shifts continuously. Permanent leadership in such a situation is not only impossible but is usually detrimental to survival. Leaders cannot lead if they have no followers close at hand.

Yet Native Californians, in their sedentary communities, had elaborate social hierarchies. There were hereditary (and full-time) chiefs, who could be quite powerful. In some cases native societies had differentiated into rudimentary class systems, with hereditary rulers, bureaucrats and artisans, poor or lower classes, and sometimes slave or outcast groups (Bean and Lawton 1973; Bean 1974). Among the Chumash of the Santa Barbara coast, political and economic differentiation

207

reached their most complex form. Wealthy people owned part of the means of production: large boats made of sewn planks, used for fishing in the Santa Barbara Channel and for trade with villages on offshore islands (a sea voyage of more than 32 kilometers each way). The economy was at least partially monetized, with food, goods, and services exchangeable for shell beads made in standard forms. Chiefs owned stores of goods, which they could distribute at ceremonies or trade with other chiefs. Clusters of villages were integrated into political units, of which one village was considered the capital. Villages distributed across large areas were integrated into ceremonial confederations, and food was shared on ritual occasions (thereby integrating the religious system into the economic one) (Harrington 1942; Landberg 1965; Brown 1967; L. King 1969; C. King 1971; Blackburn 1974).

It is a matter of both anthropological and ecological importance to understand how Native California came to be so unusually populous and highly organized. The answer seems to lie in the management of the California landscape—for manage it Native Californians did, to a degree perhaps unequaled by hunter–gatherers elsewhere. Manipulation of animal populations and plant communities created an aggregate effect across the landscape. The California landscape was continuously manipulated to create a mosaic of patches that favored early successional stages and thereby species of greatest human value.

The foremost manipulative tool was fire, the use of which was definitively studied by Henry Lewis (1973). By one estimate, at least 35 groups of Native Californians used fire to increase the yield of seed-bearing plants, 33 used fire to drive game, and 22 used it to stimulate the growth of wild tobacco. The burning strategy apparently was to set small, spot fires, which would greatly increase the number of edges and the overall diversity of the landscape system.

California vegetation, in its Mediterranean-like climate, responds well to fire. In the oak woodlands of the Sierra Nevada and Coast Range foothills, fire maintains a balance between trees and grasses, keeps shrub vegetation in check, and favors young shoots on which deer browse. California's coniferous forests, now supporting heavy fuels and subject to disastrous crown fires, formerly burned only lightly. Frequent fires moved swiftly through light ground cover, so that tree density and other fuels never grew to the levels seen today. Forests were open and park-like, and in the Sierra Nevada John Muir described a mosaic of even-aged stands. Some who study these forests have long suspected that native burning helped to maintain these conditions.

Chaparral, a dense, shrubby, sclerophyllous cover, was especially important to Native Californians, and they used fire to improve its productivity. Many species not evident in late-stage chaparral appear after fire, and several species in chaparral can germinate only after their seeds are scorched. Fire destroys the aerial portions of chaparral, and brush sprouts return rapidly with growth that browsing animals find highly nutritious. Deer browsing in recently burned areas experience ovulation rates more than twice those in late-stage chaparral, and their density per square kilometer increases by more than 30 percent. For coastal black-tailed deer, the ratio of fawns to does grows 37 percent in burned chaparral (Taber 1953:184). Smaller game species also respond favorably. The beneficial effect diminishes a few years after burning, and in time a patch must be reburned. Thus in prehistory California chaparral lands formed a mosaic, with patches ranging from recently burned, with herbaceous vegetation and much game, through mid-successional stages with decreasing game, to dense chaparral needing to be burned again.

Fires occur on their own, of course, but the Native Californians both augmented the natural fire regime and departed from it. In the coastal coniferous forest, for example, lightning is less intense than in the Sierra Nevada, and lightning fires are rare. Yet in the northern Coast Range the redwoods in historic times showed clear evidence of burning at a frequency that is difficult to explain without human agency. Grasslands often were burned immediately after the summer seed harvest, causing great consternation to the Spanish, who wanted to graze cattle on the same grasses. Primary burning occurred in the fall, when vegetation was dry. Autumn rains following these fires provided for rapid growth of new vegetation. This practice augmented the natural fire regime, for lightning fires occur primarily in summer and fall. In some areas, though, the natives also practiced a secondary spring burn, which was a departure from the natural fire pattern.

Native Californians manipulated their environment in other ways as well. In a famous case the Paiute of Owen's Valley irrigated natural stands of seed-bearing grasses—surely one of history's only examples of irrigating without agriculture. Elsewhere seeds of native plants were sowed or broadcast, shrubs or small trees were transplanted to new locations, plants were pruned or coppiced, plant communities were weeded and tilled, and water diversion structures were built to limit erosion (Blackburn and Anderson 1993:19). Basket weavers set fires in stands of the plants that they used, to augment the production of long, straight shoots without lateral branches. In other areas fire was used to help produce straight shafts for arrows. Areas were both burned and tilled to produce bulbs, corms, and tubers. Fern patches were pruned to regenerate each year (Anderson 1993). Sedges (also used in basketmaking) were enhanced by loosening the soil and removing weeds, rocks, branches, and other roots (Peri and Patterson 1993).

Thus the prehistoric California landscape was in part a human artifact and for several thousand years had increasingly been so. The remarkable population densities of California foragers were achieved by a combination of systematically manipulating vegetation to increase the yields of foods and other resources, and using complex ritual and social systems to even out food distribution (Bean and Lawton 1973; Bean 1974). At the time of European contact, aboriginal California was composed of hundreds of independent villages, each largely constrained within its territory and forced to use that territory intensively. Native Californians did this so well that, as some scholars have noted, agriculture as we understand it was unnecessary (Bean and Lawton 1973). Like the Hanunóo (chapter 2), Native Californians acted as the context for the successional cycle, manipulating natural ecosystem processes to sustain themselves. This exemplifies what we consider the supply-side approach. Native Californians achieved sustainable ecological management, and sustainable societies, by using just enough human intervention to start a cascade of ecological processes that produced the desired result.

Although California is an exemplary illustration of landscape domestication, it is far from unique and indeed may differ from other cases only in the degree of manipulation. Hunter–gatherers have commonly modified their environments. Fire was used by almost all foragers of what is now the western United States. It was used in western Canada and is used to this day by aboriginal Australians. Hunter–gatherers probably modified most landscapes of the globe well before they domesticated plants and animals. Thus domestication in human history seems to have been applied first to ecological communities and landscapes (as discussed here), then to ecological communities and populations (conventional domestication), and finally to the creation of fully agricultural landscapes and ecosystems.

Maya Landscapes. In the climactic battle of the film *Star Wars*, the evil empire's ultimate weapon closes ponderously on the rebels' base. As a rebel scout stands watch we see beyond him a tropical rain forest stretching to the horizon. Here and there the forest canopy is broken by a gleaming white pyramid, ruinous yet still standing to a magnificent height. It is a vanished civilization, gone eons before the era of spaceships depicted in the film.

The background footage is of the lowlands of Guatemala, and the pyramids are temples of the Maya Civilization (ca. A.D. 200–1000) (Fig. 4.12). The scene is compelling, and the poignancy of a civilization lost in the jungle fits the popular image of the Maya. For a long time archaeological writing reinforced this image, depicting the Maya as peaceful, low-density forest dwellers, dependent on slash-and-burn agriculture (swidden), who for unfathomable reasons built magnificent ceremonial centers and staffed them with priests interested only in calendars (the Maya one being the most accurate of its day) and ritual.

As we have learned more about the Maya, however, this idyllic view had to be discarded. At the maximum extent of Maya civilization its landscape looked nothing like the film scene. We now know that the Maya lowlands were so densely populated that by about 2000 B.C. the area was deforested. The Maya supported themselves by intensive farming often involving hydroagricultural engineering. Their "ceremonial centers" were actually cities, which formed alliances with other cities or made war on them. And although their elites were assuredly interested in calendars and ceremonialism, the information they carved in stone concerned primarily dynasties, royal marriages, and domination. All the while Maya society developed and collapsed in the context of a manipulated landscape. It would not be amiss to say that Maya society and the Maya landscape coevolved, nor to say that the sustainability of Maya society depended on sustaining the agricultural landscape.

The Maya lowlands were one of the most densely populated areas of the preindustrial world, with an average of 200 people per square kilometer. Swidden agriculture could not have supported even one-

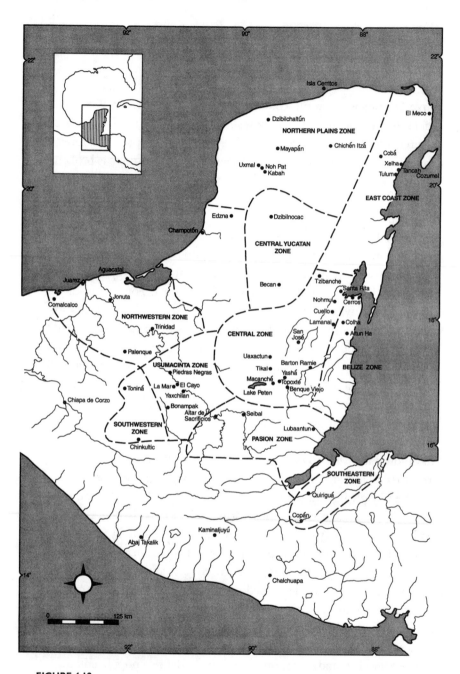

FIGURE 4.12

The Maya area and its major cities.

third this density (Culbert 1973, 1988). Tikal, one of the major Maya cities, had a population estimated at nearly 50,000, which is roughly the same as ancient Sumerian cities (Haviland 1970). In the southern part of the area known today as Quintana Roo, there are stretches of terrain 40 to 50 kilometers long in which there is no gap of more than 100 meters between Maya structures (Willey 1981).

To support so many people, the southern lowland Maya transformed their landscape. One technique they used (once the forests had been cut) was to channel and store water. Although the area today is a high tropical forest, precipitation varies seasonally, and many areas have little surface water. The Maya accordingly built canals, dams, reservoirs, and small wells, and in Yucatán they modified the large limestone sinkholes known as *cenotes*. At Edzna, a site in west Yucatán, the Maya built a water control system consisting of a moat, canal, and reservoir complex. It was designed to collect and store rainwater for agriculture, human consumption, and defense. The volume of earth that was moved for this project made it a construction as great as the Pyramids of the Sun and the Moon—great, famous temples at Teotihuacán in the Valley of Mexico. There are 180 kilometers of canals along the Rio Candelaria. They represent at least 500,000 days of labor to move (entirely with human workers) 10 million cubic meters of earth. This is 10 times the volume of the Pyramid of the Sun (Matheny 1976, 1978).

Hillsides across the lowlands often were terraced for planting. Terracing directs water flow, traps silt, and creates cultivable land. Archaeological surveys have encountered hundreds of thousands of these features, often with related stone works. They are spread across more than 11,000 square kilometers (Adams 1981; Turner 1974, 1978, 1979; Wilken 1971; Willey 1980). Much of the southern lowlands (the area known as the Petén) also contains walled fields (Turner 1974; Wiseman 1983), a type of feature compatible only with intensive, sustained cultivation.

Perhaps the most interesting agricultural features of the lowlands are systems of raised fields and associated canals. These were built to supply dry, cultivable land in areas otherwise inundated. Raised fields were built within lakes and lagoons and at river margins, but they were built primarily within *bajos* (swamps), thereby reclaiming land that could be highly productive. Modified and intensively cultivated swamps may have become the most valuable land. Not surprisingly, the larger cities tend to be located near swamps. Tikal has the most swampland nearby of any major city. By building raised fields in such areas, the Maya could have maintained a moist, root-level environment, fertilized fields with rich, organic muck from canal bottoms, and perhaps raised fish as well. Tended carefully, such fields could have supported continuous cropping.

The extent of these systems is difficult to discern on the ground. Not only has the forest regrown, but the amount of transformed land is truly large. Most of the managed landscape has been detected from aerial photographs and radar mapping. By one estimate, between 1250 and 2500 square kilometers of the lowlands were transformed into canals and raised fields. In contrast, the famed *chinampas* (lacustrine raised fields) built by the Aztecs in the Valley of Mexico, which fed their city of Tenochtitlán, covered only about 120 square kilometers (Adams 1980, 1983; Puleston 1977; Siemens 1982; Turner 1974, 1978).

Clearly the Maya had greatly transformed the southern lowland landscape. By the first millennium A.D. the once-ubiquitous rain forest was so reduced that the Maya would have had to maintain woodlots for firewood and palm fronds (Wiseman 1983). The landscape was temporarily domesticated, transformed into water control systems, terraces, walled plots, canals, and raised fields. With the use of these facilities the energy of the lowlands was channeled to produce a simplified biota consisting of maize, squash, avocado, cacao, cotton, root crops, and fruit-bearing trees. The Maya themselves became the context for the lowland landscape, while the landscape and its productive capacity set the constraints within which Maya society developed. What emerged was one of history's most fascinating and enigmatic societies.

If one could have seen the southern Maya lowlands 1500 years ago, a conservation biologist might have perceived a degraded landscape with few species, greatly diminished ecological complexity, and little canopy. To an archaeologist, though, these simplified food webs were the foundation of a diverse and complex society. Ranks of bureaucrats, artisans, elites, and paramount rulers held sway in cities whose centers were built of gleaming limestone. Their sense of time was based on astronomical observations so precise that they were not equaled in Europe or China for centuries to come. The Maya counted time in cycles stretching eons into the past and developed a system of writing that has only recently been deciphered. Their art is greatly admired today and no doubt always will be. All of this complexity was based on the energy flow made available from their managed landscape.

This energy flow generated problems as surely as it produced food. Directing energy along new pathways is always a transformative act. Changes are set in motion that, in human systems, may take generations to play out. One consequence of the development of complex societies in the Maya area is that the cities apparently were often at war. Warfare is a prominent theme in Maya art, with many scenes of domination and execution of captives. There are great earthen fortifications at such sites as Becan, Calakmul, Oxpemul, and Tikal. The defensive works at Tikal include a moat and earthwork 9.5 kilometers long. Warfare in turn selected for the development of centralized political authority (as prolonged warfare, or the threat of it, always does). Maya cities formed alliances, which they cemented by royal marriages. Lesser cities were subordinated to greater ones, although the balance between political centralization and autonomy always fluctuated. A state's position in the hierarchy of cities could rise or fall, and it seems that wars often occurred one level below the top—a result of lesser cities jockeying for position among themselves or trying to free themselves from the domination of a place such as Tikal (Webster 1976, 1977; Sanders 1981; Marcus 1993:147–148).

Political centralization and warfare coevolved with the Maya agricultural landscape. Although cities were provided with moats and earthworks, the Maya lacked the logistical capability for sieges or prolonged warfare. An enemy's support base could be weakened, however, by carrying away many captives or destroying the agricultural production system. The transformation of the lowlands from a rain forest with shifting human production to an engineered agricultural landscape with fixed production assets completely changed the potential for Maya warfare. Raised fields and terraces, as well as those who tended them, were tempting targets. Being immovable, such assets were highly vulnerable. At the same time, stationary agricultural features were both easier to defend than if they had been dispersed and more worthwhile defending. In the absence of a managed agricultural landscape, Maya warfare and politics could not have developed as they did.

The Maya landscape, with its magnificent cities, imposed a high cost on the support population. Great efforts were needed to transform the forest, keep the land from reverting to forest, and keep the soil fertile; to support nonproducing classes of bureaucrats, priests, artisans, and rulers; and to build and maintain the great cities. Nor were the labor demands constant; over time they actually grew. The construction of monuments increased over the years. Across the southern lowlands, 60 percent of all dated monuments were built in a period of 69 years, between A.D. 687 and 756 (Marcus 1976:17, 19). Each time a temple was rebuilt it was rebuilt over its predecessor. Thus the volume of material, amount of sculpture, and necessary labor grew with each reconstruction.

All this effort took its toll on the support population. At Tikal, by about A.D. 1, the elites began to grow taller than the peasants, and within a few centuries the difference averaged 7 centimeters. That elites should have enjoyed better nutrition during childhood is not surprising. By the Late Classic (ca. A.D. 600–800), however, the stature of adult males actually dropped among both elites and peasants (Haviland 1967). The nutritional value of the Tikal diet had apparently declined for both rulers and ruled. At the site of Altar de Sacrificios there was a high incidence of malnutrition and parasitic infections. Childhood growth was interrupted by famine or disease. Many people suffered from anemia, pathologic lesions, and scurvy. The stature of adult males declined in the first few centuries A.D., and life expectancy dropped in the Late Classic (Saul 1972, 1973). Not surprisingly, the population of the southern lowlands, which had grown during the first several centuries A.D., leveled off between 600 and 830 and thereafter grew no more (Culbert 1988). As noted later in this chapter, such population trends often foreshadow collapse (Tainter 1999).

The magnificence of late Maya cities clearly came at the expense of the peasant population and ultimately at the cost of the entire system's sustainability. It was a classic case of diminishing returns to complexity (chapters 2 and 3). As the complexity of the political system and the cities grew through time, the costs of the system, in human labor, grew also. Yet it is clear that the marginal return to these investments was declining by the Late Classic. Ever greater investments in military preparedness, monumental construction, and agricultural intensification brought no proportionate return in the well-being of most of the population. On the contrary, as the demands on the peasant population increased, the benefits to that population actually declined. This was particularly the case in the seventh and eighth centuries, when a great increase in laborious monumental construction fell on a stressed and weakened population that was no longer growing (Tainter 1988:175).

The consequences were predictable. One by one the Maya polities began to collapse, and their cities were progressively abandoned (figs. 4.13 and 4.14). In A.D. 790, 19 centers erected dated monuments. In 810, 12 did so. In 830 there were only 3. And the last monument with a full calendrical inscription was built in 889 (Culbert 1974). At the same time, for reasons not fully understood, the southern lowlands experienced one of history's most wrenching demographic disasters. In the course of a century or so the population declined by 1 to 2.5 million people and fell ultimately to a fraction of its Late Classic peak (Adams 1973; Culbert 1974). The relations of constraint in the landscape simultaneously reversed. There were not enough Maya left to control the landscape, nor did those remaining need to farm as they once had. Intensive agriculture was abandoned, and the forests regrew. The landscape once again became the context for the Maya, who reverted to the slash-and-burn agriculture practiced by their ancestors three millennia earlier.

It seems, then, that the capacity of the lowland high tropical forest of Guatemala and surrounding areas to support dense populations and the features called civilization varied inversely with criteria of sustainability (such as species diversity) that conservation biologists might develop. Cultural complexity in this region was based on an engineered landscape with low species diversity, short food chains, and diversion of energy from plant to human biomass. In the long run, however, although the lowlands landscape could seemingly support dense human populations, it could not sustain the complex political systems that large populations often need. Although it took 3000 years or more, the forested landscape eventually reasserted itself as the context for the human population.

The Landscape of Roman Greece. The history of Greece, from the last half-millennium B.C. to the first few centuries A.D., illustrates how

217

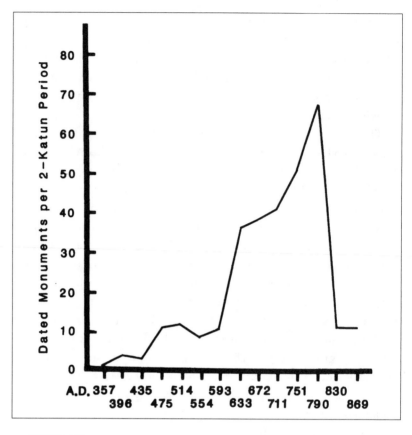

FIGURE 4.13

Construction of dated monuments at Classic Maya sites. One katun is approximately 20 years.

Source: Erickson (1973:151).

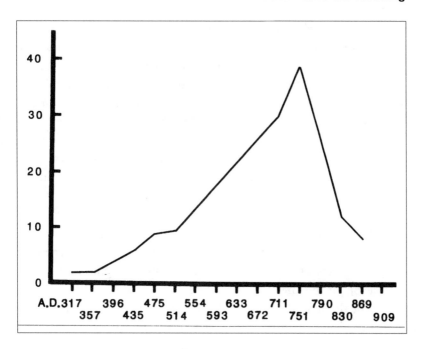

FIGURE 4.14

Occupation of Classic Maya centers. One katun is approximately 20 years.

Source: Erickson (1975:40).

landscapes evolve under social and political influences that may be far removed. Landscape generally is perceived by human actors as a local phenomenon, and the sustainability of a landscape in a particular configuration is viewed at the same scale. Yet the landscapes of Greece were increasingly incorporated into hierarchical systems at first the regional and then the international levels. Political, military, and demographic developments far removed in time and space interacted with local processes to cause Greek landscapes to change in directions that ancient writers described unfavorably.

As early as the second century B.C. Polybius complained, "In our times the whole of Greece has suffered a shortage of children and hence a general decrease of the population, and in consequence some cities have become deserted and agricultural production has declined"

(Polybius 1979:537). The observation was echoed by authors of the first and second centuries A.D. Strabo observed that once-noted cities were in his day deserted. Plutarch, in the late first or early second century A.D., complained that all of Greece could barely muster the number of soldiers formerly sent out by one city. Dio Chrysostom, about A.D. 100, observed that in part of southern Euboea two-thirds of the land was neglected, and large landowners could not find enough men to work their estates (summarized in Alcock 1993). We noted earlier that by the second century A.D., Pausanias felt it worthwhile to remark on the presence of trees whenever he encountered them.

These observations are made particularly compelling by the fact that the authors were all Greeks, who might have been expected to portray their cultural homeland in a favorable light. The image they have given us is a Greek landscape that was underpopulated, untilled, and devoid of the forests that had once covered its mountains.

Although these descriptions are compelling, they are also to be approached with caution. In the absence of actual census data over time, it is difficult to disentangle such observations from prevailing attitudes toward population. In ancient states, power, wealth, and prestige were substantially a function of population—rural agricultural population. The ability of the Greeks to defeat the Persian invasions of 490 and 480 B.C. was remarkable not just because Darius and Xerxes were wealthy in precious metals but also because Persian rulers could call upon masses of Mesopotamian peasants to supply their armies with food and men. Similarly, the success of the Romans had much to do with the enormous number of peasants available in a united Italy. A large peasant population was fundamental to a strong, prosperous state.

There was thus a moral dimension to ancient observations on population. We noted in chapter 2 that to ancient writers, the social and biophysical worlds were intertwined: The decline of one stood as a surrogate for the weakening of the other. An unpopulated, uncultivated landscape was an unmitigated evil, signaling military weakness and political insignificance. Moreover, a change in land use from agricultural to pastoral was considered a retrogression (Luigi Gallo, cited in Alcock 1993:27). The landscape desirable to sustain was one of small farmers, and Greece was such a landscape no longer. Greece's subservient political position and its unpopulated, uncultivated landscape both stood in contrast to its glorious past. Thus Roman-era travelers often found Greece better in the telling than in the viewing. To what degree, then, do the ancient observations reflect the true condition of the Greek landscape, rather than a moral parable about the loss of Greek freedom?

Fortunately, there are archaeological data bearing on this question, which have been carefully assembled and analyzed by Susan Alcock (1993). In recent decades several archaeological projects in Greece have attempted systematically to find and enumerate changes in settlement through time. These projects have included in their observations not only the great cities and temples—until recently the core of Classical archaeology—but the whole range of settlements down to very small ones. Much of the spectrum of rural settlement has been captured in these studies. Thanks to these surveys we do indeed have census data for ancient Greece, of an indirect yet telling kind.

One of the watersheds of Greek history was the campaigns of Alexander the Great (336–323 B.C.). In the wake of his conquests the whole of the Near East was opened to Greek settlement. The Classic period of Socrates, Plato, and the city-state gave way to the Hellenistic period (ca. late fourth through late first century B.C.), when large Greek-ruled kingdoms were founded in the former territory of the Persian Empire.

The archaeological data indicate that it was in the Hellenistic era that the rural Greek population began to decline. This began around the second century B.C. and lasted well into the early Roman period. Beginning at the same time, many sites appear to have been abandoned, suggesting changes in land tenure. There is noticeably a decline in the very smallest sites, especially in the early Roman era. This seems particularly to reflect the loss or abandonment of farms by small cultivators.

So Pausanias and his fellow observers, though they might have tried to make a moral point, were basically describing true conditions. Yet these conditions were not exclusively demographic. They were conditions that culminated many generations of political conflict, both local and international. Although today we think of Greece as the birthplace of western democracy, in fact democracy in Greece was never universal or assured. The history of Classical and Hellenistic Greece was, city-by-city, one of continuous struggle between democracy and oligarchy. Democratic and oligarchic forces often fought for political control. Whichever side was temporarily victorious would banish its opponents, or they would flee in fear of their lives. These exiles would immediately regroup and plot their return, sometimes in collaboration with other cities. Even Athens endured periods when its democracy was interrupted by oligarchic rule.

In the Roman period the competitive cycle came to an end. The oligarchs won. The Romans, like all imperialists, ruled native populations through their own leaders. They favored native aristocrats as their local agents and placed friendly ones in power wherever they found them. Where elites didn't exist in a native social system, the Romans created them (just as Spanish, French, British, and Americans created "chiefs" among American Indians who previously lacked them). Thus in the Roman period the oligarchs triumphed, and their triumph is reflected in the Greek archaeological record. Apparently the early Roman era (first to third centuries A.D.) was when most small holdings were abandoned. The remaining sites increased in size and display evidence of affluence. Large villas were established, some of which contained monumental structures on a scale not previously built in rural areas.

Small farmers in antiquity often lost their lands through debt. From the third century B.C. on Greece experienced periodic revolutionary outbursts in which calls were made for redistribution of land and cancellation of debts. The Roman conquest apparently exacerbated this economic problem: Removal of booty to Rome and demands for war reparations forced a liquidity crisis that would have fallen most heavily on small farmers. In such a crisis the wealthy, having diverse sources of income, often thrive. The urban poor become clients of wealthy patrons and are able at least to survive. The middle class of small landholders fares worst, for they cannot earn money. They cannot repay their debts and so lose their lands. The estates of the wealthy grow larger, while former small landholders either work for them or become part of the urban poor. Greece was not alone in this experience. Similar transformations, and social unrest, occurred also in Italy and elsewhere around the Mediterranean.

There is probably no simple reason why the rural population declined. Many able-bodied men were no doubt drawn to the eastern Mediterranean, where they could find employment as mercenaries in the armies of the great Hellenistic kingdoms. There they were paid in

silver or gold, which surely was more attractive than farming labori-
ously for a pittance. In the last two centuries B.C. a few areas had much
of their population taken as slaves by the Romans (150,000 are said to
have been taken from Epirus in 167 B.C.) or lost population from other
effects of wars fought on Greek soil. Rome also was a magnet to those
with skills, initiative, and a willingness to emigrate. Many rural dwellers
no doubt moved to the cities, Greece being one of the most urbanized
parts of the empire. Yet peasant farmers are prodigious producers of
children, and none of these factors explains why the population did not
recover. The apparent reason is that once small farms were lost to debt
or taxation, the oligarchs thereafter controlled the countryside.

The transformation in ownership had enormous consequences for
the Greek landscape. The abandonment of land was not entirely deleterious, for archaeological surveys show that it was marginal lands that
went out of production. The better lands continued in use, and aristocrats even complained of having insufficient labor to work them. The
intensity of use, however, changed decidedly. In subsistence economies,
the intensity of land use varies with the size of the holding and the status of its owner. Small landholders generally work with small margins
of surplus. They work their land intensively and need extra hands at
peak labor periods. In part to supply this need they have many children,
but of course the burden of feeding them further strains the subsistence
margin. When there is drought or a frost, or the locusts come, they fall
into debt or have to sell children into servitude. This peasant landscape
was the energetic basis of ancient civilization and was the ideal to which
Polybius and Dio Chrysostom compared Greece of their day.

Large-scale proprietors, on the other hand, use their land more
extensively and may even neglect parts of it. The sheer extent of some
holdings dictates that more distant parts be used in ways that demand
less labor. Land can be allowed to lie fallow longer or more often. Lean
years, rather than threatening the wealthy, are actually a time to make
great profits by slowly releasing grain into markets. Therefore the
threat of lean years does not force the wealthy to cultivate intensively.
Aristocrats have the freedom to cultivate in response to market
demands. In Classical antiquity they often planted vineyards and raised
livestock for urban markets. They indulged also in using their land to
compete with their peers, raising prize horses and cattle.

Thus the Greek landscape, following the victory of the oligarchs, evolved from intensive to extensive use. Marginal lands went out of cultivation. Tilled fields were converted to pasture. In many areas Mediterranean shrub vegetation would have regrown, giving the land an unkempt appearance. The landscape that Greek intellectuals preferred to sustain (with good reason) was now changed in ways not easily reversed.

The cities were also part of this landscape. While maintaining their country estates, aristocrats in the early Roman Empire spent much time in the cities. Both social life and an arena for competitive display were found there. Cities were expected not only to pay taxes to the empire but also to support their own functions. This obligation fell to the wealthy, who used the opportunity to outdo each other in largesse. Much of the public architecture still to be seen in ancient Mediterranean cities—theaters, temples, public markets, and the like—was privately built. Aristocrats paid for games and on occasion even supplied food and water. The money to do this came largely from landed estates. Ironically, the magnificent cities of antiquity that we admire today were made possible by the same rural depopulation that ancient authors decried.

In the third century A.D. this began to change. With the increasing pressures of this period and the rising cost of government and defense (chapter 3), civic service became a burden. Many oligarchs fled their now-obligatory duties to live in self-sufficient villas. Cities and their services necessarily decayed. The Greek landscape seems at the same time to have been repopulated. The number of archaeological sites increases, with rural activity in some areas returning nearly to the level of the Classical period (fifth to fourth centuries B.C.). Marginal lands were again brought under cultivation. Yet the social and economic relations of production were quite changed. Now the imperial government *required* that land be cultivated and that taxes on it be paid. To ensure sufficient agricultural labor, peasants were bound to their plots. When they fell into debt or could not pay their taxes, they became tenants of the great estates. The Greek landscape of the fourth through sixth centuries A.D. was again intensively used, but under social and economic conditions that presaged the Middle Ages.

The story of the ancient Greek landscape is one of changing contexts. These contexts were always social and political but varied in their scale. The emergence of city-states set a context in which the rural landscape had to be one of small landholders farming intensively. In the era of Classical Greece, land and demography constrained each other through the reciprocal relations of productive requirements (labor) and productive capacity. Later writers idealized this condition, and rightly so. It was the energetic basis of much that we admire in ancient Greece: art, architecture, literature, science, mathematics, democracy, and civic life. Developments far afield brought this system to an end. As political units grew in scale, the context for the Greek landscape shifted first to the entire eastern Mediterranean, then to the Roman Empire, then to the pressures exerted inexorably on the empire from central Europe and Persia. The Greek landscape in the Roman era was constrained by the entire known world and much that lay beyond it. As the Greeks lost their political freedom they also lost control of their landscape and their demography. Throughout all this period the landscape system was sustained in a biophysical sense: The land produced, and most of the time people had enough to eat. Yet Dio Chrysostom and Pausanias clearly felt that the Greek landscape had not been sustained to their day. As a social construct indeed it had not, for regulation of the landscape was no longer within the power of city-states. The social conception of sustainability remained constant even as the scale of the landscape context changed.

Implications of Landscapes in a Human Context

Clearly we cannot analyze or manage landscapes as closed loops of pure biophysical processes. Many of the earth's landscapes, perhaps even most, have been used, tended, managed, domesticated, or even built by humans. Even when human use drops in intensity, landscapes may convey the imprint of that use for centuries to millennia (fig. 4.15).

Recognizing that social–biophysical interaction shapes landscapes does complicate managing for sustainability on the landscape criterion. In each of the cases discussed, the landscape that we inherit today is the product not only of past land use but also of changing values and economies. In prehistoric New Mexico, hunter–gatherers preferred lands that today are considered unproductive. When Europeans arrived, these marginal lands were available for cattle raising because

FIGURE 4.15

Lingering effects of past land use. A prehistoric agricultural field (*left*) along the Rio del Oso, northern New Mexico, although abandoned for about 600 years, shows higher grass density than the area outside the field (*right*).

Photograph by J. Tainter.

Puebloan farmers preferred much the same landforms as Euro-American cultivators: rich river valleys. In California, hunter–gatherers supported cultural complexity by deploying ecological subsidies to diversify the landscape. The demise of the hunter–gatherer way of life has resulted here in ecosystem simplification as landscapes lost part of their patchiness. The Maya, in contrast, supported even greater cultural complexity through ecological simplification, and with the Maya collapse the landscape again diversified. No simple formula can depict human–landscape interactions, which must be understood case by case.

Greece shows how landscapes change in response to distant processes of which local people may be unaware. The triumph of the Greek oligarchs illustrates how landscape diversity varies positively with scale and inversely with human density. A landscape of small farmers was ecologically simple, with small plots, limited variety of crops, short energy chains, and little unused land. As the oligarchs acquired land and built great estates, they used land in more diverse ways. Land was used extensively, in ways requiring less labor. Market production of commodities such as wine or olive oil replaced subsistence production of annual crops. Lands were managed to lie fallow longer, and some areas were allowed to revert to whatever vegetation would grow without human intervention. Thus those who own or manage larger pieces of land have more options than do small landholders, and can manage for greater diversity.

Policy Implications on Landscapes

Whereas the lay public has few opinions about ecosystem dynamics or any other intangible ecological consideration, the tangibility of landscapes invites firmly held views about their proper form and management. In Britain there is much public outcry about the removal of hedgerows. The combining of fields is a business decision about economy of scale, with larger fields being more economical. True, the English landscape is characterized by small fields and charming hedgerows, a place of Constable landscape paintings. The irony is the public dismay in Elizabethan times, the period when the enclosures occurred and the lines of the hedges were established. Landscape is one of the most political of ecological criteria.

The battle over those original enclosures of 400 years ago is still relevant in modern times when there is intrusion into public common lands. The few public common lands that survived the Tudor and Parliamentary enclosures are jealously defended today. Epping Forest reaches down toward London and is bounded by the North Circular Road beyond the East End. In places the forest crosses the road to survive as a sad little strip of trees between the highway and the fence of Whipps Cross Hospital. Despite the hospital serving the east quadrant of London with about a thousand beds, the entrance to the main road through which ambulances would arrive was a single lane that cut economically across the narrow strip of trees (fig. 4.16). This bottleneck meant that an emergency vehicle going into the hospital would block one coming out and vice versa. Making the entrance a rational two lanes involved widening the single lane by just a rod for a length of less than 30 yards. This meant cutting down only the odd tree in the tired

227

FIGURE 4.16

The entrance of Whipps Cross Hospital from Whipps Cross Road, now two-way but until recently a single lane. A narrow strip of trees was championed for years, taking precedence over ambulances being able to pass each other into and out of the biggest hospital in east London.

Photograph T. Allen.

little strip of forest. In deference to a centuries-old issue, this minor encroachment took years of heated negotiations for the permission to be granted by the guardians of the public forest. Landscapes and policy make a volatile and not always logical mixture.

Policy at a national level has great impacts on the landscape, but it is not clear that there are concomitant effects according to other criteria. Under Earl Butts, secretary of agriculture for President Nixon, policy encouraging fence row to fence row planting removed hedges and changed many landscapes for at least half a lifetime. A decade later national conservation programs put much private farmland in conservation status for a decade. Now those lands are coming due for reassessment by their owners, and yet another shift in landscape character is expected.

National governments and international agencies often are unaware of local agricultural traditions and may harm them in their ignorance. Farmers in Burgundy, France, draw on millennia of experience with climate and local conditions. They have used this knowledge to farm sustainably. Yet as the French government accedes to international demands for open markets, traditional systems are being destroyed. The French government must now contend with tractor parades of angry farmers through Paris. In Burgundy the consequences are more serious: To compete in international markets these farmers must expand their scale of production. Hedgerows are being destroyed to allow for larger fields, and centuries-old conservation practices are being abandoned (Crumley 2000).

Although it is clear that changes occur on the landscape with national policy, there is little formal data collection that allows prediction of the influence of these policies on the material system viewed on other ecological criteria. The effect of the expiration of conservation status on water quality, an ecosystem consideration, is not known. The landscape sometimes can be used as a surrogate for all ecological aspects of the system, but the prevailing incompletely integrated approach to ecology leads to uncertainty. Fragmented approaches to environmental management mean that the extent to which the landscape is a good surrogate is unknown. A more fully integrated approach to ecology and management needs to be a high priority.

Landscape indicators are facile environmental indicators, whereas signals carried by other criteria, such as a process functional ecosystem, are poorly understood and often overlooked. For instance, let us contrast an ecosystem signal with a landscape signal in two fish and a bird example.

Example 1: The inability of some game fish to reproduce in the Great Lakes necessitates stocking these species every year. The problem emerges as an ecosystem process problem involving persistent toxic accumulation with effects on fertility directly and on eggs and embryos (Mac 1988). Economically important as the infertility of these fish is, the root of the problem was not well understood until very recently. If the fish had been an obvious landscape component, we submit that their problem would have been understood quickly. Furthermore, they would have been used as lead indicators of ecological dysfunction rather than recognized later as part of the continuing problem.

Example 2: Contrast this with the use of the Endangered Species Act in blocking and directing environmental impact in the context of perceived ecological dysfunction. On the face of it, the act seems to be about populations and organisms, but it has played a much larger role when used according to the landscape criterion. The act often is invoked in a landscape setting. Let us hasten to add that landscape does not exclude the habitat of fish. For instance, the snail darter challenged human activity on the landscape criterion more than on any other ecological criterion. A dam changes the landscape in obvious ways, and the endangered fish is seen as occupying a piece of the landscape, albeit a known stream section (U.S. Fish and Wildlife Service 1982; Etnier and Starnes 1993). The act often is used to stand in the way of changes in land use that can be expected to press a species to local extinction. Again the essential tangibility of landscapes allows them to play a critical role in conservation and remedial action.

Example 3: Contrast the ecosystem example of the Great Lakes with that of logging in the Pacific Northwest of the United States. In the Cascade Mountains, logging is first seen as a landscape issue, with great square tracts of clearcut assaulting even the most inattentive observer at any scale. The spotted owl is an endangered species. On the face of it, the bird might be in danger because of a large number of factors, including bioaccumulation of persistent toxic chemicals, like those that interfere with the Great Lakes fish. After all, the bird is a carnivore at the top of a long food chain. However, its retrenched position is seen primarily as a result of a landscape influence, the clearcutting of great tracts of old-growth timber (Carey, Horton, and Biswell 1992; Bart 1995). The spotted owl is the darling of conservationists because it has been translated into landscape terms and is a charismatic surrogate for old-growth ecosystems. We do not intend to deny the validity of the conservationist arguments but rather to point out that it is the landscape issue that is tangible enough to cause government action that affects thousands of lives and millions of dollars of resources. The contentious debates in the Pacific Northwest do not revolve around ecosystem or community issues but rather turn on landscape arguments. Policy can be turned by landscapes even when equally important ecosystem issues are ignored.

places, and *place* is a landscape term. Landscapes can unlock ecological understanding particularly because they invite a narrative treatment.

Narratives have a unique property of providing a serial arrangement for events that are very differently scaled. A powerful device in science is the dimensionless number, and it has some of the same properties as a narrative, not because it tells a story but because it has special scaling properties. In dimensionless numbers, all the units cancel out to give an essential set of relationships. For instance, drag is expressed in units that give an exponential increase in drag as the wind increases. However, some objects, such as flags or plants, change shape in the wind, and this has an effect on the drag that is unique to the deformation that occurs for the object in question. To see the effect of the deformation on drag, engineers cleverly divide the drag expected from the profile offered by the object in no wind by the actual measured drag. Because drag expected is divided by drag measured, the units of drag disappear in what is called the drag coefficient. For flags, the drag coefficient rises as the wind gets stronger because the fluttering intercepts the air to an ever-increasing degree. Plants, however, begin by increasing their drag coefficient in a light wind, but in a strong wind they bend and streamline against the wind, avoiding the worst effects of the blow (Vogel 1988). The dimensionless number of drag coefficient captures what would otherwise be a very difficult concept to grasp. Linking this back to the qualities of narrative, they remove the particular scaled properties of the components of the story. Stories consider events, and events are instantaneous in time. The order of events is not changed by physical size of the actor involved in the event. Large, slow things such as a 10-year drought may be put immediately before a large, quick event such as an earthquake that is followed by a small event involving a single person at an instant. Dimensionless numbers remove the quirks of the units in a set of quantified relationships. Narratives have a similar power in their ability to make equivalent otherwise disparate relationships, and equivalence allows seriation.

The power of narratives comes from their consonance with the human experience and the way humans remember and reflect on their experience. This is illustrated in the apocryphal story of the man who wanted to know whether the most powerful computer in the world thought and was conscious like a human being. A reasonable thing to do was to ask the computer itself. The computer replied, "That reminds me of a story." Enough said. In a large and messy situation, it is particularly satisfying to have some defining event that appears to offer the explanation. Sometimes there is a meteor that ends an age, like that of the dinosaurs, but for the most part large-scale systems appear to have only vague causality. One reason might be at the root of the way human cognition works. To understand how narrative informs us about ecological sustainability, we digress briefly into the field of environmental perception.

The two polar theories of learning and understanding are those of Piaget at one end and Gibson at the other. Gibson (1979) suggested that the physical environment is the source of knowledge, and we experience it and learn. Piaget's (1963) view is that understanding comes from doing and interacting with the environment. A Piagetian view is supported by the case of a boy with immune deficiency who therefore had to spend his life in a bubble, separate from infectious humans outside. He could see outside the window but appeared to have a peculiar and not very helpful conception of distance. The notions of near and far were in important ways beyond him, despite repeated explanations from people outside his bubble. One day an unusual fog made more distant things disappear, and this appeared to help his understanding a lot (Murphy and Vogel 1985). Without an ability to interact with their environment, humans in general do not learn very well. The vague causality that we apply to most aspects of most large systems probably comes from our limited capacity to interact with them.

Of all the types of large-scale ecological systems, landscapes are those about which we learn most easily. The explanation is that we can interact with landscapes by passing through them by this path or that. The summary of the experience can be easily captured in a narrative about place, as when Aldo Leopold told stories about places in his riveting *A Sand County Almanac* (fig. 4.17). Piaget's view is further supported by the commonplace experience of not learning one's way around a new town by being a passenger; it really helps to drive oneself, or at least take charge of the journey by being the navigator. Thus landscapes function at spatiotemporal scales commensurate with those of large ecosystems, biomes, or communities. Therefore, large landscapes resonate with the processes of less tangible but also large

FIGURE 4.17

The ecological narrator. Aldo Leopold with his wife, two daughters, and son outside "the shack," the setting for stories in his *Sand County Almanac*. The shack was the farmhouse on the land that the Leopold family worked to make wild again. Leopold remains unsurpassed as a scientific narrator.

Photograph reproduced with permission of the State Historical Society of Wisconsin, from photograph lot 3358, negative number WHi(X3)43179.

types of ecological systems. Landscapes therefore offer us a unique window on ecosystem, biome, and community processes.

One of the few ways to interact directly with ecosystems and communities is to make them small by resorting to microcosms. Piaget would not be surprised that microcosms help ecological understanding because they are manipulable, and we can learn through interaction with them. Notice that the notion of sustainability in microcosms is easy to understand. They remain stable and survive, or they do not.

The small size of microcosms takes the vagueness out of their sustainability, but the importance of insights coming from microcosms is limited by their scalar characteristics. Microcosms can be useful, but as indicators of sustainability microcosms can often be trivial.

In contrast to microcosms, sustainability of large systems is rarely trivial. Ecosystem, biome, and community sustainability in large systems will always be kept at a distance from our understanding because large-scale manipulation and human interaction with such systems are rarely available. When we address large systems, it is sometimes difficult to be explicit in circumscribing the system to be sustained. That is why the window of the landscape criterion is so important to the issue of sustainability at large. The landscape criterion often can be used to pick up signals from ecological systems viewed from less tangible criteria. The Hubbard Brook whole watershed manipulations (Likens et al. 1970) are not only unique, but they offer us a rare understanding of ecosystems. By making whole watersheds the units of manipulation, the Hubbard Brook ecosystem was made tangible, and the intangible flux of mineral nutrients was captured in the experimental data.

In summary, landscapes are interesting cases for studies in sustainability in their own right. On the other hand, they are something of a skeleton key to open understanding of sustainability on other ecological criteria. The landscape criterion will sit more comfortably on some systems than others, but it is always a powerful organizing principle that can be brought to bear on sustainability across the spectrum of ecology and resource management.

THE POPULATION

Populations bear a straightforward relationship to both organisms and landscapes, and so they capture some of the tangibility of those criteria. It is easier to move discussions of abstractions forward with tangible examples. The relationship of populations to organisms and landscapes is part of most working definitions of populations. By definition, populations are collections of organisms. Furthermore, populations are collections in which spatial contiguity usually is an organizing principle. The organisms of a population often can be circumscribed on a landscape in a way that is not available to ecosystem or community conceptions. Individuals often are assigned to a population because of their proximity. A spatial characterization of a population may be used as a surrogate for other characters shared by individual members, such as the genetic homogeneity that arises from interbreeding within a local area. If organisms are physically close together, then differences in their environments are likely to be so small that they can be discounted. Proximate organisms often share at least a recent history. Relative placement in space often is so central an organizing principle for populations that it is taken for granted and overlooked. Sustainability is a diffuse issue that needs all the help it can get from examples made tangible by their occupying a particular space.

Sustainable Populations

The population lends itself particularly well to a treatment of sustainability. Here sustainability takes two forms: species preservation and biodiversity.

First, the population criterion is manifested in efforts to preserve particular species. Unlike many other criteria, when populations are unsustainable, there is a particular time at which the last member of a species succumbs. The death of the last passenger pigeon was recorded as the last female dropped off her perch in the Cincinnati Zoo. The passenger pigeon is a striking example because its flocks darkened the skies in the nineteenth century. Apparently large accumulated capital is not security enough for systems that find themselves unsustainable. Despite the example of the passenger pigeon, often species that receive attention in the context of sustainability are those that are rare. This invites questions such as, "If rare species are in danger, do we want to try to make them common?" Probably not, but why?

The second major population issue in sustainability is biodiversity. The critical point here is that biodiversity puts little emphasis on the emergent structures that allow species coexistence, namely the organization of populations in community structure. Biodiversity is a depauperate notion that sums species in any order in a mere collection. Biodiversity makes little explicit reference to elaboration of organization.

Considering evenness, diversity is seen as greater when the equitable numbers of individuals in each species increase the chance of any individual of a given species next encountering a member of another species. Note that evenness assumes random encounter between individuals, and randomization is one standard way of making all things equal. This leveling again denies or ignores any organization there might be. There is a large literature that attempts to relate diversity to stability, starting with Hutchinson (1959), who supposed that increased diversity increased stability. A proper presentation of scaling issues and diversity has only emerged very recently (Peterson, Allen, and Holling 1998). May (1974) suggested, from arguments of system connectedness, that under certain conditions systems might be expected to become unstable with the addition of more species. Lack of success in relating diversity to stability has been instructive in that the predictions of decreasing stability with increasing diversity depend on assumptions of random or full connection between species. Because there are many examples of diverse and stable systems, May (1974) pointed out that the assumptions of random connection appear untenable. Thus diversity itself is more a bookkeeping consideration than any reference to functional sustainability. Even so, biodiversity is a touchstone term, and it needs to be unpacked as a facet of, if not sustainability itself, then a contemporary discussion of it.

Populations are adept devices for some aspects of the problem of sustainability, but population is an incompetent notion for others. The advantages of the population criterion are precision and consistency. The limitations of populations as an ecological device pertain to situations in which heterogeneity of environment and organismal response to it become unavoidable. The population criterion has achieved undeniable success and predictability in situations in which homogeneity within the population is a workable assumption. More than most criteria, the population criterion invites misapplication, arising from overconfidence, because of successful applications in some settings, such as in epidemiology. In situations where heterogeneity is unavoidable, the patterns seen in populations are principally an artifact defined into the study by invoking a criterion that is too narrow in its focus (Allen and Hoekstra 1992). Inappropriate use of the population criterion is reminiscent of the drunk looking under the street lamp because that is where he can see, although he lost his cuff-links some distance from the light. Invoking populations as the unit of study when the system is heterogeneous is more focused on the device for illumination than on the problem at hand. Such can be the case when competitive exclusion, niches, and optimality are invoked in the context of sustainability.

A unit in an observation scheme is raised for a particular purpose, and when the purpose changes, the scientist is advised to contemplate changing the units. One might well think of changing even the general criteria for observation. Consider the idea of a flock of birds and its relationship to sustainability. If breeding as it relates to sustainability were the important issue, then the heterogeneity within a multispecies flock would make it an unworkable entity. It would be better to decompose the flock into its constituent species and reach outside the flock to include other local members of the various species. Thus the population as a flock is abandoned in this case in favor of a population circumscribed for other purposes. Of course the members of the flock continue to flock together, but that is not relevant to the breeding that occurs within each species. Alternatively, the argument might be that flocking with members of another species allows increased survival. In that case multispecies flocking becomes relevant to the survival processes in the functioning of the population defined by breeding within one species. At that point one might work some sort of mapping between the multispecies flock, as a population, and single-species populations. A new upper-level population with members from different species that is based on coevolution between species probably is at odds with the facts. The point is that populations, like every other type of ecological entity, must be defined for each specific purpose. Nature itself does not tell the observer what is the correct definition, for the burden of responsibility for asserting a definition rests on the one who devises the question that invokes this as opposed to that population. The essential tangibility of populations encourages a naive realist tone in discussions of populations, which manifests itself in inflexible definitions, usually consisting of single species. Precision in definition that characterizes populations brings with it the need for flexibility as new definitions create new populations ad hoc.

Sustainability in Aquatic Populations

The positive contribution of chaos to discourses on sustainability hardly ever turns on a technical application of some nonlinear equation. Nevertheless, chaos is very helpful as a metaphor. Chaos tells us that just because a situation is extremely complicated as to the states found in ecological data, the emergent pattern may nevertheless have a simple underlying generator. Chaos is very reassuring. It tells us that sustainability need not be hopelessly lost just because some aspects of the situation are hugely complicated. Chaos says that no matter how complicated an issue of sustainability may appear, there may still be a solution that is simple enough to be tractable. We do not know how often that is the case, but it does deny the option of immediate acquiescence, which ecologists facing complexity had always done before chaos theory. "It is too complicated so let's not try" is disallowed.

The fishery off the Northeast Coast of North America probably is best described as chaotic (Acheson, Wilson, and Steneck 1998). Of course chaos applies to deterministic equations, and we have no hope of ever knowing the equation that characterizes the fishery, if such could exist. Even so, the tendency to increase the proportion of fish with rapid population growth rates can be expected to create a fishery that moves unpredictably, even if we knew most of its important parameters. Once growth rates invite description of the system as chaotic, even if we could run the history of the material system over again, it would not turn out the same way twice. There are ample data to suggest that the ecosystem is supplying roughly the same biomass of fish (Apollonio 2002), despite the systematic destruction of species, one at a time, in the order of the most desirable. The fishery off Maine has been worked all the way down to dogfish, with the desirable haddock and cod long gone. The experiment based on predictions of orthodox population models has been a disaster.

The population model for fisheries turns on medium-sized fish growing fastest. Small fish have little mass as the base on which to grow, and large fish do not translate food into increased mass. The most efficient time to turn a live fish into a human resource is when it has reached a size just beyond the period of fastest growth, the point when the transfer of food into body mass begins to decline (fig. 4.18).

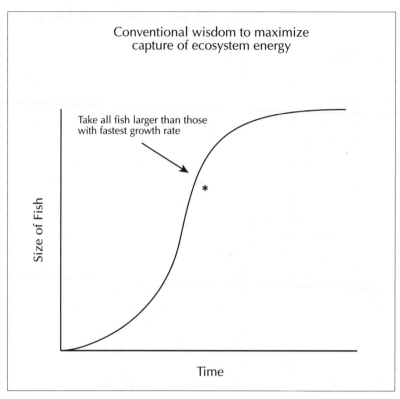

Conventional wisdom to maximize
capture of ecosystem energy

Take all fish larger than those
with fastest growth rate

*

Size of Fish

Time

A

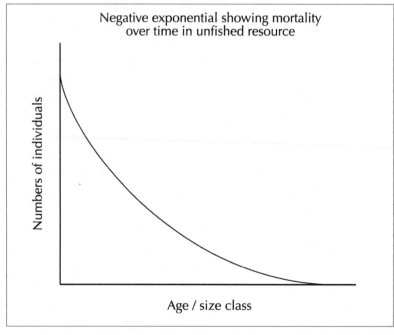

Negative exponential showing mortality
over time in unfished resource

Numbers of individuals

Age / size class

B

This is the logic of an equivalent terrestrial model that says any forest growing slower than the rate of monetary depreciation should be cut. Clearly there should be a limit on fishing small fish, for they have a large potential to grow and capture resources that can be channeled into humans if only we wait. We disagree with the conventional model when it says that old fish should be fished out of existence because they take resources away from fish of a size that could turn those resources into fish harvest faster. Clearly one does not want to harvest too many fish before they are significantly reproductive.

With 150 years of this experiment behind us, we can be confident that the population model just described does not preserve the fishery. As a species is fished, it is devastated. Therefore, it is time to invoke some other population model before there is literally nothing of value to lose. Population models, like the conventional fishery argument, are most commonly driven from below. That is, the assumption is that the population is produced by processes from below and inside the population. It is a reductionist approach, in the parlance of Rosen (2000).

As an alternative, let us try a strategy that works top-down, imposing constraints that are not derived from the aggregate behavior of the lower level. In this argument, management derives from a need to replace a missing context with some sort of surrogate. For example, modern forests often exist today as mosaic patches rather than the matrix of forest that was there before significant human influence. Therefore, one needs to identify what services would have been offered to the managed remnant by the missing context, if it were there. Notice that if one did indeed offer those services, then the orphaned patch would behave as if the missing context was in place. Accordingly, without any need for ecologists to deal with the low-level intricacies of detailed function within populations in the management

FIGURE 4.18 (*opposite page*)

Fish growth and death. (A) Fish grow fastest and so convert food into fish biomass most efficiently as middle-sized adults (the asterisk corresponds to the same size as in figure 4.19a). (B) The unfished population declines in numbers by size or age on a negative exponential path.

unit, the forest patch would return to healthy function. In a top-down model, the managed unit would subsidize the management effort by achieving full functionality.

In case of the Gulf of Maine fishery, is there a missing context? Context can be considered as a constraint. Constraints give the constancy against which predictions are made. The temptation is to assert some simple tangible entity as the constraint, without showing how the constraint works. Plausible narratives as to how such constraints work will not suffice because there is always some other story that can be erected instead, with no indication as to how to choose which narrative deserves privilege. Plausibility is the Achilles heel of biological models in general, and population models are no exception. We offer the big fish in the Gulf of Maine as a surrogate of the missing constraint, not as the actual limiting factor itself. We do not want to rely on plausible narratives as to how big fish might work as a surrogate. The alternative strategy therefore is to fish the middle-sized fish as hard as you like but allow a small number to become big. Explicitly, do not take the large old fish.

There is evidence that large fish lay eggs of particularly high quality. The narrow genetic base that characterizes particular cohorts of fish implies that the entire cohort comes from a very small number of individual fish, the few large specimens. Apparently, high-quality eggs give individuals that survive the rigors of growing and surviving much better odds of reproductive success than do the mass of lower-quality eggs coming from moderate-sized parents.

Constraints not only impose limits on lower levels in the system, they also offer a context that protects. Large fish are more resilient to changes in temperature and food supply. Their large body mass has a favorable surface-to-volume ratio to shut out the environment. Their large mass provides resources to bridge times of deprivation, and their physiology means that they use less of what they have. Therefore large fish offer a reliable supply of eggs for the next generation, even after a hard year.

The appeal of this alternative model is that it invites translation into more workable regulations than the prevailing model that is the basis of extant regulations. The model that relies on large fish invites a number of feasible management actions. A total ban is easy to impose, particularly if it is intended only as a window large enough to create a cohort of big fish. Explicitly rearing large fish as opposed to flooding the waters with hatchery fingerlings is a possibility. Large fish survive in places that have been difficult to fish until recently. However, sonar allows fishing in shallow rocky areas inside barriers. Once the large fish are reestablished, simply banning some modern technologies might be sufficient to keep them from being fished. A short total ban on all fishing, followed by continuing long-term restrictions in certain small areas, should be sufficient to achieve a turnaround in the fortunes of the fishery.

Following the caveat that the big fish are not in themselves the constraint but are a surrogate for that constraint, it is probably important to work with a simulation model that turns on some simple, preferably mathematical parameter. In such a model, the mathematics of system behavior is transparent, even if the rich scenario of fish survival is ignored. A negative exponential survivorship curve reveals the underlying sizes of fish and their numbers in the system were it not fished. The consequences of different fishing regimes may be superimposed on that negative exponential (fig. 4.19). Notice that this curve is modified in a very particular way by the intrusions of the modern fishery. Basically, the abscissa is raised so that the larger growth classes are eliminated. The model we propose would take care not to remove the largest age and size classes. The more intensive the fishery, the more it would bite into the middle age classes. All that is needed to simulate such action is a stronger negative exponential or power function on the survivorship curve. With just one term modified to achieve the desired effect in the model, one is unlikely to be misled by artifacts from hidden, unexpected mathematical ramifications. While plausible scenarios might be reassuring to the natural historian, implementing the details of a plausible scenario in a model is an invitation to self-deception. There are too many turns in the story to model them all in any general way. The approach we support relies on the more general and powerful notion of constraint, as opposed to the narrower concept of limiting factors with their explicit rate-limiting effects on the outcomes.

There are two precedents for this model of preserving the large, old fish. First, there are data from freshwater, where replacement of big fish has been associated with improvements in total system function. Kitchell and Crowder (1986) asserted that the reason for the improvement in water clarity depths in Lake Michigan is the new fishery in

243

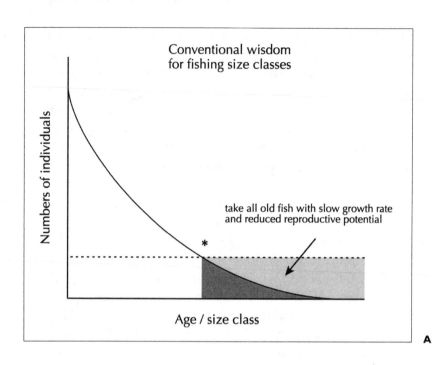

Conventional wisdom
for fishing size classes

Numbers of individuals

Age / size class

take all old fish with slow growth rate
and reduced reproductive potential

*

A

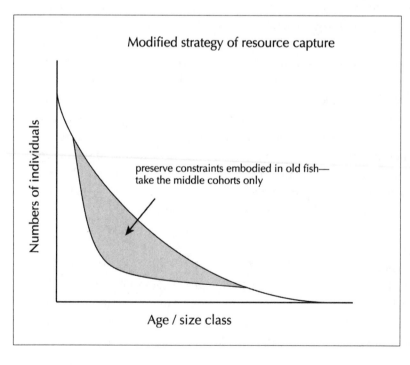

Modified strategy of resource capture

Numbers of individuals

Age / size class

preserve constraints embodied in old fish—
take the middle cohorts only

B

large sport salmonids. Second, the Maine lobster fishery continues to be heavily harvested but remains in good condition. The nature of the fishing gear for lobster, lobster pots, lets the small ones out and does not let the big lobster enter. Accordingly, there are both a lower and an upper limit on the size of lobsters taken. It is encouraging for managers that there is still a thriving lobster fishery in the neighboring states and provinces, even in the absence of appropriate regulation of fishing gear or legal upper size limits for lobsters there. On the face of it, it appears that those lobster fisheries are supported by the efficacy of the Maine lobster fishing regulation.

Sustainability and Human Populations

In social as in biophysical systems, managing by the population criterion means that the system has already become unsustainable. As population declined in the later Roman Empire, for example, the state responded by regulating the population of workers. Shortages of labor in agriculture, industry, the military, and the civil service were met by freezing workers into hereditary occupations and by conscription. Shortages of agricultural labor meant that landowners offered the least fit of their workers for military service. City councils were held collectively liable for the taxes on deserted lands. In response, many urban elites moved to country estates, leaving city services to decay or be assumed by bishops. The main attempts to manage the problem at a system level were periodic government efforts to aid poor children and orphans and to give tax remissions to parents with large numbers of children. These were fine efforts but did not have enough demographic

FIGURE 4.19 (*opposite page*)

Contrasting models of fishery management. (A) The prevailing management strategem is to allow fish to achieve middle size when growth is maximal, and then fish them out, along with all other larger fish. The strategem amounts to raising the abscissa on figure 4.18b to the size of fish at the asterisk on figure 4.18a. (B) The preferred strategem is to fish hard on the middle-sized fish but preserve the context of the population by leaving the large, old fish.

effect. As discussed in chapter 3, the great system-level problem—the crushing costliness of complexity—was something the beleaguered government could do nothing about.

Human statistics under the population criterion often have foreshadowed societal unsustainability. In the Western Chou dynasty of China (1122–771 B.C.) and the southern Lowland Classic Maya (ca. A.D. 250–1000), as well as in the Abbasid Caliphate and the Roman Empire, a leveling or decline of population generations before collapse signaled that the peasant population was stressed so much that it failed to replicate itself (Tainter 1999). Peasants are well known for prodigious reproduction. A peasant population unable to perpetuate itself is a certain indicator of systemic instability, which may manifest itself eventually in a political collapse.

Modern Conservation Biology

Sustainability and conservation have been a consideration for some time but with a rather different posture than prevails now. There are giants in the literature such as Aldo Leopold, for whom these issues were a rich collage of community, landscape, ecosystem, and population biology. By contrast, the modern self-conscious field of conservation biology is dominated by the population criterion. This invites questions as to the origins of the modern population-based conservation and the desirability of leaving such important issues in the hands of a specialist group. Whence comes their sense of ownership? So complete is the ownership of conservation that population biologists specializing in conservation often appear unaware that they are taking a distinctive population posture. In modern conservation biology, species and genetic foci, two hallmarks of a population posture, are almost ubiquitous. Conservation is seen by a majority as preserving species because their local or global extinction would be an irretrievable loss of particular genetic configurations. The loss of species as members of communities or as pieces of ecosystem function is not the main issue, and community and ecosystem implications often are raised as token, secondary justifications for genetic, population-based arguments.

Part of this expansion of population biology into conservation comes from genuine desires to do something about an ugly situation. Part of it may come from a need to justify funding in a political and fiscal climate that demands accountability. Given the need to appear useful, applied population biology amounts to epidemiology and conservation. Part of the sense of ownership of conservation may come from the fact that the population criterion gives one of the simplest and therefore most widely accessible accounts of sustainability. One of the clearest cases of an ecological situation that has been obviously unsustainable is when a species goes extinct. The population criterion does indeed own that manifestation of nonsustainability, and its practitioners play it for all it is worth.

Generalizing from one species to many, devotees of the population criterion champion biodiversity. In attempts to aggrandize a first-order notion, biodiversity has taken on other meanings. For some, biodiversity includes aspects of the ecosystem criterion, but rarely is this operationalized or even taken beyond hand waving. The diversity is expanded to include a diversity of process and function. Under such expansion of definition, biodiversity can mean anything and is likely to take on many and confusing meanings. Here, we will limit biodiversity to mean diversity of species, and in doing so we hope to cut the Gordian knot. That being said, we now take issue with the general usefulness of diversity as an ecological measurement. Diversity is a very easy thing to measure. True, there are some complications beyond S, species number, but evenness and like considerations add few complications. Various sorts of diversity are recognized: diversity within the plot, diversity across samples, and diversity across large regions, such as continents or faunal realms. We submit that the interest in diversity is a reflection of its ease of measurement as opposed to large ecological meaning. To put it bluntly, diversity is a measurement waiting for a concept.

Little distinction is made between the diversity of a functional whole as opposed to merely a collection whose only characteristic is that all the bits are in the same place at a given time. The diversity of coherent units is worthy of study, whereas mere collections are more a reflection that the data are at hand. The difficulty is separating systems whose diversity is a reflection of some biologically meaningful whole from mere collections that are caught in a data collection protocol, with little else to connect them. Preston (1962) did make that distinction on an empirical basis, cleaving the diversity of "islands" from mere sampling units, but he receives only lip service nowadays. Co-occurrence at a place is an interesting matter under the landscape criterion, and that is where the significance lies for much that is studied under the rubric of diversity. However, the literature of diversity generally does not distinguish between functional multispecies wholes, as opposed to mere collections. The essential problem with diversity is the isolation of the parts; the relationship between them is unspoken, as is consonant with a reductionist approach. Diversity as the anchor of an ecological investigation may not refer to a functioning system. If there is an organized whole, diversity indices often are a messy and generally uninformative expression of the system.

In other chapters of this volume we make the distinction between systems that are complex and those that are merely complicated. Increases in diversity are increases only in structure that complicate systems. Diversity does not relate to system reorganization in any straightforward fashion with regard to stability or complexification. The act of counting species in a list excludes elaboration of organization. Even adding the complication of relative numbers in species explicitly does not help. Diversity indices capture only elaboration of structure, with no reference to functionality beyond a general expectation that more parts means more organization. Sometimes diversity may indicate a larger organization, but there is little theory that suggests when or why. Structural elaboration is independent of organizational elaboration, but almost all of the literature of diversity sweeps that issue under the rug.

Hierarchical Structure in Populations: Metapopulations

Early work on metapopulations goes back to Levins (1970) and to Sewell Wright (1970) writing on demes on adaptive landscapes. Andrewartha and Birch (1954) gave the first explicit statement on metapopulations. The collection of populations over time is called, in modern parlance, a metapopulation. Arising from a need to deal with system complexity, the notion of metapopulations has emerged in conservation biology. Much as the new population-based conservation biology pays inadequate attention to the large heritage of conservation of decades earlier, the literature of metapopulations in conservation is very self-contained. Many of its principles were laid out years before in hierarchy theory (Simon 1962) and the ecology of complex systems. That being said, metapopulations are a worthy member of the family of complex ecological entities. Metapopulation models are the hierarchy theory of the population criterion, although they are not as new an idea as their proponents suggest. More than a quarter of a century ago, Loucks and colleagues (Loucks 1970) identified upper-level stability in a mosaic of habitats turning over, each patch at a different stage in the disturbance cycle.

Metapopulations can be a useful concept in matters of sustainability. Some species are impressive in the size and density of their populations. However, there does not appear to be safety in numbers. If processes that reduce numbers in such dense and large populations are inexorable, then even these common species can disappear. The once ubiquitous American chestnut fell the victim of chestnut blight in about the same period as the extinction of the passenger pigeon. On the other hand, many species persist as uncommon members of the flora but as metapopulations (McEachern, Bowles, and Pavlovic 1994). It depends very much on the grain of the discourse, but there are some scales at which many species appear to work less as persistent large populations and more as isolated populations, some of which become extinct. However, local extinctions are replaced in the larger scheme of things by the establishment of new local populations elsewhere. It is not clear that the scale at which metapopulation is an apt description is also a scale where critical constraints operate. However, in some cases, the species persists importantly as a set of local populations winking in and out of existence by a process of invasion and extinction.

Allen and Hoekstra (1992) observed that management often is the process of compensating for a missing context. When a rare species needs help from management, it is often exactly because the local population is fine, if vulnerable, but the metapopulation has been removed from the larger context. The difficulty here is that it is a perfectly natural part of the biology of the species for local populations to become extinct, including the last one. Management practice includes shortening the time to reestablishment at the local site, such that there is functionally no time between periods of site occupancy; local extinction is met immediately with contrived reestablishment.

Because breeding within populations is one of the central characteristics that make a collection of organisms a population in the first place, populations biologists often see sustainability in genetic terms (e.g., many contributions in Temple 1978). True, each local population in a metapopulation is likely to have distinctive genetics, which might indeed be a matter of local selection and adaptation to local conditions. Distinctive local genetics may also be just the persistence of a natural founder effect. In prairie restoration circles, there is much criticism of nurseries that provide seeds of native species from out of state. It is our opinion that such concern for sustainability of local genetic integrity is overstated, and we raise two arguments that address the issue.

First, genetic integrity is a local matter. In local population biology, extinction snuffs out such local variation. We stated at the outset that sustainability must be set in a realistic and explicit time frame. Failing to recognize the ephemeral nature of local genetic integrity is a failure to be realistic and to acknowledge an appropriate time frame. Second, genetic integrity is easily recreated by microevolution, which appears fully capable of replacing with a facsimile what is lost in local extinction. At worst there is replacement with alien genetic strains. True, entire species can be genetically overwhelmed, as in the case of the hermaphrodite species of hawthorn that has recently disappeared in the Pennine uplands into a genetic ocean of the common weedy *Crataegus monogyna*. Loss of a whole species globally is a real loss, but it does not provide an excuse for being a genetic purist at the level of more narrow taxa such as demes. Simberloff (1988) pointed to a literature where inbreeding depression cannot be demonstrated, despite conventional expectations.

There probably is one situation in which the concern of genetic integrity is well founded. It is almost certainly very important in local cultivar varieties, much more so than in plants that are not usually sustained by human husbandry. The importance of genetics in cultivars comes not from esoteric issues of population biologists but rather from the uses to which humans put local cultivars. In agronomy, genetics is a very practical matter of germlines to press breeding programs forward. Local genetic integrity is preserved in primitive crop lines as they persist as genetic entities. Zimmerer (1998) validly invoked a metapopulation model for local land races of potatoes. Fields are planted or not, and so populations of the land races wink on and off, with the metapopulation surviving as the crop of a region. Zimmerer rejected what he called adaptationist models, where local varieties are selected in a local condition. In chapter 1 we cited Simberloff's skepticism on this matter. Too often, metapopulation theory is invoked in inappropriate situations where there is a stable source population. The metapopulation model reduces to a trivial condition if local populations are reliably fed from a mother population that is stable. In agronomic genetics, the concern is indeed for a large source population in the first world and not a matter of sustaining a fragmented metapopulation. Agronomic genetics is a matter of long-term health and productivity of the huge, homogeneous populations of crops on which present human well-being clearly depends. Some crops are precisely in the same danger as the passenger pigeon and the American chestnut. We need to sustain the metapopulations of indigenous varieties of cultivars exactly because the crops that feed the large bulk of humanity are not metapopulations.

THE COMMUNITY

A critical issue in the debate on sustainability is the definition of a reference state against which sustainability is to be measured. The realist default setting often is a vaguely defined pristine condition that pertained before the anthropogenic stresses of today. The community criterion is the ideal vehicle to debunk the pristine system as the reference for ultimate sustainability. The community is the system type most often implied in the mythical pristine system, so the community conception is ideal for disabusing ourselves of an ecological Garden of Eden. Although pristine nature before the coming of our species is conjectured to be sustainable, it is not a suitable reference point for models of humanly sustained systems. As noted in chapter 1, if Greek and Roman smelting produced hemispheric levels of atmospheric lead not seen again until the Industrial Revolution (Hong et al. 1994), then there have been no pristine systems for a long time, and they cannot be recreated. Anyway, we would not want to revert to a pristine condition as the reference state even if we could. Human influence cannot be undone and, if somehow removed, will again come to the fore. As Herakleitos might have put it, one cannot step into the same forest twice.

The more critical problem with the pristine system as the benchmark for sustainability is that it is at best vaguely defined and so offers an inadequate baseline. Lack of definition is anathema to the approach we recommend; feeling confused and overwhelmed comes exactly from a lack of definition of the bounds of the system and the goals of management. Our whole strategy is to define and achieve: define what it would mean to achieve sustainability and then move on to achieve that situation. Embracing pristine nature as a surrogate for goals of a program for sustainability is defeatist and irresponsible.

In more positive terms, community restoration is one of the important tools in implementing notions of sustainability. Community restoration is more than a means of sustaining some foundering systems; it is also a device for recapturing intellectual rigor. Community restoration offers a concrete challenge to the state of the art of theory. If ecologists really understand a community, such as a tallgrass prairie, then they ought to be able to create one.

In a much more contrived circumstance than wild land community restoration, farmers have a wealth of experience in creating stable species complexes, as in a hay field. And a community is at some level a stable species complex. There are insights to be gained into the nature of the community concept by exploring the limits of the success of farmers in creating a good hay field community. Agroecosystems press the limits of community by reducing the system to a very few species. However, there is advantage in viewing pine plantations as monotypic communities rather than as populations. Agrarian systems have been extremely helpful in advancing plant demographic studies in the style of John Harper (1977), and it is time they were used in a community setting. Robert Rosen (1989) recommends stretching a concept until it fails. What then has to be done to make it work again gives insights both for the old situation, in which the concept functioned well, and for the new situation, in which it failed. Stretching the community concept into the realm of the farm is insightful and is extremely pertinent to sustainability of humanly important systems.

Community as Opposed to Population

To highlight the distinctive character of communities with regard to sustainability, we now contrast the sustainability of communities with that of populations. As a stark contrast to the population criterion, the community criterion invokes entities with heterogeneous parts. In fact, maintaining that heterogeneity is central to the challenge of sustainability under the community criterion. Loss of sustainability in communities may involve degeneration to the homogeneity that pertains under the population criterion, as the work of Roughgarden (Roughgarden and Pacala 1989) shows. Roughgarden's achievements press the whole intellectual framework to its limits. Community ecologists had been intuiting for some time that communities were not reducible to populations; understanding all the populations in the community in population terms would not solve the issues that the notion of community raises. But it takes someone from the population side of the argument to press the population paradigm to its limit, to make the point in terms that might gain general acceptance.

Roughgarden's work was an investigation of community structure using lizards in the Lesser Antilles, the chain of islands that circles southward in the Caribbean close to South America. Although the original intention of the work was not a comparison of populations and communities, nor an investigation into sustainability, it emerged as the classic example that addresses both those issues. The plan at the outset was to look at incipient communities so that a generalization could be made toward the more complex systems that are studied by most community ecologists. Roughgarden's was the conventional agenda of a population biologist.

The system appeared to cross the threshold from populations to minimal communities in that some islands had one lizard species on them, and others had two. If there was one species alone, then it was always close to what is presumably a climatic optimum size of 55 millimeters in length. If there were two lizards on an island, then one was about 100 millimeters long, and the other species was 45 millimeters, smaller than the presumed climatic optimum size. At first glance it looks as if indeed the two-lizard condition was the beginning of a community. The larger lizard would appear to be the later arrival, and the decrease in size of the original species by 10 millimeters would indicate an accommodation to having a partner. The technical name for the accommodation is *character displacement*, the argument being that the decrease in size is selected by giving advantage to smaller lizards that avoid competition with the larger lizard.

On further investigation it emerged that there is no community at all. There appears to be character displacement occurring, but it is not part of the emergence of a community. At some places on the islands there are local fossil deposits that correspond to roughly a thousand years of occupancy of favorite sites. The sizes of teeth at different levels in these deposits indicates changes in tooth size in a way that suggests oscillations in the size of lizards occupying the site over the time of the brief fossil record. Another telling point is that if there are two lizards and one is restricted in its range, it is the small species that appears to be retreating in a losing battle up toward the peaks of the mountains. From these and other lines of evidence, it appears that the two-lizard islands are not emergent communities but are populations caught in the act of competitive exclusion. No community is in process of emerging; rather, naked populations are failing to accommodate sufficiently.

Roughgarden's scenario is that isolated lizard species, one to an island, have evolved to the optimum size. In communities, physical limits are only a minor part of selection, and so often no species is selected to be climatically optimal. On larger islands, such as Hispaniola, competitive exclusion cannot reach its conclusion because the competitive regime is muddled by a wide range of environmental variation and a large number of species as players. One of these larger, less than optimal species from Hispaniola, or an equivalent island, arrives on one of the small islands of the Lesser Antilles. It then outcompetes the smaller resident, which responds by character displacement, but to no avail. The smaller species is driven to extinction, as the competitive exclusion principle might predict in such a constrained setting. Competitive exclusion is not seen often in nature, perhaps because the evidence is lost in extinctions past, in the opinion of Ives (1998). On Roughgarden's islands, the victor is selected to become smaller to accommodate to the optimal size for the climate. Thus one is not dealing with a community but with naked populations that have been excised from their communities on some larger island. In the absence of heterogeneity and community constraints on small islands, population properties have a rare opportunity to express themselves unmasked, in a natural setting.

In communities one is interested in why the loser in exchanges between populations does not lose it all. The loser is the phenomenon of interest, not the winner. John Harper (1967, 1977) came close to expressing that position, although in the final analysis he cast his findings in population terms. The study in question turns on the relative abundance of five poppy species. The phenomenon of poppies in wheat fields is nowhere near as common as it was. At the end of the nineteenth century, Impressionist painters were daubing golden yellow for corn on canvases but adding splotches of bright red poppies. In World War I, some of the fiercest fighting was in the poppy fields of Flanders. But all that was before the general application of broadleaf weed killers. Harper was lucky enough to catch the tail end of English wheat fields in their full natural beauty. He noted that there were five British poppy species, one common, *Papaver rhoeas*, and the others less so (fig. 4.20). Almost all fields contained at least

FIGURE 4.20

Poppies, a common weed in wheat fields before the coming of broadleaf weedkillers.

Photograph by R. Mitchell.

two species, one of which was often *P. rhoeas*. Sometimes all five species were present.

Counter to first expectations, the less common species as a group were represented by at least one of their number as regularly as the common species. First expectations do not take into account the fact that being present in a site is fairly independent of the numbers of individuals one would expect to see at a site occupied by a given species. Metapopulation considerations do not give clues as to density at a site because invasion and equilibration of numbers are separate processes at different levels. A density other than zero is constrained by successful invasion in the first place.

Clatworthy and Harper (reported in Harper 1977) experimented on four of the five species. Harper used the device of the de Wit replacement series experiment. It is an intuitive design, which has its critics but is still insightful. The de Wit design sows species together in different densities and compares results with species grown alone. In some of the experiments *P. rhoeas* was sown as the rare partner, and the other species, less common in nature, was made artificially dominant. The artificially common species in the experiments interfered with themselves more than with the *P. rhoeas*. Harper's interpretation was that an ability to compete comes from a capacity for interference. However, as it wins the competition, that very same increased interference is most often directed at members of the strong competitor's own species. There is a negative feedback, and success is self-contained. The pressure to deal with within-species competition is relaxed for the losing population as it encounters less of its own. Individuals of the rare species therefore can afford to interfere with their neighbors to a greater extent because their immediate neighbors usually are not of their own species. Selection therefore is toward greater interference for the rare species and toward less interference for the common species. Harper argues for a long-term seesaw, where the capacity for greater interference is alternately selected for and against: for when the species is rare, against when the species is common.

Occasionally, the deWit replacement has been used on animals. Anthony Seaton conceived of using wild-type fruit flies against a mutant line in a replacement series to see whether there was evolution of competition strategies. What made Seaton's work so distinctive is the way he put constraints on the system that resemble community limitations, thus allowing him to look at population dynamics in a more meaningful setting. The critical constraint was that he stepped in at each generation to reset the balance of the outcome of competition. Each new generation was "seeded" into the experimental tests, always at a level independent of the numbers generated in the previous competition experiment.

He raised flies in pint milk bottles on a base of agar nutrient medium. In his protocol the pure cultures were established by 12 fertilized females. The yield was the subsequent numbers of flies in the next generation in the milk bottle. He started the mixed cultures with just six wild type and six mutants. It is a tiresome protocol to manage because one has to remove the females for the next round of the experiment while they are still virgins because the different types of fly would readily interbreed. As he set up each new generation, he short-circuited the process of competitive exclusion, thus allowing interesting quasicommunity processes to work. He raised wild-type and mutant flies either naive to the other type or exposed to the other type for five generations. Then in the replacement series he pitted the naive strains of each type against naive members of the other type, naive wild type against exposed mutants, exposed wild type against naive mutant lines, and exposed members of each line against each other, giving four experimental regimes.

The results of this experiment have implications for sustainability. The critical measurement is the number of flies produced as a per capita increase over the original 12 flies in pure culture or 6 flies of each strain in mixed cultures. Results showed that naive lines from each strain suffered from competition with the other respective strain. Per capita increase was greater for both naive strains in pure culture than for the naive line mixtures. In other tests, exposed lines outcompeted naive lines from the other strain. In the final test, Seaton mixed exposed lines. In terms of per capita reproductive success, both strains benefitted from the presence of the other strain. Thus the mixture of exposed strains led to the highest densities of flies in the next generation. It emerged that the exposed lines had evolved to fill separate niches.

Although a milk bottle with agar in the bottom might seem a homogeneous environment to a casual human observer, the flies who have to live in there make fine distinctions. Apparently, the agar at the edge of the plate, near the side of the bottle, dries out to a degree, whereas the agar in the middle, away from the glass, remains sloppy. Over the five generations of being raised with the other strain, the exposed strains appear to have evolved to use only one part of the agar, either the dry edge or the moist middle. Each strain evolved a special preference that kept it out of the way of the other strain. The naive wild-type and mutant lines together blundered into the other type in their resource capture and so felt more between-type competition. Minimizing competitive interaction appears to be a device that leads to fecundity and was therefore selected.

Seaton's fly experiments cast a shadow across the utility of population work that depends on the exquisite homogeneity of experience of individuals. Seaton's work indicates that even the most apparently homogeneous settings are heterogeneous for the organisms that are adapted to live in them. Furthermore, if the environment starts as appearing homogeneous, then microevolution can easily make it functionally heterogeneous in as few as five generations. The implications of this for the usefulness for sustainability of the population style of investigation are discouraging. There is necessarily a long time line associated with sustainability—that is, long relative to the speed of the processes under investigation in most population studies. Very few generations can change the functionality of the environmental setting, even if it is the same physical place in essentially the same physical condition. If one studies the system for such a short time that environmental heterogeneity has no time to manifest itself, one might as well use the organism criterion and ignore populations. Alternatively, if one studies the system for a time long enough to be relevant for sustainability studies, then the populations will change the way they read even the same environment. The problem is compounded in changing environments. The assumptions of homogeneity that are needed in most working definitions of populations are likely to be violated in ways that are important.

Changing significance of the environment is the matter that is raised by Harper's poppy work. The common species occupy the landscape at a different scale from that of the rare species. The term *mean area* refers to the average size of the area surrounding an individual without another individual present. Given a population of a specific density, some individuals have more unoccupied area around them, but others have less than the mean area. The mean area of occupancy of an individual belonging to the common species is small. The common species occupies the space square foot by square foot. The mean area occupied by a member within the rare species is much larger, each plant occupying a space hundreds of times larger. Of course its area of occupancy does have crowded into it plants of other species, particularly members of the common species, but here we are concerned only with landscape occupancy of the rare species. There is much more environmental heterogeneity in areas the size of the mean area of the rare species than in the smaller mean area of the individuals of the common species. In an area that is heterogeneous, the rare species can regularly find a place where it emerges as the anomalous winner in the process of resource capture. Heterogeneity over space and time places significant limits on the utility of the population approach to sustainability. In contrast to the population criterion, the community concept has no trouble dealing with changing environments over both time and space and so is generally a more useful device for dealing with issues of sustainability.

Forest Stand Simulators: Community–Population Hybrids

Closer to our conception of the community criterion but still occupying an intermediate condition between population and community work are forest stand simulators. In population biology they fall under the rubric of individual-based models. Modeling all the trees in a plot, as does the family of models derived from Botkin's JABOWA model, depends on massive computational power. In Shugart's FORET model, hundreds of trees are tracked individually in the simulation of approximately one tenth of a hectare of forest.

Shugart has had success in simulating forests around the world by just reparameterizing the species in the FORET stand simulator to the new species in the new region to be simulated. The theory that underlies these stand simulators is that the organizing principle for forested vegetation is that resources come from above. No assumptions are ever simply correct, so one should not be distracted by the fact that certain assumptions in the stand simulators are patently false. For example, there is no spatial information as to the placement of trees in the plot.

As far as the models are concerned, all the trees grow from one spot in the center of the unit. The simulation is of a certain area, but that is only implied by the amount of biomass that is allowed to grow in the area; there is not an explicit area with trees explicitly placed within it.

The critical relationships in the FORET class of models view individual trees from two levels (Shugart, West, and Emanuel 1981). First, the trees are autonomous photosynthesizing entities, with their leaves held modeled as a disc at a certain height. The leaves are exposed to light that is calculated explicitly for each individual, depending on the amount of foliage at a greater height. Second, the leaves on the individual tree are part of a canopy of leaves that starts at the exact height of the tree in question. No distinction is made between the contribution of the leaves on this, the shortest tree in the canopy, and those belonging to the tallest tree. Thus the material canopy is modeled as a set of canopies, each one starting at a height corresponding to that of one of the trees. All trees are subjected to the shade of the canopy that corresponds to the canopy as calculated to include the single next higher tree. There are probabilities for no growth, death, or for the amount of growth, all depending on the amount of light the individual is modeled to receive.

In the model the nonlinear architecture is implied in a feedback that works so that the larger the tree, the greater its chance of growing even larger. Conversely, failure to grow allows other individuals to grow past the slow-growing trees and suppress them further. Each species has its own probability distributions, shade-tolerant species being assigned a greater probability of survival and growth than a shade-intolerant species for a given low-light regime. The parameterization of species is done by simple but remarkably effective characterizations. There are two principal sources of information. One is range maps for the species that define its permissible larger climatic context. The other is knowledge derived from foresters with extensive field experience, the conventional wisdom of a manager sitting on a stump. This gives the local information as to how a given species responds to local conditions within a stand.

Much of the work on stand simulators through the late 1970s and early 1980s pressed computation of the day to the limit. Even restricted computationally in this way, at least one form of the model showed hysteresis, a character of nonlinear models of complex systems. Although this may seem an esoteric point, it has implications for sustainability. If the environment of the upland forest for east Tennessee is changed to favor the beech or is changed to favor its competitor, the tulip poplar, the change in the population of beech is asymmetric. Shugart et al. (1981) showed a graph of temperature in degree-days where the track to and from tulip poplar dominance differs depending on whether degree-days are increasing or decreasing. Alternatively, this indicates that the lower altitude limit for beech on the Great Smoky Mountains is different depending on whether beech is encroaching or retreating. The implications for sustainability are that it is easier for a declining population to hold its own than for it to reestablish. It therefore appears more important to expend resources holding on to the desired cover for sites than it is to expend those same resources attempting to mitigate loss of desired populations and communities after they have gone. Note that this recommendation comes from a minimal model, but it is one that inspires confidence by virtue of displaying convincing richness and surprising accuracy compared with field data.

Dynamics of the General Community Model

Allen (1998a) laid out our general community model that indicates how other subdisciplines of ecology might lend assistance in investigating concrete examples within the community conception. To capture community dynamics, our general model of community uses the metaphor of a folded surface. Note that we do not feel the need to identify some cubic equation that generates the fold explicitly. Often the ultimate contribution of mathematical formalities is a justification for metaphor and new conception. Others may want to use catastrophe theory (Thom 1976) explicitly, to discuss the discontinuities of species composition to which our folded surface applies; we do not.

The folded surface occurs on a plot between some integrative measure of the physical environment, often moisture, and some general measure of the state of the vegetation. There is a general trend in the relationship between the physical environment and the measure of the state of the vegetation. However, that trend is complicated by the relationship becoming so nonlinear in the middle of the curve that the surface showing the trend folds back on itself to give two stable surfaces in the region of the fold. The folded region has a top and bottom surface, one for each of two possible stable communities for that environmental condition.

As equilibria, the top and bottom surfaces of the fold are discontinuous in the dimension of the vegetation measure. The states on the surface inside the fold, the piece that connects the top and bottom surfaces, is equilibrial but unstable, thus separating two stable conditions. Discontinuity in the state of the vegetation indicates the separation between the two or more types of community that are the stable options for a given state of the physical environment. It is possible to find vegetation in a given environment not on the curve itself, but such vegetation would change so as to reach the upper or lower surface of the fold in time (fig. 4.21). Disturbance may explain why the vegetation has been moved from the surface. Vegetation positioned at any point on the surface has a special and particularly informative property. That is, it exists at equilibrium with regard to woody biomass or proportion of woody species, or whatever is being used to characterize the vegetation.

Separate community types occur on the surface because of stable plant strategies that feed back to reinforce themselves. The strategies could be fire adaptation on the part of grasses as opposed to shading strategies of trees. The vegetation is characterized on the ordinate as some general vegetation measure, such as biomass or percentage of species with a large woody form at maturity. So when we talk about vegetation showing some sort of equilibrium, it will almost never be equilibrium of species composition. The general measures of vegetation are chosen to distinguish between plant community strategies in a given comparison. For instance, woody biomass is high in woodlands reinforced by shading strategies, but it is low in grasslands that reinforce themselves with fire (fig. 4.22).

In different regions of the folded surface the relationships between environment and vegetation have explanations coming from various ecological subdisciplines. At environmental extremes away from the region of the fold, there is only one stable vegetation strategy. In the case of fire-adapted grasses against forest, at the dry end of the physical environment axis, grassland is the only possible vegetation for phys-

Tendency for vegetation type to persist

Zone where there are two environmentally feasible vegetation types

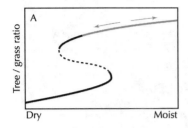

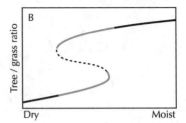

Critical historical environmental change

Disturbance

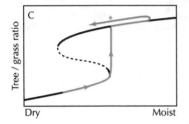

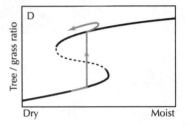

FIGURE 4.21

A model for community stability and change. The ordinate is some measure of the state of the vegetation, and the abscissa is the state of the physical environment. The upper and lower surfaces of the fold represent stable communities. In a given environment, disturbed vegetation returns to one or other of the surfaces of the fold. At extreme environments, beyond the folded region, there is only one stable option. The vegetation tends to stay on the surface on which it resides, but disturbance or environmental change into extreme conditions can move the system from one community type to another.

After Allen (1998a).

iological reasons. The region of the fold itself occurs in the middle of the physical environmental gradient. There the vegetational consequences of a given physical environment become ambiguous. Details of history, disturbance, and feedbacks between species make a physiological explanation inadequate and cause the ecologist to invoke explanations coming from other ecological subdisciplines.

A

B

FIGURE 4.22

Strategies in conflict. Madison, Wisconsin, is in the prairie–forest border region where strategems of being a tree work as well as the stratagem of grass and fire. (A) Oak woodland abuts tallgrass prairie, indicating the edge of last fire. The Green Prairie at the University of Wisconsin is burned regularly, so woody vegetation is absent. (B) In the absence of fire the oak grubs encroach the prairie, but the boundary of the strategy switch is still clear at the point of the last big fire into the woodland, per-haps a century ago.

Photographs by T. Allen.

As to the stable plant strategies, we identify only a limited number at this level of analysis.

* *Shading by trees.* This strategy requires relatively high rainfall to support the necessary biomass. Trees cannot grow and reproduce effectively in regions with extended summer drought.
* *Low nitrogen and fire adaptation.* This is the strategy of the dominants of the tallgrass prairies. David Tilman and David Wedin (1991; Wedin 1990) showed that these grasses keep nitrogen low by drawing down nitrogen to exclude competitors. Fire drives nitrogen into the air, and grazing is discouraged beyond only a few weeks. All this makes the positive feedback of low-nitrogen-encouraging plants that drive nitrogen lower.
* *High nitrogen and grazing.* By adapting to grazing, herbaceous plants keep out woody material. Relying on a nitrogen-rich soil, they crowd out the low-nitrogen strategists. This community is the meadow (fig. 4.23).

FIGURE 4.23

In Avebury Circle. European grazed systems exemplify the high-nitrogen strategy.

Photograph by T. Allen.

* *Dry out the soil to reduce fuel.* Here shrubs suck all the water out of the soil. Reduced ground cover in isolated clumps minimizes the risk of fire (fig. 4.24).
* *Fire-adapted shrubs.* In the dry far West of the United States, shrub vegetation with volatile oils burns hot every 30 years or so. Herbs are crowded out, and trees are killed by drought.
* *The desert strategy.* In very dry regions, succulent perennials dominate bare soil. The alternative life form grows very fast, to complete the entire life cycle on the water of a single hard rain.

Within any of these general strategies are substrategies that often relate to manipulating fire frequency one way or another. Around the globe there are other strategies for being a tree, shrub, or herb community. In the ideal climate for one strategy, others are often prohibited. The physiological ecologist explains where are the edges, beyond which this or that strategy is excluded. In terms of the folded surface,

FIGURE 4.24

Desert shrub vegetation. Here in Utah, shrubs dry out the soil between plants, thus keeping soil bare and fire at bay.

Photograph by T. Allen.

the physiological ecologist can tell community ecologists where is the edge of the zone of the fold, beyond which physical limits cause just one strategy to rule. There, disturbance and history play a limited role. Disturbance only takes the vegetation away from the default configuration for a time.

In environments in the zone of the fold itself, where alternative strategies are viable, physiological ecology is unlikely to be a powerful predictor. In this zone of the fold, history and disturbance offer reliable explanations. Although one does find intermediate patches of vegetation between, say, forest and grassland, they are less common than other vegetational configurations. Intermediates are unstable, and local historical events determine which stable strategy prevails. The unmistakable patterns of association on which the community concept turns arise from shared strategies between species generating a positive feedback, whereby the strategy alters the environment to suit itself. Earlier we referred to positive feedbacks of the low-nitrogen grass strategy, but it is only one example among many. Each strategy has its own positive feedbacks that lead to its own success. For instance, trees generate shade and high humidity, and even increase local rainfall, to create an environment where tree seedlings thrive and fire is suppressed.

The predominance of one strategy in a vegetation leads to the persistence of members using that lifestyle at the expense of those using other lifestyles and life forms. Disturbance takes the community composition over the divide to allow another strategy to work its feedbacks. In a series of dry years in the prairie–forest border region, grasses press the trees back with fire as their weapon. In a series of catastrophically wet years, fire may be so suppressed as to allow the forest to encroach.

The exact form of the upper and lower surfaces of the fold is determined by the happenstance of which species are present as players in the local wars between strategies. The explanation here comes from paleoecologists who can say why there is a given mix in a given area. They, with climatologists and specialists in long-term ecology, can indicate where the players will move next and at what speed.

The community is not generally well considered as a coevolved entity. However, this does not deny that there has been macroevolution that generates the players and the strategy to which they subscribe. This is, however, convergent evolution, not coevolution. Here Eric Knox (Knox and Palmer 1998) unraveled evolution of life forms and their attendant strategies for tropical mountain lobelias. In the lowlands of tropical Africa, lobelia exists as species of herbaceous plants. At various altitudes up widely scattered mountains across four countries, there is a trend to species with larger tree form at higher altitudes (fig. 4.25). The species at particular altitudes are so similar that a reasonable hypothesis is movement from one mountain to the next and the splitting of species to give a series of closely related sisters. In general that is not true. It appears that as land literally rose to high altitudes, species evolved to adapt to that condition each time it happened. Thus there is adaptive evolution in the emergence of given strategies of life form, even though the stable communities that emerge are not the result of coevolution. Communities are stable configurations of ready-made species. They are not coevolved complexes but rather consist of convergently evolved species whose important aspects that characterize communities predate the communities in which they are currently found.

Having asserted that communities are not principally ordered by coevolution, it is important to identify the lesser extent to which coevolution does lightly shade community form. The palynological record of Britain indicates that the British flora and its associated communities differ between the interglacial periods in which they occur. When the ice comes off Britain early, much ice still persists elsewhere, so the sea level is low and the English Channel is dry land. In that case hazel marches into the open habitat, along with other species, and takes its place as a major vegetational component for that interglacial. On the other hand, if the ice comes off Britain late, the sea level is high, and the English Channel is full of water before there is an ice-free substrate for colonization in Britain (Sparks and West 1972; West 1970). Hazel has large nuts that do not disperse well, so it is left behind. It does colonize later, but as a rare species.

The remarkable phenomenon is that if it makes a late start, hazel remains a minor component of the British flora for the many thousands of years of that interglacial. There must be some reason that hazel cannot recover from a marginal start. It appears that the initial

A

B

FIGURE 4.25

Macroevolution and communities. Communities are less a matter of microevolutionary adaptation, for they consist of ready-made units with commensurate strategies. Within a genus there may be macroevolutionary evolution that produces changes in strategies. Here mountain lobelias at different altitudes have evolved to produce different life forms, from (A) shrubs to (B) trees. Macroevolution offers the ready-made bits of communities.

Photographs by Martin Burd.

conditions leave a permanent mark on community structure for the entire interglacial. This, we surmise, is a matter of founder effects and microevolution whereby later species stand genetically in the shadow of the first arrivals. Jim Drake (1992) has shown assembly rules and founder effects in aquatic microfauna and flora in glass bottles, but one might reasonably question their relevance to large-scale terrestrial systems, such as the flora of Britain for 70,000 years. However, hazel would indicate that coevolution and assembly rules are indeed relevant to forest communities covering thousands of square of miles for millennia. It is in this way that population biologists can contribute importantly to the resistance to deformation of the folded surface in our community model.

Taking the Community Model Through Scale Changes

The community model just proposed applies at several scales. Watt (1947) reported a series of local patterns of cyclical replacement, such as bands of heather and bear berry on Scottish hillsides. So even at the scale of meters there can be alternate states of vegetation. Shugart's FORET gap phase systems operate over 0.1-hectare models patterns of vegetation mosaic reported in principle by Watt (1947). These patches also indicate small-scale alternate states. At a larger scale McCune's study of vegetation in valleys in the Bitterroots showed patterns of homogeneity at the scale of a kilometer or so but great heterogeneity across the 50 kilometers of the whole range (McCune and Allen 1985). Each valley in the Bitterroot range represents alternative states. Whittaker's (1956) Great Smoky Mountain vegetation can be discussed readily at several scales (fig. 4.26). His pine-dominated communities on the low ridges represent a self-reinforcing state that, once established, keeps deciduous species at bay. By now it should be no surprise to the reader that we intend to apply a model that uses the metaphor of a folded surface at several scales.

Scale here has two distinct aspects to it. First is the standard spatiotemporal scaling, where *large scale* means great tracts of vegetation over long periods of time. Second is scale as it applies to the number of types of things included in the system, such that a large-scale community is one that is defined by a stable configuration of a large num-

FIGURE 4.26

Communities of the Smoky Mountains. This same view occurs in Whittaker's classic paper on vegetation in the Smoky Mountains. This photograph, taken in winter, shows evergreens, pine on the ridges, and hemlock in the high valleys. The whole is a community, as are the pine and hemlock subsystems in their own right.

Photograph by T. Allen.

ber of species. The two types of scaled differences may well run in parallel in that larger areas are likely to include more species. However, this is not necessarily the case, for there may be great tracts of species-poor, homogeneous vegetation to contrast with remarkably local heterogeneous vegetation situated in the same general region. An example here might be in a heavily grazed regional grassland in the Yorkshire Dales, contrasted with the fine-textured environment of limestone pavements in the same region, where cracks vegetated by many rare species dissect a very local area. A second example, in the mountains of Snowdonia in north Wales, would be the tracts of homogeneous hillside manicured by sheep contrasted with hanging, isolated valleys in which *Lloydia* and other rare endemics present a more diverse system (fig. 4.27). A longer temporal window is needed

271

to study these local hanging gardens because the rare endemics survived the last ice age, whereas the grassland matrix won out over wooded only in the last 2000 years. Thus a local area can be more diverse and require a long time line. Therefore precision and flexibility in scaling are crucial for sustaining community structures, even in one locale.

The type of scaling related to diversity is less straightforward and so warrants some further explanation. Let us choose one of the conceptually more comfortable systems where the more diverse metacom-

FIGURE 4.27

Remnant vegetation. There is a series of hanging valleys on the Devil's Kitchen Cliffs in Cwm Idwal, Snowdonia. The lowest Cwm was plucked from the mountain by the glacier of the last ice age, but some valleys remained above the ice and are hanging garden valleys because the ice sheared them off at the bottom. The cliff below protects such valleys from grazing. Hanging Garden Gully is one of the short, truncated gulleys coming from the far ridge, center left in the picture. There *Lloydia* survives, protected as a patch of remnant vegetation in a blanket of modern vegetation in the foreground. The blanket of heath and narrow-leaved grasses arises from sheep farming.

Photograph by T. Allen.

munity is also of larger area, made up of the local community types: Whittaker's Great Smoky Mountain vegetation (Whittaker 1956). We noted that different parts of the mountain, such as low ridges or high valleys, are dominated by different major species. There is no reason to consider the vegetation across the entire mountain as somehow the vegetational community that eclipses the lesser types that occur on characteristic parts of the mountain. It is perfectly reasonable to consider the pine-dominated vegetation on the low ridges as a community in its own right. Of course, it represents a subset of the entire mountain vegetation but can be considered a community, should the research or conservation be suitably focused on these more local considerations.

As with most changes of scale, the more local community type merits its own patterns of explanation that may well not fit into those of the higher-level, more diverse type of community that comprises the vegetation of the entire Smoky Mountain system. Questions asked of the upper-level system usually entail a different set of considerations, if even the same question were directed at the lower-level system. The processes of positive feedback that lead to the regular occurrence of certain associations will be different at the scale that includes all types of vegetation on the range of mountains, compared with the local vegetation on just the low ridges. Consider the larger community of the entire Smoky Mountains as a basin of attraction for the vegetation as a whole. That basin has within it multiple local stability points at the center of local regions of attraction. The explanation as to why one local attractor wins out over another, say heath as opposed to grassy bald, does not offer an explanation of the larger basin of attraction that makes the vegetation of the entire region what it is. Different scales of community analysis, even within one region, usually warrant different explanations that invoke different processes.

We emphasized that the stable configurations of species association are linked to strategies. These strategies employ positive feedback whereby the state of the vegetation is attracted to either the top or bottom surface of the fold, the upper or lower community. Closer attention to the dynamics of the system on only the upper or lower community generally will reveal a less than fully homogeneous system, one with its own discontinuities. The difference between the grass and forest strategies turns on a set of considerations different from those that distinguish between different tactics within the forest strategy. At their own level of analysis, within-forest tactics become strategies, albeit strategies working at a lower level. Let us translate this conception into an extension of the folded surface model.

The first-order fold might be that plotted between a wet–dry abscissa and an ordinate that is some surrogate for the forest–grass distinction, such as the amount of woody material or leaf area index. A closer look at the forest surface of the grass–forest fold on some other surrogate for community disjunction reveals that the forest side of the fold is itself folded in a different dimension, one that distinguishes between tactics that lead to one type of forested vegetation as opposed to another forest type. To see these lesser folds, one might need to change the ordinate and abscissa to some other environmental gradient, such as altitude, against some other surrogate for the vegetational differences, such as proportion of deciduous species. This extension of the folded surface model makes it scale-independent, a requirement for any workable formal model.

Increasing the dimensionality of the folded surface has additional advantages. Disjunctions that give the discontinuity of the fold in some circumstances might give way to continuous changes in other situations. The classic example here would be savannas that give continuity between prairies and woodlands. Forest often abuts prairie, with a sharp line of demarcation only a few meters wide. On the other hand, a far from uncommon condition occurs wherein large trees present an open canopy, with prairie species occurring as an understory. The trees in this savanna vegetation are always fire adapted, with thick fire-resistant cork not only on the main bole, but also on the twigs. If we extend the distinct fold of the prairie–forest discontinuity onto a third axis that accounts for the degree of fire adaptedness of the woody vegetation, then the fold emerges as just part of a pleat in three dimensions. Thus crossing the folded part of the pleat gives a discontinuous change between forest and prairie, whereas it is also possible to move continuously from forest to prairie, with a gradually thinning canopy, by traversing the unfolded region beyond the base of the pleat, at the tree fire-adapted end of the third axis.

Implications for Sustainability

The community criterion probably is the most challenging for addressing sustainability. The question is, "What exactly does one want to sustain?" Under the population criterion, it is clear that a certain species is the structure at the center of the effort, and so deciding what to sustain under that criterion is simple; finding ways to sustain it may be a different matter. Under the landscape criterion, the desired condition and the thing to be sustained is almost as obvious as it is for populations. Even ecosystems, abstract as they are, nevertheless offer some clear choices as to desired conditions, such as a sustained nutrient status. Furthermore, ecosystem boundaries can be rationally drawn around tangible entities such as watersheds. By contrast, the community offers no such easy decisions as to what is the structure to be sustained.

The difficulty is the deep intangibility of community structure, and the richness of the dynamics associated with that structure, once it has been identified. In a previous book, Allen and Hoekstra (1992) pointed to the spatiotemporal mismatches between community members. Members of different species occupy the landscape at different scales. In this tension between organism and landscape, some community members occupy sites fleetingly, whereas others persist longer. Some species move around the landscape so that they are absent from many sites for a long time, whereas others are only occasionally absent from just about all sites. If both are community members, then the community as a collection of species in a place falls apart. That is why it is inappropriate to assert that the thing to be sustained is a species list at a site.

In *Toward a Unified Ecology* (Allen and Hoekstra 1992), we preferred to emphasize the community as a set of interacting periodicities of site occupancy, and we still find that conception useful for many purposes. The structure of the community, the thing to be sustained, therefore has embedded in it rich dynamical interaction. These dynamics are different from the structure behaving, for the dynamical interactions are the structure itself. We cast community structure as a rich wave interference pattern. The boundaries of the structure and its essential character incorporate species coming and going as part of the pattern that is the community. Wave interference patterns change radically with the addition, removal, or change in frequency of behavior of even a single component. Even so, we do see undeniable, reliable patterns that draw some ecologists to become specialists in community ecology. On the positive side, it seems that some patterns at some scales offer a coherent structure with which we might address community sustainability. On the down side, it is clear that something as static and arbitrary as a chosen collection of species at a rigidly defined site is not coherent. Such a capriciously asserted community is not sustainable and is not worth much investment of effort.

So much for the community as tension between the landscape and organism criteria. Allen (1998a) pointed out another tension between two criteria—populations and biomes—with the community concept caught again in the middle. At one extreme, populations are entirely species-focused and entail biological feedbacks such as growth and density dependence. At the other extreme is the biome, a criterion under which species specificity is almost entirely relaxed. In the biome conception, life form and physiognomy of the dominant organisms, not species composition, characterize the system. Unlike populations, biomes reflect directly the consequences of the physical environment; biomes, unlike populations or communities, are environmentally determinable. The difficulty with the community criterion is that, by most definitions, there is an insistence on retaining the species as identifiers while seeking to deal simultaneously with a large number of species, each with its own environmental tolerances. Environmental determinism fails in communities because the complex of species responds in a complex fashion to the environment. Life form in biomes is constrained by physical limits and so is environmentally determined. By contrast, species composition, so central to communities, has a many-to-many mapping to the environment. A given community occurs in a range of environments, whereas a given environmental condition can be the setting of a range of communities (fig. 4.28).

Both the population and biome are worthy criteria for directing efforts toward sustainability. The community criterion is also an appropriate organizing principle but should not be expected to work on the

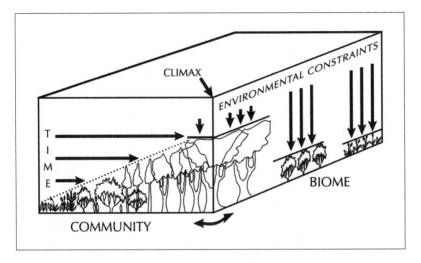

FIGURE 4.28

Communities versus biomes. The community and biome become the same thing when environmental determinism works. In succession, biotic determinism pertains because the outcome depends on biological input to the stand. Only when climatic climax is achieved does the vegetational system press against physical constraints. If there were more or less rain, then the trees would be taller and denser, or smaller, or absent. By contrast to a community, something characterized as a biome is always environmentally constrained. That is what makes the life form of the dominant plants recognizable: physiological limits interacting with physical environment. Only at climax do biome and community coincide with regard to environmental determinism.

same principles or use the same methods as populations, with their coevolution, or biomes, with their convergent evolution of members. If one insists on seeing communities as evolved entities, then they emerge as middle number systems (Weinberg 1975), where coevolution, convergent evolution, or both types of evolution pertain. When constraints switch in this manner between interlocked population processes as opposed to physical limits, prediction is not possible. In a middle number system one cannot predict, even with experimentation; for instance, the streams of bubbles will arise in a glass of champagne differently every time and cannot be predicted. Prediction fails in the face of switching constraints. However, it is the question that is the problem, not the material system. One can predict, from the inter-action of carbon dioxide partial pressure and ambient atmospheric pressure, whether a particular bottle of champagne will fizz. Turning the notion of middle number system on the community concept, one cannot make predictions as to particular communities from the inter-action of different sorts of evolution. Coevolution and convergent evo-lution offer the community different classes of constraint. The failure to predict is because the presence or absence of many species, and therefore the state of the community, depends on happenstance of the details of the case-specific interactions. The unpredictable aspects of champagne bubbles depend on the interaction of many imperfections on the inside of the glass and the turbulent flow as the wine is poured. In communities, the unpredictable interactions are between popula-tion processes and environmental variation, where constraints that might engender prediction switch back and forth (fig. 4.29).

Even so, there are other questions about communities that can be answered. To make community predictions, one identifies constraints that are reliable. These constraints must apply consistently enough to limit the possible outcomes to a workable number. The simple under-lying structure pertains to the feedback processes that are associated with the strategies at the core of the particular community at hand. The message is that communities are a criterion in their own right. Community conceptions might be inept at answering population and biome questions, but they do offer predictions, and therefore the potential for sustainable management, with respect to their own ques-tions. Think of communities as a link or conceptual filter between biotic processes of populations and the physical constraints that pre-dict biomes but not as a predictor of the effects of either.

There is a typology of problems that was formalized in a problem-solving method by Chen (1973) some 30 years ago. The essential prob-lem of ecological sustainability of communities can be captured in that typology. In its most general terms, a problem arises because of a mis-match between the state of the system and the condition desired for it. If we start with the state matching the desired condition, then there is no problem. A problem emerges when either the state of the system or one's desires for it change. Type 1 problems occur when the state of the system no longer matches the desired condition. Type 2 problems arise

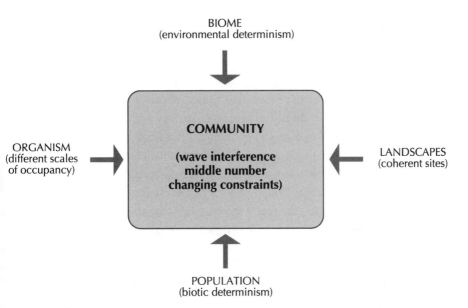

BIOME
(environmental determinism)

COMMUNITY

(wave interference
middle number
changing constraints)

ORGANISM
(different scales
of occupancy)

LANDSCAPES
(coherent sites)

POPULATION
(biotic determinism)

FIGURE 4.29

Community as a point of tension in the middle. Community concepts present the ecologist with a set of switching constraints. The switches of constraint lie on both a landscape–organism line of tension and an environmental determinism–biotic determinism tension. The latter is the difference between the population and biome conceptions.

when the desires for the system change, leaving the system in a state that is no longer desirable. Type 3 problems occur when both the state of the system and the desires for its condition change. Sustainability of communities is not a matter of achieving one state and keeping the system in that condition. Communities are dynamical, and so the desire for sustainability involves a system under perpetual change around a basin of attraction defined by the outcome of the shared species strategies. Desired population dynamics will butt up against changing environmental constraints. The state of a community up against a desired set of environmental constraints will be moved by population dynamics according to what the constraints allow. In its very nature, sustainability of communities is a type 3 problem, the most difficult sort of problem to manage. There is always change in the system, and it is likely that our desires for it change also.

Sustainability under the community criterion must depend on the feedback processes that revolve around the shared strategies of community members, such as the shared strategy of shading out competitors used by the tree life form. However, there is an added complication, which is that even a small change in the scale of the system under consideration can radically redefine the strategy and its attendant feedbacks. In terms of our model of the folded surface, the order of folding is crucial. Sustainability might be considered in terms of preserving a generalized forest cover, by aligning management with the shading strategy. On the other hand, that is very different from sustainability as it is engineered inside the forest strategy for types of forest, such as the types of forests that occur in the Great Smoky Mountains.

Deciding and identifying what is to be sustained is the critical problem in community sustainability. We want the answer to the question, "What is it we wish to sustain?" The central issue in sustainability under the community criterion is scaling and thus bounding the community in question. As we stated at the beginning of this section, a capricious bounding of the community system as a particular species collection is pointless because it amounts to very expensive gardening to mimic nature. It is essential to identify the feedback processes that give the stable, strategy-determined configuration. If one can identify enough of the feedback process, one can co-opt it as an ally in achieving sustainability. Efforts directed toward sustainability of communities that are not guided by such feedbacks will have to fight them. Instead of being an ally, the undetermined feedbacks will act subversively. They will be an untiring enemy that eventually will overcome the misdirected effort to sustain a capriciously chosen condition.

In our first presentation of the model using the folded surface, we pointed out that beyond the region of the fold itself, physiological ecological insights are particularly helpful (fig. 4.30). In the zone of the fold, disturbance and landscape ecology are likely to be of greater assistance. The form of the fold with respect to the surrogate measure on the ordinate can be understood and predicted into the future by palynological, climate change, and long-term ecological research. The autecology of the players in the stable configuration can be understood in communities in macroevolutionary terms. Meanwhile the persistence of founder effects is a microevolutionary issue. We identi-

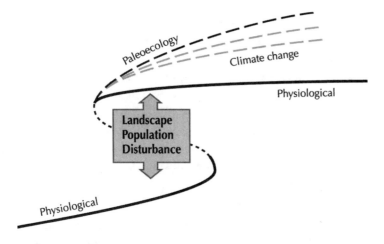

FIGURE 4.30

Community insights from other ecologies. Different subdisciplines of ecology give insights into community processes, depending on where in the environment/vegetation state space the system resides. Whereas physiological ecology informs the community ecologist at environmental extremes, disturbance ecology, landscape ecology, and population biology aid in understanding communities in intermediate conditions where multiple stable states are possible. Paleoecology gives references as to the historically pertinent components of communities in a given region.

fied how each of the sister disciplines of ecology can play its role in unraveling the forest–grassland interface. As the scale of an effort to achieve sustainability is changed to consider different smaller-scale communities inside the generalized forest, the relative importance of these ancillary ecological sciences is likely to change.

Local community differences, such as those between the local vegetation types on the Great Smoky Mountains, would have a different mix of explanatory principles than would a conception of the community system as the whole national park. Two scales of analysis would rely more or less heavily on various sister subdisciplines across ecology. For example, the mix of species in a community occupying a small kettle hole filled with water might be principally explained by founder effects. In locally defined systems, almost the entire universe of discourse might be occupied by the zone of the fold where alternative communities are an option. In such situations, physiological limits lose their explanatory power. By contrast, the general arrangement of dominant species at different altitudes in the Great Smoky Mountains may well be explained by physiological limits set by life form differences. Scale of explanation changes everything in community sustainability.

Sustaining communities is particularly challenging. The principal difficulty in sustaining a system under other criteria is getting the system to respond in the long term to give the desired management outcome. Community sustainability turns on other considerations. Under other criteria, it is usually fairly clear what must be sustained and what is the behavior of the system in such a condition. In communities, the major obstacle is defining an appropriate entity to be sustained. Sustainability is meaningful only if it is in line with the processes of positive feedback that draw species composition from one configuration to another. The distinction between positive and negative feedbacks here is nontrivial. The positive feedbacks that move the system away from a prior configuration are the beginning of a negative feedback that leads to a sustained existence in the new configuration. The community entity to be sustained lies within a basin of attraction. The difficulty is in identifying the basin of attraction and getting the material system to begin moving in that direction. The good news is that, once one has correctly identified a basin of attraction and has defined it as desirable, sustainability comes as a matter of course. Thus achieving sustainability is the challenge under other criteria, whereas under the community criterion the critical step is defining the desired condition adequately, one in which natural processes hold the system in place.

CONCLUSION

By cleaving criteria for observation away from matters of scale, we expect to be able to unravel the issue of sustainability more completely than heretofore. The organizing principle of type of ecology as separate from scale was first used as the framework for Allen and Hoekstra's *Toward a Unified Ecology*. The order in which the ecological criteria have been addressed is different here because our purpose here is different. The tangibility of organisms and landscapes gives them a special place because it allows those criteria to offer lead indicators of lack of sustainability. Population is again primary here because it amounts to a combination of the organism and landscape criterion. In the overview provided in this chapter we have not dealt with the ecosystem criterion because it receives its own treatment in its own chapter (chapter 6). Similarly, biomes are considered elsewhere in a section on physically large systems (chapter 5). The material in this chapter has laid out the ecological underpinning to the issue of sustainability. This chapter ties down the biogeophysical aspects of sustainability as we weave that facet of sustainability into the larger treatment in the book as a whole.

5

Biomes and the Biosphere

The biosphere as a criterion for observation is fundamentally different from other ecological criteria. Unlike the criteria considered in other chapters, the biosphere has a certain explicit scale. The difference between the biosphere as an ecological criterion and others considered in this volume is that there are exemplars of ecosystems, landscapes, and the other criteria, but there is only one biosphere (fig. 5.1). The definite article in "the biosphere" is more definite for the biosphere than for, say, "the ecosystem." The biosphere and the ecosystem are not just different things; they belong to distinct logical types. The former represents a material thing in particular, whereas the latter is a general concept that does not specify a particular material ecosystem. The same sort of distinction can be drawn between the biosphere and the organism. Whereas a particular example of an organism exists at a certain scale, the concept of organism is independent of scale. Because there is only one biosphere, it is in itself the definitive example and so can be assigned a scale in the same way that a scale can be assigned to a particular organism. Biospherics becomes altogether more interesting when the character of "the biosphere" is assigned to other entities. At that point, it is not the biosphere anymore but is the generalizable concept of a biosphere. Prob-

FIGURE 5.1

The biosphere as an exemplar. There is only one biosphere, and it has a particular size to it. Although particular organisms also have a size, the organism as a level of organization has no particular scale. It is the singularity of the biosphere that gives biosphere ecology its definite size.

Photograph by NASA.

ably the definitive character of *a* biosphere, as opposed to *the* biosphere, is that a biosphere would be an ecological entity that is relatively closed to matter but significantly open to energy. Part of this chapter addresses the generalized criterion of a biospherical point of view, and another part of it considers the biosphere at the scale of the entire globe. The difference between biosphere and other ecological criteria is that the biosphere refers first to a particular entity and only

285

second as a way of looking at a material system. It is the other way around for other criteria.

As ecological systems go, the biosphere is large. There is a convention that biomes too should be large, although not as large as the biosphere. In *Toward a Unified Ecology,* Allen and Hoekstra argued against such a prescription of scale for the biome, pointing to the advantages of keeping criteria for ecological observation and models scale-independent. The argument there turned on models as being useful exactly because all else significant is equal except scale. In the case of a model airplane, the "all else significant but scale" is clear: fuselage, wings, tail, nose, and so on. Only the size of a model airplane is significantly different from the full-size prototype (fig. 5.2). Although in many cases it

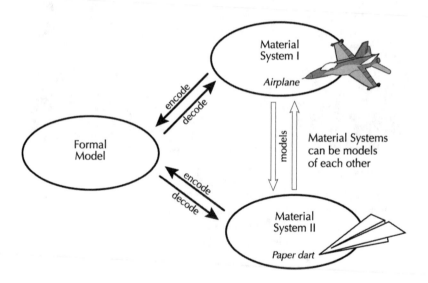

FIGURE 5.2

Formal models and scaling. A model airplane has all the critical features of the full-size machine except size. The model and the prototype both yield to description through the laws of aerodynamics, which are statements of relative scaling and so are in themselves scale-independent. When a formal model, such as the laws of aerodynamics, applies to two models such that both can be encoded and decoded to and from the formal model, then both material systems become models of each other. Not only is the model airplane a model but so is the full-size plane a model for the model.

is less intuitively obvious, the same applies to all models, including the general model of a biome. Biome is a model in the sense that everything that warrants assignment to the class *biome* has a set of significant characteristics that it shares with all biomes and that can be expressed as relative scaling considerations. The differences between biomes are found in the particular scaled values assigned to those considered significant. Therefore, if scale is mentioned in the criterion that defines the biome, the concept of biome loses much of its power as a model. This is because much of the scaling of the characteristics of the specific case of the thing modeled is preempted by biomes having a prescribed scale by definition.

In this chapter we revisit the issue of biomes independent of scale so that the distinctive point of view that is involved in investigating biomes can be generalized as far as possible. Whereas the singularity of the biosphere gives a certain logic to assigning a particular scale to the biosphere, the biome is not a singular entity. There are examples of biomes, whereas the biosphere is not an example; it just is in itself the biosphere. There is therefore much less justification in insisting on a scale for biomes than there is in assigning a scale to the biosphere.

All that being said, it does make sense to look for commonalities between ecological situations that share being large. At the end of Allen's first book on hierarchical structure, *Hierarchy: Perspectives for Ecological Complexity*, Allen and Starr (1982) considered various protocols for investigating issues in the context of scale. One strategy is to keep one particular data set and to perform scaling operations on it. Different scales reveal different aspects of the biology captured in the data. An alternative strategy is to investigate the particular properties of the scale that reveal a class of phenomena. An example here is allometric scaling of animals of different size. Knute Schmidt-Nielsen (1984) teased apart several different aspects of animal functioning taking this approach. Another strategy highlighted by Allen and Starr is to look at a range of entities that share certain particular scales, that exist at a particular size. Accordingly, part of this chapter is concerned with the commonalities of large ecological systems.

Certain characteristics emerge when very large or very small entities are compared with other entities that share their respective size. For example, gravity is the constraint on large animals, whereas surface tension figures large in the existence of small creatures. In ecological systems, physical constraints as to what physical material can do determine which features of a system are important in giving the system its integrity. For example, only certain phenomena have a large enough scope to hold large systems together. Whereas genetics can be a powerful explanatory device in ecological systems that range in size from viruses to clones of trees, genetics is a less promising tool for very large systems. Conversely, atmospheric considerations are more useful in explaining the ecology of regions than they are in explaining the temperature of leaves, a notoriously difficult thing even to measure, let alone explain (Allen, Havlicek, and Norman 2001). We address large and small biospherics and biomes. However, there is a certain focus in this chapter on ecological systems that are particularly large because they share certain general characteristics, such as the critical role of atmospherics in holding them together.

The thrust of this chapter is akin to long-term ecological research, an emerging specialty in ecology. Note that in long-term research, there is no prescribed approach when it comes to defining the entity under investigation. The focus could be on long-term species composition, which would entail a palynological analysis. Alternatively, the central concern might be for long-term sustainability of ecosystem function, as when Parton and his colleagues modeled carbon in three compartments (Parton et al. 1987). Despite net carbon loss, the two short-term compartments do not change significantly because they are held in the context of the slow compartment with the longest relaxation time. The various Long Term Ecological Research (LTER) sites funded by the National Science Foundation each have a style of investigation that can be only generally circumscribed by any one subdiscipline in ecology. The Sevilleta site in New Mexico focuses on landscape considerations, whereas the Cedar Creek site in Minnesota takes a nutrient-cycling approach with a focus on populations that determine nutrient status. With such differences in equivalently long-term research sites, it is clear that long-term research can use any of the subdisciplinary approaches. In the same spirit, large-scale ecology, of which long-term research is but one facet, can use any of a number of ecological types in its search for general principles that apply at large scales.

THE BIOME CRITERION

Sometimes, large-scale entities may indeed be best cast as biomes, necessitating the particular point of view and methodology that come with the biome model. The biome is distinctive in that the life form of plants, not plant species, defines a particular biome. In biomes, physical environmental determinism is successful as an explanatory principle. Whereas the species composition of communities cannot be predicted by environment, the life form of vegetation appears to be climatically determined. The life form that characterizes vegetation physiognomy of biomes does have a physical explanation understood in terms of physiological requirements. Another distinction between biomes and communities turns on the inclusion of animals and plants together. Species considered as members of communities generally belong to one kingdom, the other kingdoms being biotic environment. By contrast, a biome would not be complete without explicit inclusion of its animals to groom the vegetation.

BIOMES AND CLIMATE CHANGE

Given the distinction between the community concept and that of the biome, it is hardly surprising that sustainability of biomes raises different issues from community sustainability. Indeed, sustainability does not sit easily as an intellectual device for communities, whereas a large portion of the entire concern with ecological sustainability turns on biome considerations. Climate change and its effect on sustainability are a matter of shifting biomes. Under significant climate change, some biomes cease to be sustainable where they were before, whereas other biomes become a possibility in new regions. For instance, many of the contemporary closed forest types of North America could not have been sustained at the levels of carbon dioxide that existed at the beginning of this interglacial. As a result, the closed canopies of boreal forest of today did not exist 15,000 years ago, and in their place there were conifer savannas with significant grass cover between trees.

Climate from short to long term does show change that is greater the longer the time window of concern. There are cycles over decades, of which the Dust Bowl in the 1930s was one. When that cycle returns, it does not always manifest itself in such dramatic terms, but the cycling of the water level in the Great Lakes every decade probably reflects the return of dry times in that same cyclic process. Over longer time periods there are other cycles, such as the Little Ice Age that ended in the last century and that at its deepest allowed skating on the River Thames, something that Allen, a native of London, has never seen and cannot easily imagine. In yet longer cycles occur true glacial periods. Recent data from the Greenland ice cap indicate, counter to intuition, that at the beginning of the last ice age it took about 4 years to shift from interglacial to glacial (Alley et al. 1993; Fairbanks 1993). In that case, the rapidity of the shift probably turns on the feedback of winter snow never melting, causing a critical change in albedo, causing significant cooling that kept the winter snows all summer every year. So there are threshold effects that can cause grand changes in climate in short order, but in general longer-term changes in climate are gradual when seen at the scale of years, decades, or even centuries. Climate is a large-scale consideration and therefore is likely to change continuously when measured at a human time scale. A case in point is the record summer heat in the 1980s in the North American Midwest but a record-breaking cool summer in 1996. It seemed as if the predicted global warming had arrived, only to manifest a false start, if it is indeed about to occur.

That being said, even in the face of continuity of climatic change, the vegetation may not read the change as continuous. The dominant organisms that give biomes their physiognomy have particular life histories that have bottlenecks influenced by climate. Sometimes the bottlenecks pertain to just one species, but one that is so central to a biome that its limitations become boundaries for the biome.

A case in point of discontinuous response of one species to climate would be the Gambel oak, which reaches its limit in northern Utah (Neilson and Wullstein 1983, 1986). There drought and length of the growing season defined by temperature squeeze seedlings out of a window for growth and survival. North of that window, mountainsides are wooded with trees that have a narrow genetic base. Almost all trees are of vegetative origin, not grown from a seed. Fire would permanently remove the trees from those northern sites. The drought and growing season change continuously over the landscape, but the effect is discrete once the limit is reached. The absence of seedling recruitment occurs at a line on the ground.

Sometimes continuous change in climate may influence not just one species but the viability of all species with the dominant life form for the biome in question. Whereas different species have particular requirements and exhibit unique timing of critical events such as germination, there are limits to the possession of the very life form that defines a biome. At the low level of separate species, there may be differences in germination between species, but at a higher level there may be unity of the response of a life form shared by several species. At the lower level, tree species may exhibit unique requirements in soil microsites for successful germination, even if a species shares a biome, or even a narrow community type, with other species that have critically different germination requirements. For example, the hemlock *Tsuga canadensis* occurs commonly with *Acer saccharum* in central Wisconsin sugar maple forests. In the forest one often sees hemlocks growing in straight rows, even if the stand was seeded naturally. The reason is that hemlock germinates particularly well on dead logs, which rot away, leaving the seedlings and eventually trees in a file (personal observation). Sugar maple appears not to have that same preference, and one never sees them growing in rows except when they have been planted that way silviculturally. At the higher level, despite those low-level ecological differences, these two tree species both require mesic sites, and like all large trees that occur in Wisconsin, they cannot survive more than a limited period with the soil water essentially absent in the root zone. The strategy of being a tree involves support of a large biomass that competes for sites by means of the strategy of shading other would-be occupants of a site. That strategy works well but requires a certain minimal amount of soil water through the growing season (Neilson 1986, 1987a, 1987b). The prairie forest border region of Wisconsin gives way to a sea of prairie in Iowa at the very point where soil water limits the strategy of being a tree. Thus not only are particular species likely to show discrete response to continuous changes in climate, but so will entire vegetation types, should the limit be one on life form.

It is limits in viability of given life forms that determine the limits of biomes. Climate usually varies continuously in space, but the response manifested by the vegetation is a sudden change of biomes when the limit of the prevalent life form strategy is reached. The situation is a little complicated by historical accident in that the exact position of the line of forest on the ground is determined by the last fire if the forest is retreating. However, in a drying climate, the fire merely makes visible the situation that prevailed before the fire, namely that the forest is outside its range and has no recruits with which to recapture the site decades later. At the other edge of the forest–prairie border region, on the wet side of transition, fire occurs too, but it is only part of a continuing tussle back and forth in a zone where occupancy by either prairie or forest is equivocal. At the dry edge of a forest–prairie border region in a drying climate, the ouster of forest by fire is permanent until a reversal of climate, perhaps millennia later. Neilson has data to suggest that there is wholesale switching across the entire Great Basin of desert grassland to desert shrubland when climate changes press one or another biome type to the wall (Neilson 1986, 1987b).

Thus lack of sustainability of a biome may manifest itself remarkably quickly, even though the change in climate is gradual. Once one has defined a sustainability goal—what to sustain, for whom, for how long, and at what cost (chapter 1)—then sustainability becomes, like pregnancy, not a matter of degree. "Almost pregnant" has no meaning. Sustainability therefore is a rate-independent notion, a step function, and this is particularly clear in the sharp boundaries around biomes and their sudden appearance and disappearance in a given region. The vegetation in a biome on the losing end of climate change may be a perfect example of the physiognomy of that type of biome. Even so, if the climatic threshold has been crossed, such that the processes of renewal and regeneration are compromised, the system is nonsustainable. It is only a matter of waiting until the physiognomy of the vegetation undergoes radical change.

By contrast, as long as there is a local habitat where the central species can reproduce, relics can persist for millennia. In Wisconsin there are relics of the boreal forest hundreds of miles to the south of the present position of that biome. Wisconsin has a tension zone where a large number of northern species have their southern range, and the southern species have their northern range. In the cold air drainage at microsites in the Baraboo Hills, there are isolated stands of hemlock, a boreal and mixed conifer hardwood component, well into the southern half of the state. Curtis (1959) observed hemlock pollen in southern Wisconsin bogs, indicating an ancient southern presence (although he suggested a postglacial southern migration route instead of relic status, with which that observation is also consistent). The critical issue appears to be that there has been successful reproduction in the special local climate regime. The great increase in the deer population has all but stopped reproduction in recent years, so these relics may be gone in a few hundred years, even though they have survived there since the ice retreated at the end of the last glaciation. Charles Canham (1978) calculated that at any time, 20 percent of hemlock forests are recovering from a blowdown within the previous 250 years. Therefore, everywhere in the state is blown flat on the order of once a millennium. As long as there is reproduction, such disturbance is part of the normal course of things. But in the absence of reproduction, relics will be lost.

SUSTAINABILITY OF AGRICULTURAL SYSTEMS AS BIOMES

The population of our species is at such numbers that it not only influences ecological systems that are actively groomed but also affects vegetation toward which attention is not particularly directed. The role of animals in biomes is to groom the vegetation and to move materials such as propagules or diseases that give the biome in question integrity. The human creature has a peculiar effect on dispersal of propagules. Unlike the random walks of animals moving seeds on their fur, we are sentient and purposeful in moving some propagules. Even when we are unintentional in moving invading species, human activities move materials at unprecedented rates over unlikely large distances. The movement of noncultivated species under climate change is a difficult matter. A lot depends on where species are when the climate thresholds are crossed. The situation is further complicated by unpredictable human intervention. Whereas tree species with large seeds that have poor dispersal might take millennia to move across a state in the primeval condition, humans in the modern realm might move those same species such distances in a day.

293

Agricultural systems are aptly considered a manifestation of a biome as much as an agroecosystem. With cultivars, long-distance dispersal is the rule. Accordingly, should an agrobiome suffer critical climate change, humans can move it to some other viable location through response to market pressures alone. However, just as there is no place for some biomes of the past in the modern world climate regime, agrobiomes on the move in response to climate may also find no place to go. For example, should climate change move the climate of the United States corn belt up into Canada, the pertinent human response would not be a simple matter. A move north would be onto the thin soils of the Canadian Shield. Corn needs more than 4 inches of acidic soil to grow. The present corn belt is a matter of blowing 20 feet of loess into Illinois at the end of the last glaciation to make the modern topsoil, as much as it is contemporary warm humid summer days and a reliable growing season into October.

Climate greatly influences soil development, and most biomes have a characteristic soil regime. In agrobiomes soil is also characteristic of the type of agrobiome, but it is often maintained only through human effort. More than that, there are processes inherent in agrobiomes that would destroy the soil characteristics, and even the entire soil altogether. The human effort directed at agrobiome sustainability often is directed toward ameliorating processes that destroy soil. Agricultural systems are unusual as biomes in that there is normally a significant export of material directed outside the region of the agrobiome. There are two critical types of material export: mineral soil nutrients and carbon. In cropping systems, the export of mineral nutrients is balanced with anthropogenic imports, as best the humans can manage. Farmers in the Midwest know to plow under cornstalks and straw rather than to use them always for food and bedding. A significant part of manuring is the return of soiled bedding materials along with the animal waste that makes up the manure. Another practice that increases soil carbon is green manuring, in which the farmer plants a crop of winter rye with the intent of plowing it under in the spring. Soil structure, which depends on soil carbon, is as important as mineral status, for mineral and carbon status are linked.

The capacity of agricultural systems to function as biomes is indicated by the work of David Tilman and David Wedin (1991). They have been able to explain the capacity of agricultural systems to resist the return of the native biome in the North American prairies, even after the agriculture is abandoned. Native prairie bunchgrasses, such as big bluestem, use a strategy of reducing the soil nitrogen to a level at which only they can compete for the site. It is a classic example of Tilman's (1988) model of the winner of competition setting a positive feedback to draw down resources so as to keep winning the competition. It is a positive feedback that uses root scavenging, fire, and grazing manipulation. The bunchgrasses encourage hot fires and discourage grazing. The fires recycle mineral nutrients, with the exception of nitrogen, which is driven off in the flames. The bunchgrasses also discourage grazing by offering very poor forage, except for a period of about 6 weeks. Thus the native grazers are encouraged to graze off the woody material, which would overtop the grasses if given the chance. At the same time, the animals are not encouraged to stay long enough to increase soil nitrogen by urination and manuring. The native dominants of the tallgrass prairie biome work in a positive feedback to increase their hold on the vegetation (fig. 5.3). Agriculture sets the feedback going in the other direction to keep out the native vegetation of the biome.

As with all positive feedbacks, when the process of success leading to greater success of bunchgrasses is reversed, the reversal keeps mov-

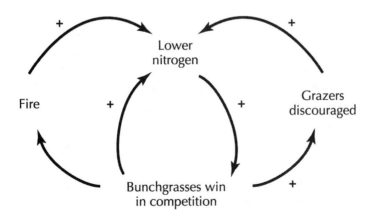

FIGURE 5.3

The feedback system that keeps the bunchgrasses dominant.

ing to an ever more complete exclusion of native grasses. It is the same feedback as that of bunchgrass success but working in the other direction. Only a few years of agriculture is enough not only to remove the bunchgrasses but also to keep them out. As agriculture increases soil nitrogen, a feedback is set that keeps soil nitrogen high. High nitrogen encourages vigorous growth of Eurasian weeds. In high-nitrogen regimes, the weeds can crowd out the native bunchgrasses. They are economical in their consumption of nitrogen, thus leaving soil nitrogen high. The bunchgrasses indulge in luxury consumption of nitrogen as an interference strategy directed at other life forms. Once established, the Eurasian weeds encourage long-term grazing by offering good forage as long as there is aboveground biomass. The grazing stresses any surviving bunchgrasses while also raising soil nitrogen levels through constant urination and manuring. As a result, the strategy of the Eurasian weeds keeps a high-nitrogen, heavy-grazing scheme in place, holding the site with a tenacity equal to that of the native biome dominated by bunchgrasses. Thus agrobiomes have an integrity equal to that of prairies.

In grazing systems, there is an import of mineral nutrients in the form of animal feed. To an extent, this balances the export in the livestock that are shipped out to urban centers. A problem here is distribution of mineral nutrients in that grass-fed beef is finished with grain in feedlots. Thus the loss of minerals is from the whole system, whereas the commensurate imports are into local foci. The more significant loss in grazing systems is carbon, a part of the budget of material that is often ignored. For quite a long time this net carbon loss is fed by fast processes that supply carbon to the soil. The loss is not immediately noticeable because the fast processes of carbon supply draw from a slow-moving carbon store (Parton et al. 1987). In the unexploited biome of the shortgrass prairies, that slow compartment works over millennia. With even prudent western grazing practices, that slow compartment is being significantly drawn down.

Holling pointed to fire as a fast-moving process that holds together biomes. Animals, such as spruce budworm in the boreal forest, can also move at rates that are commensurate with biome integrity across regions (Holling 1986). Rapid movement of materials across the landscape is of greater importance in agricultural regions than in other biome types. In managed forest systems, the context of wildfire is managed with prescribed burns. In agricultural systems, fire is suppressed and animals are fenced. Whereas fire moves fast across unpopulated landscapes, the decision to suppress fire moves even faster than fire, and so it gives landscapes heavily populated by humans an integrity that is at least as great as the integrity of biomes where fire can run its course. Domesticated animals generally are not allowed to move across the landscape, except when they are actively transported. Thus, whereas animals such as ungulates have played a primary role in organizing past biomes, they are often reduced to a secondary effect of human movement. The agent that gives modern biomes their integrity is not fire and primary consumers per se but the human animal as it manipulates the agents that gave past biomes integrity.

The fast-moving processes in agrobiomes that give them integrity are driven by commerce. When primary consumers and fire influenced biomes directly to give them integrity, there was an element of positive feedback. Both wildfire and animal population numbers are driven by positive feedback forces. Flame spreads and animals breed, although both are constrained by negative feedback in the mid-term as fuels and feed are temporarily used up. Now commerce is more widespread and responsive than the processes that pertained to biomes when human presence was marginal. It is the positive feedback properties of commerce that allow it to dominate ecological situations that are touched by human hands. Hunting for subsistence has its own built-in constraints on consumption as the hunter becomes sated, but hunting for commercial gain overrides those constraints. Positive feedback occurs wherein successful trade leads to yet greater success in expanded trade. Beaver and alligators were once trapped intensively to support international commerce, to the point of near extinction. As fashions changed (in the case of beaver) or conservation measures were enacted (for alligators), the species rebounded to the extent that people who occupy the animals' former habitats now consider them nuisances. Commerce involving pelts has been replaced by commerce in land for housing. Positive feedback loops follow one after the other as the circumstances of commerce change.

In most systems that involve rapid movement, there is a countercurrent flow, some sort of a recycling of the flux. A biological example might be the arteries and veins in the limbs of wolves; they lie close together so that the warm blood from the core is cooled by, and warms, the cool blood coming from the extremities. In human-designed systems an example might be traffic flow, either as opposing traffic on a freeway or paired one-way streets in urban centers. Commerce generates such rapid movement of materials and biota in agricultural systems that it needs countercurrent flows of an ecological type to achieve balance of material flows. The difficulty is that commerce dominates agricultural systems, but the obvious countercurrent processes that are needed to set the context of commerce are not particularly commercial. The processes that hold agrobiomes together appear not to be as well designed as those of systems with a smaller human presence.

A broken loop occurs in the patterns of resource use in the western rangelands of the United States. Commerce moves livestock from the western United States to slaughter and consumption in urban markets. There is no countercurrent to close the cycle to balance the flux to market. Carbon more than mineral nutrients is the critical consideration. Although rangelands do respond remarkably to mineral supplements, particularly nitrogen, the long-term issue for sustainability appears to be carbon loss from the soils. Nitrogen is a natural limit to rangeland production in that there is less response to unusually high water input if nitrogen is left at natural low levels. Much of the nitrogen removed in consuming forage is redeposited by the animal on the range, and much of the loss by removal of the animal to slaughter is matched by deposition. Nitrogen limits forage and therefore also limits stocking rates. This negative feedback keeps nitrogen status under control. The real problem is carbon (Parton et al. 1987). The low rainfall of the region leads to light, sandy soils. The capture of resources from the western rangelands appears to be lowering long-term carbon status. This is important in that low carbon leads to soil that is ready to blow. The disaster waiting to happen might remove the soil, essentially taking with it the entire agrobiome.

A solution would be to devise some means whereby organic waste from cities could supplement the diminishing carbon of western rangeland soils (White, Loftin, and Aguilar 1997). When commerce can drive both sides of the countercurrent, new markets and patterns of human resource use arise spontaneously. For example, the commerce that takes sugar and tropical fruit in ships from Hawaii leads to empty ships looking for cargo in the Pacific ports of North America. The loop was closed by shipping foods, furniture, and other materials to the burgeoning economy and population of Hawaii. Whether or not increasing the population of Hawaii is a good thing, it illustrates how an unlikely situation can emerge de novo if there is potential for countercurrent. On the face of it, Hawaii is too isolated to supply a consumer society with the most well-developed health care system in the United States, but countercurrent flows have a way of creating the unlikely through positive feedback. It is by no means obvious as to how to put the links in place for a countercurrent flow of food from the West balanced by sludge from the cities. It will at first require an elaboration of the system of municipal waste processing, linking them to commercial transport systems, and sludge dissemination systems at the source of the livestock.

What makes the situation hopeful is that, unlike elaboration of government that costs and stimulates resistance, the positive feedback that can arise in countercurrent systems might cause the changes for the good to be self-sustaining. Contrast this hopeful scenario with elaboration of exploitation of resources by regulation. For example, despite extensive and changing government regulation, the effort exerted by fishermen in achieving the catch is almost impossible to control. Limit the number of lobster pots, and suddenly their size increases. The innate force in the system is the intelligence and experience of lobstermen, who will get the best living they can. That force works against any attempts of government to regulate fishing effort. Very few lobsters make it through the 5-year window wherein they are big enough to keep yet small enough to catch. The government program that buys part of the catch for release is self-defeating. It only feeds back to increase the incentive to put more effort into capturing the resource. By contrast, an effective countercurrent system for sending sludge to the dry lands of the West would offer commercial opportunities at both ends of the pipe. The power of market forces resides in the intrinsic positive feedback properties of commerce. The weakness of government regulation is its intrinsic static nature, which comes from negative feedbacks. Of course, not all positive feedback processes that govern commercial exploitation of natural resources are forces for good. However, engineered commercial forces for good are likely to be more powerful than regulations that aim to achieve the same end. The positive feedbacks of commerce will create elaborate infrastructure, but it is more likely to be a cost-effective infrastructure than that which emerges from regulation. It is more efficient to use the positive feedback forces of commerce to outcompete undesirable aspects of commercial activity than it is to attempt constraint by regulation alone. Regulation is best used as a catalyst to put dynamic forces in play rather than as a direct constraint.

LACK OF SUSTAINABILITY IN PALEOBIOMES

The classic manifestation of an unsustained biome was the loss of the North American megafauna, as mammoths and many other species disappeared at the end of the last glaciation. This appears to have been a good dress rehearsal for modern global warming. Furthermore, there was more than one extinction, so we are not trapped inside a single scenario with no ability to tease the general characteristics from the unique happenings of just one extinction. There was an extinction of equal size in the late Pliocene, 5 million years ago. The most recent extinction has invited speculation about the possible role of human hunters. The Pliocene offers a parallel scenario, unadulterated by a convenient narrative about human predators hunting beasts to extinction. These ideas, and those that follow in this subsection, come from and are central to the dissertation research of Thomas Brandner at the University of Wisconsin. To lay out the final arguments for the relationships between modern sustainability and megafauna extinctions, we must take the discussion through notions of system instability, modern grazing and predation systems, and climatic limitation of life form. Extended as it is, the thread of the following argument does indeed lead back to megafauna extinctions.

Underpinning Brandner's work are the ideas on faunal lumpiness of C. S. Holling (1992). When Holling arranged the animals of a region in order of increasing size of the adult, most animals were close in size to those adjacent in the series. However, there were gaps in the sizes of animals, such that animals adjacent in the series were of significantly different sizes. There appear to be prohibited sizes for animals, for there are no examples of creatures in those critical sizes in a given region. Holling found that the gaps between the lumps are not universal but appear to be characteristic of particular regions. They can also be expected to be characteristic of particular times. The megafauna extinctions can be cast as the disqualification of a lump of a certain size, very large animals, by the climate change at the end of the Ice Age.

Also pertinent here is the scaling work of Rosen (1989), as addressed by Bruce Milne et al. (1992). In an analysis of the van der Waals equation, Rosen made the distinction between systems that are different and those that are dissimilar. A gas at different pressures and temperatures is only different, whereas a gas and the same material in liquid form are dissimilar. Systems that are so different as to be dissimilar are under different constraints. The constraints on the relationship between gas particles are discretely different from those on the relationship between liquid particles, making liquids and gases dissimilar.

That discrete difference, the dissimilarity between gases and liquids, is manifested as a fold in the response surface of the van der Waals equation (fig. 5.4). The change of constraints manifests itself as an instability. The simple gas laws use the linear equation $PV = rT$, which shows no instability. A modified pressure expression of temperature and volume appears in the nonlinear van der Waals equation (the pressure becomes pressure on the particles, not the box, and volume becomes the volume that is compressible between the particles, excluding particle volume). In the van der Waals equation, pressure, volume, and temperature manifest discontinuous behavior at the point of liquefaction. One seeks that unique instability point. In the van der Waals equation, that unique point of the equation is the triple point, where the equation has three real roots. On the response surface, it is the base of a pleat where pressure, temperature, and volume are noted as having critical values. That critical point of P_c, V_c, and T_c is different for each gas.

To deal with the instability, one normalizes the measured P, V, and T around the critical point P_c, V_c, and T_c at the base of the fold. This amounts to expressing pressure, temperature, and volume relative to the respective critical values. After the normalization, the instability disappears, and different gases can be compared with each other in the

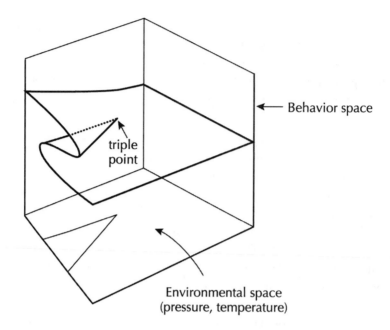

FIGURE 5.4

Gas–liquid instability. The van der Waals equation works in a two-dimensional environmental space of pressure and temperature, with an added dimension of behavior: volume. The equation is cubic and displays an instability pleat because there is a change of state from gas to liquid. The fold shows the region of instability, which is unique for each species of gas.

After Allen and Hoekstra (1992).

same terms. Milne and his colleagues looked for critical points in energetics of simulated animals of different sizes. Much as Rosen expressed gases relative to the critical triple point of the van der Waals equation, Milne expressed the behavior of each size of animal relative to the critical energetic point of animals of that size and the time of the inflection of population numbers on the move. After that normalization, the rescaled behavior of all animals of all sizes became equivalent.

For Rosen's gases, the distinction is between the instability of particular gases under pressure and the continuity of behavior in a comparison between all gases (fig. 5.5). The normalization around the triple

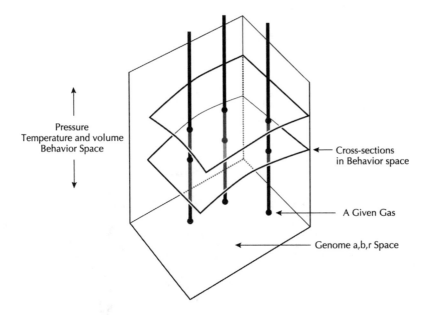

Pressure
Temperature and volume
Behavior Space

Cross-sections
in Behavior space

A Given Gas

Genome a,b,r Space

FIGURE 5.5

Stretching across instability. The pressures, volumes, and temperatures of all gases are normalized relative to the critical values at the base of the pleat in figure 5.4. A given pounds-per-square-inch of pressure becomes different for two species of gas because of differences between their critical points on the pleat, relative to which the pressure is expressed. The gas species are defined by a, the mutual attraction term; b, the volume of the gas particles; and r, the temperature related constant. In the a, b, r space on the bottom of the figure, each gas occurs as a point. Note that the dimensionless expression of the pressure, temperature, and volume gives a continuous space above, with the pleat of figure 5.4 normalized away. All gases become mere mutants of each other in a space with no instability.

After Allen and Hoekstra (1992).

point changes the discourse from a low-level phenomenon to an upper-level concern. Liquefaction of individual gases is the low-level discourse, whereas the unity across all gases after normalization is a high-level discourse. There is thus a hierarchy that considers gases at two distinct levels: one that shows instability and the other normalizing away that instability (fig. 5.6). In biology, there are multiple instabilities

303

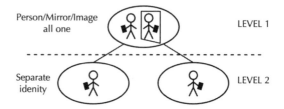

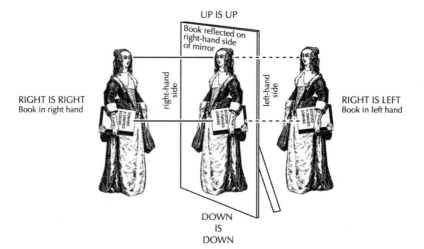

FIGURE 5.6

Changes in logical types. When the instability of the gases is normalized away to give a different level of analysis, gases are being addressed with a different level of analysis, introducing a change in logical type. A pair of logical types concerns the same issue but at another level of analysis. The image here shows how the up–down distinction belongs to a different logical type from the left–right distinction. The former addresses a shared reference to the context, whereas left and right are possessed by the person and the image separately. This explains why a mirror switches you left–right but not up–down.

From Ahl and Allen (1996), reproduced here with permission of Columbia University Press.

that give a more complicated pattern. Each instability gives a new level of discourse. As a result we see the elaborate hierarchical patterns that occur in biology from nested taxa in taxonomy to nested biochemistry and cell structure in cell biology. Unlike the behavior of gases, the behavior of biological material involves multiple instabilities and so creates hierarchies with multiple levels (fig. 5.7). Milne showed a hierarchy with two levels, like that of Rosen. His lower level showed energetic instability at one size of animal, whereas at the upper level there was unity of energetics even for animals of different sizes. Thus Milne and Rosen both dealt with rudimentary two-level hierarchies.

Because Milne considers a biological system, one could easily imagine yet higher levels. Milne's animals were of different sizes, but the range was not large (only 2 to 6 kilograms). Now consider what might happen if the range of animal size was greatly extended. Although the energetics are likely to be different but equivalent in animals of slightly different size, the energetics of animals of very different size are likely to be dissimilar and so not even equivalent. The equivalence that Milne showed would break down, with the emergence of a new level of distinction. This would entail yet another normalization, which would expose a yet higher level of equivalence.

As an example, consider the way in which large predators such as lions and cheetahs kill in roughly the same terms. Even a house cat might be considered as just a very small cat, hunting and killing in its own but equivalent fashion. By contrast, insects are so much smaller than the average mammal that there are fundamental dissimilarities that are more than a matter of scale. When things become more than a matter of scale, qualitative differences emerge. Colinvaux (1979) compared lions with ladybugs. He found that lions live alongside their prey in a slack relationship. Slack in the system at some level is crucial for survival. By contrast, insect predators are lethal and wipe out the entire population of aphids on a given plant. The instability inherent in eating the last prey item is clear. On the other hand, insect predators are about as inefficient at finding a new plant with new populations of aphids to exploit as are lions inefficient in exploiting their one population of prey. However, if one relaxes the criteria for the unit of prey, at another level of analysis lions and insects show an equivalence of lethality. Insects cannot just nibble on a population of aphids on a sin-

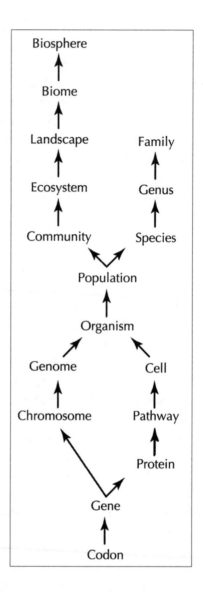

FIGURE 5.7

Instability and hierarchical levels. Taxonomic, morphologic, and bio-chemical hierarchies reflect biological processes that go unstable between levels. Whereas the gas–liquid instability separates only two levels, it is characteristic of biological and social systems to show multiple instabilities and so multiple levels. To deal with the different levels, we assign them different logical types. Whereas a dog and a cat are the same logical type, a carnivore is a different logical type. However, a carnivore is the same logical type as an herbivore.

gle plant, keeping some survivors to replace prey, and maintain the population of aphids on one plant. In rather the same way, lions cannot just nibble on a single item of prey. It is not an option to eat the front legs today and leave the rest of the giraffe to reproduce later. There is thus an equivalence between entire populations of aphids and single individuals of ungulates when one looks at them relative to their respective predators. Thus animals of only moderate scale differences prey in the same terms, whereas animals of very different sizes redefine predation relative to the other animal of very different size. In the same way that a parasite leaves its host alive, lions are parasitic on ungulate populations but predatory on ungulate individuals.

Predation of animals on plants also gives an example of critical rescaling. When animals graze on plants, several levels of analysis are needed for an understanding of various aspects of the process. There is a contentious literature on plant compensation in response to grazing; the phenomenon of overcompensation, in which plants grow more because they have been grazed, is counterintuitive and has been accepted only with reluctance. Without entering that arena of debate, consider the process of replacement of grazed material just in terms of the unit of replacement. When a forb is grazed, the level at which material is replaced is activation of buds or the continued growth of the tip of the surviving stems. Replacement is achieved by growing new stem material with new leaves that take the place of leaves that have been eaten. By contrast, grasses have leaves that grow from the base. When blades of grass are grazed, the very same leaf just grows longer. Replacement need not involve new stem growth with whole new leaves on those stems. Ungulates and grasses have coevolved so tightly that the grasses respond to grazing in a manner that is qualitatively different from that of forbs. It is a rescaling of response so that grasses are not more responsive to grazing; they are differently responsive.

Changes in morphology and life histories can be pushed only so far, whereupon qualitative differences emerge. This manifests itself in the differences between larvae and their respective adults. Amphibia live across a such a wide range of water availability that there are qualitative differences across metamorphosis. Insects can occupy such widely different habitats that larvae change through various instars continuously, only to emerge qualitatively different across the metamorphosis through chrysalis to adult. Some creatures appear to make the transition seamlessly, whereas others are more vulnerable. One of the advantages of having larval stages is a capacity to exploit qualitatively different resource bases with qualitatively different morphologies and strategies for resource capture. The danger for metamorphosis, in the context of sustainability, is that the life history as a whole presents a larger target for misfortune to hit. If all stages of the life history are obligatory, then elimination of only one crucial type of habitat will cause the species to go extinct. It seems that amphibia might be suffering from exactly that problem as their numbers decline at a global scale. Animals such as birds, which must spend different times of their lives in different biomes as far apart as the tropics and temperate zones, exhibit a similar vulnerability. Some of them are disappearing for that reason. The discrete life stages of organisms that inhabit radically different places are another manifestation of the discrete lumpiness in the size of animals, as discussed by Holling (1992). When a growing animal reaches the size that is prohibited by the top end of its lump, metamorphosis changes the rules of engagement.

Parasites are metamorphosing organisms that appear to be less vulnerable as a class. Parasites often have radically different life forms as they move from one host to another. It is noteworthy that they usually have one host that acts as a safe haven from which enormous numbers of eggs are disseminated. The other hosts bridge the gaps and may play a special role in dissemination over distance as opposed to production of a large number of disseminules. The masters of this approach are the fungi, which can not only spend most of their life cycle in one particular stage but also can reproduce within that phase. There is a class of fungi, the *Fungi imperfecti*, in which the sexual phase of the life cycle appears to have been lost altogether, the species achieving sustainability through the spore-bearing stage alone.

Trees appear to exhibit a strategy parallel to that of parasites in their life cycles. The adult tree is a safe haven for the genome and is an organism from which enormous numbers of propagules can be disseminated. The option for vegetative reproduction in plants appears to reduce their vulnerability further. Some plant species, in the same manner as the *Fungi imperfecti*, appear to be able to survive through vegetative reproduction only. Plants that hybridize sometimes lose sexual competence in the hybrid offspring. However, they often regain their sexual parts by doubling their number of chromosomes. This means that the incompatible chromosomes of the sterile hybrid now have a partner with which they can achieve successful pairing in meiosis. Meiosis is the means whereby sexual cells are achieved. Our crop plants often show successive hybridizations and subsequent doubling of chromosome numbers in a process of allopolyploidy. The widespread occurrence of allopolyploidy in plants indicates that a retrieval of sexual competence can wait a long time.

The capacity of plants to persist in the absence of an option for sexual reproduction offers some explanation for the sharp boundaries of biomes on the ground. Trees face problems when the climate for their biome begins to move out from under them. We mentioned the Gambel oak earlier in this chapter. There is a certain amount of rescaling of life history that Gambel oaks can achieve beyond which a full life cycle becomes impossible. They can germinate at different times to avoid drought at the critical stage as well as killing frost. However, there is a point in northern Utah where there is no room for further adjustment of germination strategy, and the species can reproduce only vegetatively. There are Gambel oak forests that are significantly north of the zone wherein the species can reproduce by seed. These populations are limited to sites where there are already established trees. Presumably, these sites became occupied under a different climatic regime that allowed reproduction through seeds and seed dissemination. Adjacent sites that are equally favorable but disjunct are not now available because seedlings cannot survive, even if acorns do jump the gap. Thus the northern limit of Gambel oak cannot involve metapopulations, for each population must defend its occupancy of its site in isolation from the species as a whole. In a metapopulation of Gambel oak at its northern range, populations can only wink off, not on. There are clear implications for sustainability of that species in those sites. The point of mentioning Gambel oak again in the present context is that life histories can only be rescaled so far, whereupon there is a qualitative change in the potential for sustainability.

All this can be related back to Holling's lumps in the size of animals and their relevance to the megafauna extinction. One can imagine widening Milne's scale of operation to compare not animals of somewhat different size but animals of very different sizes. Our expectation is that one would find not just differences in the scale of energetics across which translations to equivalence could be made (Milne et al. 1992). We would also find that with larger differences in size, animals could not be made simply equivalent by rescaling their energetics. There would be a change in the significance of energetics across gross size differences in animals. Holling's lumps are that higher level of dissimilarity. There is a range of scales over which an animal deals with its environment in roughly the same terms as slightly smaller or larger animals. Animals of commensurate size need to rescale their ecology only as a matter of fine tuning. That is what defines a single lump in Holling's work. However, there comes a time when the rescaling does not work any more. That is what defines the edge of a given lump. Thus Milne's simulated animals relate to Holling's (1992) lumps: His animals are inside a single lump.

Milne's energetics relate to the allometry of animals with regard to the size of the home range of animals. For very large animals, there is a need for very large home ranges. Note that for the surviving large animal of the Great Plains, the bison, a very large home range is needed. Indeed, that is one of the problems for those who would raise bison for meat. Bison are very persistent in their efforts to move on to new pastures, such that animals in small enclosures work massive fence posts out of the ground. The role of the buffalo in shaping vegetation before Euroamerican settlement is only emerging in our understanding at this time.

A change in climate tends to dissect site occupancy of plants. It is not so much that plant species cannot persist in a region, for trees live on long after their favored climate has moved away. Bunchgrasses live as long as many trees, such that sexual reproduction events are the exception rather than the rule, even when the climate appears suitable for all phases of bunchgrass life history. More than 20 years ago nutrient and watering experiments were conducted as part of the Grassland Biome Project of the International Biological Program. The vegetation was radically changed. Now the water and even the mineral nutrient agents for change have long gone. The surprise is that the vegetation remains altered, though not in the same way as the original change in vegetation (fig. 5.8). The message is that changes in climate can change vegetation, and the change does not reverse easily. Once large tracts of vegetation have been dissected, a return to continuous cover of the large landscape is slow, such that it may never happen.

The scale of occupancy of the landscape by large animals is large scale, and necessitates large, stable habitat. Climate change will dissect habitat. At first the dissection will be at a small scale. The large animals can move across the small gaps in favorable habitat for the plants they need. However, the plants under climatic stress cannot move across such gaps, just as the Gambel oak cannot bounce back from local extinction. An ability to jump gaps in the mosaic does not matter until

FIGURE 5.8

The experimental plots of the grassland group on the Colorado plains east of Fort Collins. During the International Biological Program of the 1970s the grassland biome group from Colorado State University treated this series of experimental plots with various forms of nutrient loading and added water. The differences can be seen here to persist many years later, with the double water and nutrient enrichment sites still showing patch structure and species different from the native condition.

Photograph by T. Allen.

a one-way ratchet begins to remove patches of plant occupancy. Once a favorable patch for a plant becomes unoccupied by that species or life form, reoccupation will not occur under climatic stress. What start as cracks in continuous cover of occupancy of the landscape by a given life form become barriers that stop metapopulation processes from replenishing lost site occupancy. Small cracks in the landscape turn into missing mosaic pieces. At first the large animals can move across the cracks and even around missing mosaic pieces. However, the process of change is directional, so missing mosaic pieces over time become adjacent. Animal movement is energetically costly, with no guarantee of success in the end.

The model of percolation theory may appear esoteric, abstract, and unreal for material landscapes. Percolation theory assumes that pixels in the landscape are added or removed at random. This assumption works well with regard to air in soil. Water passes through those air spaces, so percolation of water through soil can be modeled by random placement of transmitting pixels. Percolation theory applies to engineering issues, such as when expensive gold is used to transmit electricity across the surface of nonconductive material. Percolation theory tells the engineer how little gold can be used and still conduct electricity. This is again because the material that allows percolation is well modeled by random placement. But on material landscapes, types of habitat and patterns of occupancy are far from random. Corridors cause material landscapes to percolate at lower thresholds than abstract landscapes on computer screens. That being said, cracks in favorable habitat and loss of isolated mosaic pieces of vegetation together give a model that may well fit the assumptions of the neutral percolation theory landscape. The sudden disappearance of North American megafauna could be well explained by a percolation model. The lump to which mammoths belong failed to percolate across sufficiently large pieces of landscape to reach new favorable habitat. Because the lump is defined by a commensurate scale of occupancy of the landscape, all members of the lump should have disappeared at the same time. That fits the facts of the fossil record and might indicate a percolation model for megafauna extinctions.

Predation may well have a role in defining the size range of lumps, although Holling (1992) considered and rejected what he called the trophic trough hypothesis as a qualitative issue. Holling concluded that the gaps are principally a matter of textural discontinuity, such that only certain textures of resources and habitats are available. No habitat texture is available to support animals in the gaps. Predators may not cause the gaps between lumps, but there is every reason to suppose that they can influence the size of the gap between lumps. There are two strategies for prey to deal with predation. One strategy is to be big enough to make all but the old, the sick, and the young strong enough to make hunting them not worth the danger. The other strategy is to be small enough to be able to hide in cracks and hollows in rocks, soil, and trees. The first strategy is followed by mammoths and other megafauna. The other is to be small as a mouse. Some predators feed on only one general size of animal. Others prey on both rodent- and ungulate-sized animals, as when wolves in the far north switch to mice in the summer, when the large prey are absent. A reasonable hypothesis is that the predators press large and small herbivore sizes apart.

The carnivores would be more fragile than the megafauna herbivores in that the range needed by carnivores is 20 times that needed by their prey. Predation is closer to being marginal on isolated places such as actual islands and on functional resource islands formed by fragmented resources for the prey. With the carnivores gone as a result of diminished or discontinuous resources, one of the constraints that limits the lumps of herbivores would be gone. On islands such as Crete there were pygmy mammoths. On other islands rodents significantly increased in size at the end of the Pleistocene. Change in size outside ranges that we could consider normal today have occurred in many types of animals. For example, but not particularly on islands, some beaver evolved gigantism as large as 600 pounds. Presumably they were not still swimming and building dams. Small mammoths and large rats on islands indicate that, although the predators may not keep a herbivore window open between a rodent-sized lump and a large ungulate–sized lump, predators probably do influence the size of lumps by grooming some of their edges. The geographic islands where there were changes in size at the end of the interglacial were formed then by the rising sea level. Predators would have gone extinct on these islands first. With predators gone, a new size of prey is available on islands. For the larger animals, such as mammoths, the constraints of limited range on islands would put on them pressure to become smaller. Supporting evidence here would be the fluctuations in size that occur in lizards on the Lesser Antilles, as reported by Roughgarden and his colleagues. As noted, invading lizards from large islands such as Hispaniola arrive at twice the size of the natives, only to have their descendants reduce in size down to the size of the native species that the invader replaces.

In modern climate change combined with human dissection of animal habitats, evolutionary responses are almost certainly out of the question. Pygmy bison or dwarfed elephants in national parks are an irrelevant consideration. There are nevertheless implications for sustainability in a modern context coming from linking Holling's lumps and megafauna extinction. Given the intrusive human presence in modern biomes, some lumps in some regions may well be missing. Urban places have the same lumps as the surrounding environs, but with some lumps missing. The converse of missing lumps is a potential size range that is as yet unmet but could be so. In the same manner that small versions of megafauna and big versions of rodents appear on islands, the human cultural setting offers a situation that has some features in common with islands at the time of the megafauna extinction. Despite the best efforts at predator restoration, predation is not what it used to be in any modern biome. Accordingly, some size limits that used to be in place are now moot. In designing modern human biomes—that is, agrobiomes—there is an invitation to be more accepting of odd-sized livestock. Of course, the pygmy varieties of the megafauna were only hanging on for a while, but human ingenuity might well be up to sustaining unlikely-sized lumps of domesticates in an ecologically guided human biome. The process of breeding unlikely-sized animals has been in place for some time, as evidenced by the wide range of body sizes of domesticated varieties, even within a single species.

Humans themselves are also animals of a most particular size. Like the beaver, we are of the size of the carnivore lump. Our ancestors were smaller than us. An arboreal habitat, or at least an ability to take refuge in trees, made our ancestors an animal with a special habitat that moderates predation. Some ground-dwelling primates have moved to the gigantic lump, as in the case of the gorilla and the robust australopithecines. Our species in particular persists in the middle size range between rodents that hide and ungulates that are large enough to discourage predation. Our predatory characteristics may be associated with our body size. Only an animal of about our size could rely on fire: we are big enough to build campfires that do not go out but small enough that our fires usually do not become wildfires. Fire may have allowed a greater success for our species existing in the middle of a lump full of predators.

GLOBAL ECOLOGY

Global ecology is distinctive in its size, and that size causes certain types of relationship to be important. The relationships that occur in community ecology are for good reason primarily biotic. Competition might be mediated by environment, but organismal form and tolerances at the species level are always central. In ecosystem ecology, water in liquid form figures large, as does the relationship between biota and minerals. Atmospheric considerations are present in ecosystem ecology but are not entirely central. In global ecology, the atmosphere is indeed central; that is the hallmark of a global level of analysis. Whereas plant–animal interactions and fire are the only things that can move fast enough to give a biome integrity, only the atmosphere can work as the glue to hold the system together at the scale of the entire biosphere. It is the atmosphere that allows biota to relate to the planet. The oceans also receive a lot of attention in global ecology, for they too link the parts of the global system by moving and storing large amounts of material. The atmosphere and the oceans are homogeneous at small to moderate scales; they exhibit their most interesting heterogeneity at the global scale. By contrast, soils are heterogeneous at moderate to small scales and so play a role in more local ecological situations such as ecosystems, which are usually defined as much smaller than global. Variation in temperature and precipitation appear to be central in global issues. In smaller systems, temperature and precipitation often are effectively treated as just background variables.

Until recently, the preoccupation with major abiotic aspects of the global system has led to an underestimation of the effect of biota at a global scale. Darwinian evolution still has a stranglehold on almost all conception in biology and environmental sciences, and that is probably why the influence of life on events at a planetary scale has been underestimated. Our objection here is not that we see Darwinian evolution as distinctly wrong, for as we have indicated, rightness and wrongness are not the best way to characterize scientific models. It is just that Darwinian evolution, though an undeniably useful model, is more particular in its point of view than its realist adherents generally realize. "Survival of the fittest" refers to fitness as a matter not of conditioning and health but of how the organism fits into its context. Something that fits into its context survives, whereas something that does not probably will die out. In a Darwinian world, life fits into its environment and so does not unduly influence it and is almost never seen to dominate it.

Across conventional views of biology, life is a player inside the global system, not a controlling stakeholder. The central paradigm in modern biology is that genes interact with environment to produce phenotype. Once again, this is not a distinctly wrong point of view, but it is again particular. It is an elaboration of the central dogma and a holdover from the paradigm fight in the early twentieth century, between the Anglo-Saxon Darwinians and the European continental Lamarckians. The Darwinians still think it was and remains an unqualified victory for them alone. In a Darwinian world view, the organism is seen as smaller than its physical environment, the environment being beyond biological control. Of course, since ecosystem science emerged there has been some acceptance of local influence of biota on temperature, water, and minerals. But it stops at about the scale of the forest stand; beyond that size, most biological conceptions have been of life accommodating to a world over which it has little control.

James Lovelock's (1979) Gaia hypothesis challenges the conventional wisdom that life is best seen as held in the context of the physical aspects of the biosphere. In the conventional view, even the emergence of oxygen has been viewed as something that happened to happen, not as a manifestation of a global system that is systematically controlled by life. Oxygen is conventionally seen as a one-time occurrence, not as part of a process of an entire planetary biotic system. At a larger scale than the skin of the surface of our planet, life does indeed have to accommodate to external happenings. Life has had no influence on the sun, nor even on our planet seen at a scale that includes the mantle and core. Even so, life in the biosphere has enormously moderated the temperature of the entire atmosphere in the face of a 30 percent increase in solar output. At the scale of the biosphere, life is at least an equal player with the physical processes at the earth's surface, for how else could the planet's surface have been held for 3.5 billion years within such a narrow temperature range that life could have survived?

The prevalence of the environmental determinist paradigm can be seen in attitudes toward the Gaia hypothesis. The ignorant dismiss Gaia as mystical, even spiritual, ascribing to it an outlandish teleology. When the term *Gaia* is taken up by New Age mystics, allies of the original Gaia thesis as an intellectual abstraction are apologetic. Gaia is in fact a very modest statement that needs no apology. The difficulties that it suffers in gaining intellectual currency can be put down to its implicit attack on the prevalent paradigm, discussed in Darwinian terms earlier in this chapter. All that the Gaia thesis says is that for many concerns, at a global scale, life as a whole is a regulator of the temperature and composition of the atmosphere. Without purpose as expressed in interpretations of function, life makes little sense. For example, one could not study pathology without at least an implicit normative state of proper, intended function, from which disease is a departure. (This is not to say that purpose is a material aspect of the biological system independent of human observation and understanding. We are discussing epistemology, not metaphysics.) To the extent that life requires an explication of purpose (function) for an adequate understanding, purpose becomes part of an understanding of the biosphere when it is seen as regulated by life as a whole. But Gaia is no more mystical or rudely teleological than is the position of a physician who diagnoses that someone is ill.

As one moves conceptions upscale, there is not just one criterion on which to aggregate effects. Furthermore, change the criterion for aggregation, and some levels disappear, whereas others emerge as important when they were not so before. Advocates of any one criterion for aggregation may overestimate the general importance of their favorite criterion. Certainly biologists working at a biochemical level can become overly ambitious, even condescending to biologists working at levels of the whole organism and above. However, there is one large-scale issue wherein biochemical explanations are entirely in order. Whereas neither community structure nor ecosystem function is adeptly seen in biochemical terms, global ecology often has an explanation that is essentially biochemical.

Consider rain. Rain requires nucleation sites, which are largely sulfuric acid. Over continents there is plenty of sulfuric acid coming from pollution and terrestrial plants. The quandary was rain over the ocean: On the face of it there is no adequate source of sulfuric acid, so the scientific community was puzzled. However, it was realized that phytoplankton produce dimethyl sulfide and release it to the atmosphere. That appears to be the source of the rain-seeding nucleation sites over the ocean. With the warming of the oceans by as little as 1 degree, arguably an anthropogenic effect, there is a profound effect on plankton. It is not water temperature affecting plankton directly but a matter of stratification of the ocean. The warm water sits on top of the colder water, sealing off the lower layer of cold water. The oceans are mineral deserts. Paul Colinvaux (1979) surmised that the oceans should be green, not blue, except for the lack of mineral nutrition away from the coastal regions. Where there is plankton production in the ocean, it is around upwellings of cold, mineral-rich water from below. As warm water blocks that upwelling, the plankton cannot grow, so they do not provide the dimethyl sulfide, and one can argue that rain will not fall where it did before. The plankton are moving north, and we might expect some changes in the distribution of rain.

At a global scale, the issue of sustainability often is moot. Although as a whole life might be an active player in the to-and-fro of temperature and atmospheric composition, humans probably are held in the biosphere as constrained by external physical forces. Humans as a species do have effects at some level, but there is little to suggest that the things we do are other than one-time events. Humans appear at a global scale not as working partners with the biosphere but more as a blip that has produced some one-way effects. Rarely do humans have a greater long-term effect than an active period of volcanism, and we are never in the league of a good-sized meteor impact.

So, even recognizing anthropogenic climate change, humans appear as a perturbation to the global system (comparable, as noted, to a period of volcanism) but not as large-scale players in the global scheme of things. This means that there is much more slack in the system at a global scale with regard to sustainability, compared with more local systems that are better discussed as communities or ecosystems. There is bad news and good news. The bad news is that we do not know enough about the working of the biosphere as a whole to be able to predict adequately. The inability to predict operates at two levels. First, we cannot plan how to respond to change imposed on us that is unpredictable. Second, we cannot predict what will be the global effects of actions we do take and so are fairly impotent to direct the affairs of the ecological happenings at a global scale.

The good news is that there appears to be much more regulation of the planet by the world's biota at large than we had supposed. Therefore total disaster, such as a runaway greenhouse effect, appears unlikely. At the scale of biomes, or even floras and animal realms, humans affect the entire globe one piece at a time. Communities, ecosystems, and biomes of very wide extent are clearly being influenced deleteriously by the changes wrought by humans. At those spatiotemporal scales, the situation looks bleak, not just for what is left of nature but also for human systems that rely on renewable resources. Our impotence with regard to sustainability at a global scale tells us something very heartening. At the global scale, the world is still largely unscathed. A few small geological deposits of garbage, or even some slag heaps, are no match for real geological processes such as ice ages. Humans and their greatest effects stand dwarfed by processes of global regulation. Sustainability is a nonissue in systems that have escaped human intervention or interest. The primeval forests looked after themselves, as does the globe. Although humans have destroyed many parts of the earth system, at the scale of the integrated global system we are still incidental. But we should not allow this good news at the planetary scale to be used as an excuse for inaction at more local scales. We must work hard at the less-than-global levels wherein we can be influential and at which the lack of sustainability resides. There is merit to the adage to think globally but act locally.

6

Ecosystems, Energy Flows, Evolution, and Emergence

Earlier in this book we discussed historical examples of sustainable and nonsustainable systems. Economic and social processes were central to the discussion, and we have named dates, places, and historical personages as components of our narrative. We then turned to the biogeophysical aspects of sustainability, but in this chapter we maintain continuity by linking the ecology of sustainability to political sustainability. There emerge general principles that apply to both social and biogeophysical sustainability. Of the criteria for identifying the ecological foreground, the ecosystem criterion best facilitates the social and political connection to the conventional ecology of sustainability. The ecosystem conception of biogeophysical systems achieves this link by emphasizing process and flux over assemblages of tangible things. The process and flux might be nutrient cycling, whereas the tangible things, which for the moment we eschew, could be organisms identified to species. Much as there are limitations to the utility of dates and great people as cornerstones of historical happenings, there are similar limits to the utility of ecological tangibles in accounting for ecological happenings. Both great figures, such as Diocletian, and charismatic fauna, such as spotted owls, invite an emphasis on structure and event over underlying process. The emphasis on structure in biology comes from the tyranny of natural selection as an explanatory principle. We do not deny natural selection its place as a powerful model, but its intrusion into discourses where it offers little explanatory power is akin to cluttering a process-oriented view of history with gratuitous dates when kings, prime ministers, and presidents made utterances. Even though Queen Victoria gave her name to a period, her words and actions had little to do with the changes that the British wrought on the world during that time (fig. 6.1). In part because of its independence from evolutionary population biology, the ecosys-

FIGURE 6.1

A Victorian halfpenny with the queen's likeness on the head is known as a Bun halfpenny because of her hairstyle and to differentiate that portrait from the one used on coins in her later years. Although she gave her name to an era, Queen Victoria's actions as an individual, like those of all other "important people," were less important than the underlying processes that surrounded her.

Photograph by C. Lipke.

tem criterion achieves a greater balance between process and structure, and that balance will help in bringing to the fore the general principles of sustainability. This chapter justifies these contentions.

DEFINITION OF *ECOSYSTEM*

The word *ecosystem* has many meanings, some technical, others vernacular. It is important that we lay out what we mean with our use of the word *ecosystem* as opposed to all the other usages. The vernacular usage usually means something in the spirit of "that collection of ecological stuff over there." The vernacular meaning of *ecosystem* is inclusive and integrative, which we like, but it suggests that the boundaries are vague, which we do not like.

Leaving boundaries vague is an abdication of responsibility to bound the system explicitly. In our definition, we reject vagueness and substitute the notion of a defined but intangible boundary for the ecosystem. Ecosystem boundaries should be explicit, exactly because they cannot easily be drawn on the ground. In an approach to a material ecological system that turns on flux and process, there would be no movement of carbon or nitrogen up food chains without animals eating. However, the individuality and the manner in which an animal is an organism usually are unimportant in the movement of carbon and nitrogen. The organismal quality of an animal melts away when the focus is on movement of matter and energy. The parts of an ecosystem in functional terms are things like cycles, not the tangible animals through which the cycle passes, because the flow of nitrogen through animals is in parallel, not in series. Only as an exception do any organismal parameters of a nitrogen carrier affect the pathway of that element in a fashion that influences the properties of the ecosystem. Whereas a community or ecological landscape is populated by organisms, hedgerows, and other things to which one can point, the concept of ecosystem invokes pathways of matter and energy, and it is these pathways that populate an ecosystem. One might have to use tracers to follow those pathways, exactly because they cannot be seen literally. The intangibility of these ecosystem pathways presents no more difficulty than does our inability to see atoms literally while investigating chemistry. Unseen as ecosystem pathways may be, they do have spatiotemporal extents that can be measured, albeit only by technical means. The difficulty in bounding an ecosystem comes from the intangibility of its components. The measurements of the spatiotemporal extent of ecosystem pathways give an explicit boundary to the ecosystem. The ecosystem, by our definition, is bounded by the spatiotemporal extent of the cyclical pathways of material that flows around the system.

The emphasis on a certain wholeness embodied in the vernacular use of the term *ecosystem* is related to many of the technical definitions. The *-system* root is central. It means that the ecosystem is more than a collection; the parts must interact in some way for it to qualify as a system, giving a defined wholeness of function. For all the diversity among the particular possible types of ecosystem, anything called an ecosystem is identified as having parts that are connected to give emergent properties of the whole. There is a notion of integrity across usages as diverse as "forest ecosystem," "the Great Lakes Basin Ecosystem" or "cow rumen ecosystem." A forest can be considered a mere collection of trees, an undefined context in which babes get lost or to which Shakespearian characters retreat in times of banishment (fig. 6.2). On the other hand, a "forest ecosystem" emphasizes an integrity to the whole. In a first conception at an intuitive level, one looks at a "forest" from the inside, but one regards a "forest ecosystem" as a bounded entity seen in particular from outside. "The Great Lakes Basin Ecosystem" (Vallentyne 1983) is the term used by the International Joint Commission to indicate an ecological entity that explicitly includes humans as a working component. When Allen and Hoekstra made reference in our previous book to the "cow rumen ecosystem" we were emphasizing an integrity of function despite a significant throughput. Ecosystems are importantly integrated in a way that gives the parts an active, functional role.

The intellectual lineage of the ecosystem concept includes Transeau in the 1920s, although it was Sir Arthur Tansley who coined the term a decade later. Tansley (1935) recommended that the physical environment be folded in as part of the ecological entity under investigation. At the time Tansley coined the term *ecosystem,* he was merely pointing to the way in which not only is the vegetation at a site influ-

FIGURE 6.2

The meaning of forest. A hillside west of Fort Augustus in the Scottish Highlands. The locals call this patch of scrub and heather a forest, and it is known to be ancient. Thus the modern meaning of *forest* to be always land that is heavily wooded is different from *forest* in an earlier vernacular. In the past, *forest* meant only that it was a wild place, not much occupied by humans. Royal forests were not so much places where trees were as places where people were not supposed to be. Shakespeare does not refer so much to trees when his characters go off into the forest to escape society because trees are a correlate of low-density human occupation rather than an essential character of forests.

Photograph by T. Allen.

enced by the soil and local climatic considerations but the vegetation reciprocally influences the soil and the microclimatic processes. Tansley's concern was for a more inclusive view of vegetation functioning than the community conceptions that came before. His was a remarkably modest proposal that still turned on how the organisms in the community came to be where they are and in what condition. The direct connection between the physical context and the biota led Lindeman (1942) to apply the notion of energy flow in the 1940s, in a far more elaborate account than that of Transeau (1926) in simple agronomic systems. The Odums matured the ecosystem concept through the 1950s and 1960s with a diversification of the types of fluxes and processes that are considered ecosystemic (Odum 1969; Odum and Odum 1976). The general notion of an entity consisting of flux in pathways was the basis of the ecosystem modeling at Oak Ridge National Laboratory in the 1970s. These ecosystem simulations were created as part of the American International Biological Program (IBP) and represent a maturation of the ecosystem concept (Patten 1975). The state of the art was captured in the edited volumes of Bernard Patten (1971, 1972, 1975, 1976).

The technical usage of the term *ecosystem* in this book comes from the IBP simulation studies, which operationalized the concept of ecosystem as a separate type of ecological entity. In this view, the ecosystem is set apart from the community or the population. This distinction was achieved by discarding species and organisms as the parts that typically make up the whole. Although there are organisms involved in ecosystem function, the focus is on the flux of matter and energy. Even when modeling an explicit type of organism, forest floor arthropods, O'Neill (1971) did not mention species. Only occasionally did he distinguish genera. Mostly he lumped whole orders of insect, with other arthropods mentioned also only as whole orders. O'Neill focused primarily on energy and material fluxes, with a nod to some population dynamics of *Collembola*. Crickets and centipedes were graphed along with leaf litter as to intensity of radiotracers for each compartment, crickets turning from a Jiminy Cricket tangibility into a black box with radioactivity in it. In the modern ecosystem conception, organisms become conduits, connectors between short- and long-term storage compartments. In a botanical example, Sollins, Harris, and Edwards (1976) modeled the physiology of a temperate forest. Atmospheric carbon flows into leaves of trees to become part of carbon storage in tree boles, or it flows out to become soil organic matter. Thus parts of the tree, the leaves, become mere connections between the atmosphere and a storage compartment that consists not of whole trees but of only one part of the tree, the trunk. It is hard to see the tree as a whole when it is divided up in this way because the connections of carbon inside the organism as a whole are no stronger

325

than connections of that same carbon to carbon outside the organism in animals or the abiotic world. Of course there is still a material thing that corresponds to the tree, but the coherence of that entity may have no significance in a particular ecosystem conception. By modeling the flux of matter and energy around ecological systems, the computer simulations of the IBP established the ecosystem concept as something discrete from other ecological types, such as communities or populations.

Sustainability involves complex situations that we need to manage over the long term and that require sophistication as to the choice of models. Management involves prediction (chapter 2), such that a given management action has the desired outcome. Prediction requires that we have appropriate models. An appropriate model deals with the system at the right level of analysis. Of course the scientist is human and will have predilections for this or that sort of model, but effective scientists will move away from their favorite strategic models when the best solution lies in another direction. C. S. Holling (1986), aware that his favorite causes for the behavior of insect outbreaks focused on bird predators, was explicit in moving to other classes of explanation so as not to miss a more powerful explanation. In the end he returned to avian predators because nothing else looked as if it was going to work, but his open-minded inquiry allowed him to discover that squirrels were important along with the birds. Lesser scientists stick to their default model, and the most parochial scientists are not aware that there are other models than their favorites.

The default model in biology is evolution and adaptationist explanation, but as we noted in chapter 4, this is only half the story. Sometimes that model is appropriate, but when it is not, adaptationist arguments are worse than useless because it is the wrong half of biology for the question at hand. We are being literal when we say half the story because there is a duality involved here, and evolution and adaptation represent only one half of the dual. As we argued earlier, an ecosystem view of a system deals with fluxes of energy and material from a thermodynamic reference. Thermodynamic processes do not relate easily to adaptation because ecosystem fluxes happen for physical reasons remarkably independent of adaptive function. An adaptation is embedded in the structure of a biological entity, whereas flux is a matter of process, not structure. So an essential dualism arises in which thermodynamic process and adaptive function each tells half the story.

THE ESSENTIAL DICHOTOMY IN BIOLOGY

When one takes a flux-oriented view of ecological material emphasizing pathways, natural selection loses power as an explanatory principle. Discrete organisms may persist in the material system and can be seen if one looks for them in the right way, but in the ecosystem model they recede and fail to maintain their particular identity as bounded entities. As organisms dissolve in an ecosystem conception, there are no discrete entities to display evolutionary adaptation. Therefore, the principles of orthodox Darwinian natural selection also recede as explanatory devices for most ecosystem conceptions. Adaptations through the evolutionary processes may still occur, but adaptation for the most part ceases to be a relevant, predictive process in the ecosystem conception.

The rejection of natural selection as an explanatory principle may be a radical consideration for many ecologists. We do not mean to deny the significance of the loss of species and the termination of their adaptations. The point of any intellectual device is its ability to provide prediction or understanding in terms that are as simple as possible. When an ecosystem conception offers such simplification, without recourse to natural selection and adaptation, it might be tempting to include natural selection in the argument because it explains something else or because we know that the ecosystem actors are evolved entities. However, such inclusion of natural selection and its principles can complicate the science with no added value. If local or global extinction of a species removes its particular adaptations from the material ecological system, that loss probably passes unnoticed in the measurement regime that comes with casting an ecological situation as an ecosystem. Another species often takes up the slack by assuming the part of primary productivity or consumption that was vacated by the departing species. For instance, when the American chestnut proved unsustainable because of chestnut blight, that species had disappeared from the canopy, causing radical changes in forest texture and composition (Shugart, West, and Emanuel 1981). However, other species in the Great Smoky Mountains performed almost all the ecosystem functions of the chestnut. In the measurement regime that is satisfied with primary productivity as a general process, individual species appear not to perform the ecosystem function in themselves. Of course the leaves of a species that takes over the space will be different in shape and in adaptive details of photosynthesis, but in an ecosystem conception, fixed carbon is just that, fixed carbon, no matter which species fixes it. Therefore, as radical as it may seem, evolution is rarely a player in ecosystem function.

If the adaptive significance of particular organisms is often unimportant, then some other unifying principle must be responsible for the predictive power that comes with casting ecological material as an ecosystem. That unifying principle is thermodynamics. Natural selection does not deny thermodynamics and its laws, but nor does an evolutionary explanation readily relate to thermodynamics in a coherent fashion (Wicken 1980; but see Wicken 1987 for a spirited attempt to link them). The thermodynamics of living systems is not interesting if the system is seen only as closed and running down. The thermodynamics that are the complement to evolution must be concerned with the creation of new order, not how order disappears over time (Schneider and Kay 1994). Thermodynamics applied to open systems far from equilibrium does indeed indicate creative forces that complement evolution. The thermodynamics of far-from-equilibrium systems has strengths at exactly the points where evolution is weak.

THE DUALITY OF EVOLUTION AND THERMODYNAMICS

Natural selection and the thermodynamics of order are fundamentally different, not just simple alternative views that can be invoked on a whim. The difference amounts to a duality in which both models are needed to deal adequately with what appears to be one material situation. Evolution and thermodynamics together appear contradictory in that evolution generates adaptations that turn on functionality, whereas that function is executed thermodynamically by carbon and other atoms that do what they do whether or not the outcome has functionality. The tension is between the rate independence of the significance of an adaptation and the rate dependence of thermodynamic material flows. The temptation is to try to find the single unitary truth as to which approach applies. However, if the scientist plumps for one model, the phenomena associated with the other prove the chosen model wrong. It does not matter which is chosen; the other will always be needed and will be proven wrong by its complement. Equally troublesome is the strategy of trying to bring together the two models in one unified explanation. The only way out is in a duality in which both alternatives are used but are kept apart as separate descriptions. This is the principle of complementarity in physics, but it applies also in biology (Pattee 1978). There is no comfortable solution. There are aspects of evolution and thermodynamics that force the duality.

Because ecosystems are intangible things, and there are issues of duality surrounding them, it is important to remember that the ecologist deals with models, not reality per se. The hero of ecosystem science, Tansley himself (1926:685), offered some early cautionary statements about the dangers of realism.

We must always be aware of hypostasizing [reifying] abstractions, that is, of giving them an unreal substance, for it is one of the most dangerous and widespread of vices through the whole range of philosophical and scientific thought. I mean we must always be alive to the fact that our scientific concepts are obtained by "abstracting from the continuum of sense-experience," to use philosophical jargon, that is by *selecting* certain sets of phenomena from the continuum and putting them together to form a concept which we use as an apparatus to formulate and systematize thought. This we must continually do, for it is the only way in which we can think, in which science can proceed. What we should not do is to treat the concepts so formed as if they represented entities which we could deal with as we should deal, for example, with persons, instead of being, as they are, mere thought-apparatuses of strictly limited, though of essential value. . . . [Tansley continues in a footnote:] A good example of hypostasization of an abstraction, exceedingly common 40 years ago, but now happily rare, is the treatment of the process of natural selection as if it were an active force, a sort of *deus ex machina* which always and everywhere modified species and created new ones, as a breeder might do with conscious design.

Wicken (1987) highlighted the differences between models for natural selection and thermodynamic models. He emphasized that natural selection is an *external* force that evaluates function in an ecological arena, but there are also *internal* limits as to what biological systems offer for selection. Wicken pointed out that amino acids in simply chemical situations do not polymerize to form dipeptides at random but with frequencies that reflect "energetic preferences. . . . The Second Law [of thermodynamics] is obliged to operate within the particular bonding predilections of chemical elements so that only limited kinds of structures can be produced. . . . Perspectives from the physical sciences are essential to keeping the Darwinian program honest" (Wicken 1987:88). Energetic limits define which are the pertinent physical preferences that the system must accept, and these are critical in an understanding of the relationship of evolution to ecosystems.

Wicken's notions of a duality between internal and external considerations came from his identification of two functions in biology and so ecology (Wicken 1979). Rosen (2000) also emphasized internal and external considerations, and he used a distinction that occurs in linguistics. A proper description of a language requires both the internal syntax of how the language works and a semantics that relates the language to the external world of meaning. Similarly, in ecosystems, the external force serves as an *evaluative function* that gives semantic meaning to adaptations, whereas the syntax of the internal energetic limits come from a *generative function* that previously caused the emergence of the structure that has adaptations.

Wicken (1979) pointed out that any theory of evolution or adaptation must have both principles. An example of the generative function in neo-Darwinian evolution is found in random mutation of genes. Mutation generates material de novo in emergent processes such as radiation exciting of atomic bonds that settle down thermodynamically in some alternative molecular arrangement. Speciation and macroevolution show clear emergence over evaluation. Grene (1974) noted that Darwin (1964 [1859]) spoke of natural selection, an evaluative process, and barely addressed speciation, a process under the generative function. When Eldredge and Gould (1972) identified discontinuities in the fossil record of punctuated equilibria, they were addressing an aspect of evolution that comes under the generative function.

On the thermodynamic, generative side, emergence of structures de novo in the generative function occurs through thermodynamic dissipation of a gradient. Thermodynamic gradients pass from a concentration of energy or matter to a diffuse, most likely arrangement of that same energy or matter. When the second law of thermodynamics states that a closed system runs down to an entropic condition, the system does so by material or energy passing down the gradient. Steep gradients give a certain tension that creates a condition of an accident waiting to happen. When the accident occurs, it sets in motion a positive feedback that dissipates the gradient. The accident is an event often amounting to the intrusion of some structure that acts as a template for the ensuing cascade. The pattern that emerges in a thermodynamic cascade comes from the details of the initial accident. In this way every snowflake is different, although the general pattern of snowflakes occurs in their hexagonal architecture. When human societies emerge under a process of resource dissipation, each is unique because of the importance of the initial conditions that started the cultural happening.

Complementarity between dual models invokes apparent contradiction, such as the existence of an adaptive organism composed of material that is in a configuration determined by thermodynamic energetics, not adaptation. Viewing the generative function, there is a contradiction that turns on continuity and discontinuity. The new structure is discrete but must emerge in a continuous process. Furthermore, observed very closely at its boundary, the discrete structure grades into its surrounding. But as Tansley (1926) advised, we are dealing with intellectual devices in the evaluative and generative functions. Therefore the newness of the new system that has emerged under the generative function is a matter of the observer's decisions. These decisions turn on what the observer is forced or chooses to recognize. The issue of the ontological status of emergent structures—that is, whether they are new in reality in themselves—need not be of concern and is not really a scientific matter. The central issue is dealing with reliable observables, and it appears that observables regularly generate the dualities, with their inherent potential for contradiction embedded in the act of observation.

The generative and evaluative functions play off each other, and we do not want to give either function primacy. Whichever function is given privilege, there is a reciprocal dependence working in the other direction. The evaluation maps substantially to natural selection, and the grist for its mill comes from the output of the thermodynamic, generative function. Meanwhile, the process of emergence is held in the context of the survival of irregularities; that is, they survive some sort of natural selection. An emergent structure will become full blown only if it fits into its local environments well enough to survive. The emerging structure might be an embryo in biology, or in physics it might be a wrinkle in the laminar flow in a stream, developing turbulence. The uneasy mutual dependence of emergence and inspection should not cause dismay, for such shared necessity of opposites is well known in other dualities (Pattee 1978). One is the wave–particle complementarity in physics. As in the wave–particle duality, one gets in trouble only if the two models are mixed. Therefore it is best to keep the evaluative and generative functions formally separate while acknowledging that both are needed for a full account.

The evaluative and generative functions address different causalities. The inspectorate, under the rubric of Darwinian natural selection, explains the existence of particular elaborate structures. It is most often used in attempts to address issues of *what* and *why* in biology. The generative function often answers biological questions of *how*. The generative, thermodynamic function does not have a label in the scientific biological vernacular but could be called "blind process, without adaptive significance, under energetic constraints." Studies of thermodynamic function do not generally address the Darwinian questions of why such-and-such a situation exists. Instead, studies invoking thermodynamics focus on biological questions of how a particular system works, such as the functioning of a biochemical cycle. One could cast the thermodynamic part of biology as addressing efficient Aristotelian causation, whereas evolution is most adept at addressing ultimate causation. Ulanowicz (1997), in this book series, did a nice job of laying out the different Aristotelian causations in ecology. Dylan Thomas (1954) once received a book, as he described in *A Child's Christmas in Wales*, that told him everything about the wasp "except why." In a conventional view, evolution answers the question "Why?" as an ultimate Aristotelian cause.

Our view is that students of the Darwinian approach need to relax their stranglehold on biological discourse and allow the generative function in thermodynamics room to make its special contribution in tandem. The intellectual histories of thermodynamics and evolution are remarkably independent, with little evidence of much of a confluence of ideas and culture. That intellectual divide is being institutionalized as new academic departments of biology are disaggregating into cellular, molecular biology, with its thermodynamic posture, as opposed to evolutionary, population, and organismal biology. More broad-based ecology embraces not just population biology but also community, ecosystem, and landscape conceptions. The sort of ecology that the authors here embrace finds itself often betwixt and between. Ecosystems are effectively cast as principally thermodynamic but incorporating natural selection, populations, and organisms as given.

The evaluative function is studied by natural historians, usually at levels of organization of the functioning of whole organisms and organismal associations, such as predation, parasitism, mutualism, and community. The appealing richness of biology is captured in studies of the evaluative function. Biological structures take a particular form, often as tangibles, as a result of having survived inspection in natural selection and other evolutionary processes. The richness in biology seen in this light arises from the significance of the details in structure. Students of the evaluative function and its products often are concerned with adaptive significance of the patterns they detect or, on the other side of the coin, with the unexpected neutrality of distinctive characters. On the issue of neutral characters, the essential problem is that the pressures for adaptation may well have disappeared, leaving a pattern behind that has no contemporary functional significance. One cannot in fact easily distinguish between apparently meaningless characters that emerged originally as adaptations and those that were always neutral and never had significance (Koestler 1967). Past selective pressures have left a tantalizing pattern, the meaning of which can be considered only through plausible scenarios. Plausible narratives are an endemic flaw of the literature of evolutionary investigations. The best use of plausible narrative is only as an example of what might have happened among an infinite set of possibilities. Only imagination is needed to erect alternative plausible scenarios, making the original story remarkably ad hoc. Though effective within its domain, the evaluative function manifested in natural selection is weak when pushed beyond its bounds.

Entirely separate from the evaluative function and natural selection, the generative function presents the inspectorate with particular structures for selection. We still want to avoid giving primacy to either function, but we have to start somewhere in a discussion of the relationship between the two functions. The generative function creates structures. Once those structures exist, they are placed under selection by the inspectorate through the lens of the biotic and abiotic environment (Wicken 1987). The generative function is responsible for the general types of configuration that are presented to natural selection and the other evolutionary filters. The generative function causes systems to exist, whereas the evaluative function sifts through the lot, identifying those destined to persist.

A PRIMER ON THE MECHANICS OF
THERMODYNAMIC EMERGENCE

In ecosystems, the thermodynamics is captured in the fluxes of energy and matter that pass around the ecological material. Sometimes ecosystem ecologists depend on the first law of thermodynamics, the one that says everything must balance because there is conservation of energy and matter. When the ecosystem is treated as a black box, the first law is invoked. The second law of thermodynamics speaks of closed systems running down to a state of maximum entropy, where only the most likely, least distinctive arrangements of matter and energy pertain. However, a new thermodynamics is emerging that refers to open systems, such as living material, where constant inputs of resources stop the system running down. It appears that as systems are pushed away from that run-down, dead state, they create unlikely configurations that are the order that is seen in living systems. This thermodynamics of emergence is that which pertains most importantly to sustainability because it is those unlikely configurations that are to be sustained, at the cost of critical inputs that hold the system dynamically away from the dead state.

If there is any local input to a homogeneous system at equilibrium, the system moves away from equilibrium and a gradient will arise. The input may be energy or matter, and the gradient therefore can cause flows of energy, matter, or both. In the case of a heat source in a cold environment, it is a temperature gradient, with heat flowing away from the hot spot. Weak inputs to systems create shallow gradients, whereas a stronger input, say a greater heat source, causes a steeper gradient. A return to equilibrium eventually would arise spontaneously should the input cease because the energy in the system runs down the gradient until the gradient disappears. There is nothing special about energy or material flowing down the gradient, for it is something that will occur on any gradient according to a conventional statement of the second law of thermodynamics.

Steep gradients cause the emergence of structures (Wicken 1987; Schneider and Kay 1994). Along a shallow gradient, one where the top of the gradient is not far from equilibrium, the flow of material or energy is unremarkable (Allen, Tainter, and Hoekstra 1999, 2001; Allen, Tainter, Pires, and Hoekstra 2001). However, as the gradient becomes steeper, the flow begins to elaborate. Patterns elaborate spontaneously on steep gradients as structure emerges. Such emergent structures are called far-from-equilibrium structures.

It is worth going through a simple physical example of thermodynamic emergence so that the general principles can be demystified. In the case of a strong heat gradient in a fluid system, the far-from-equilibrium structure will be some sort of convection cell. Convection cells take the form of a torus, a doughnut with heat passing through the hole, away from the heat source, toward the cold end of the gradient. Schneider and Kay (1994) gave a detailed account of the energetics of convection cell emergence, based on the original experimental work. As hot material reaches the cooler end of the hole, it encounters the cold heat sink. At that point, the fluid moves centrifugally from the cold end of the hole, radiating out across the cold surface, to the perimeter of the torus. In doing so the fluid cools. Now that the fluid is cool, it begins its passage down the outside of the doughnut, back to the heat source. At the heat source the fluid is reheated and passes through the hole in the torus again, just as it did the first time. Hot fluid passes in one direction through the hole, and cooler material passes in the other direction on the outside of the doughnut.

The fluid in a convection cell is entrained in the flow inside the cell, such that the fluid in the cell is isolated from the fluid in the environment. Often more than one convection cell emerges, in which case they pack together like doughnuts crammed into the bottom of a flat box. This makes a hexagonal, honeycomb pattern, one cell per honeycomb or doughnut unit. The fluid moves rapidly around inside each cell, but there is very little mixing with the fluid of neighboring cells. Thus there are boundaries between convection cells, as there are boundaries around all structures that emerge from strong gradients. Indeed, their structural identity depends on those boundaries. A boundary encloses fast reactions within. Those fast reactions run parallel to the boundary, with only slow, weak reactions running across the boundary. It is these relative rates of reaction inside as opposed to across boundaries that give far-from-equilibrium structures their integrity and invite recognition of their structural identity.

In the example of convection cells, it is clear that the emergence of structure facilitates flow of energy down the gradient from the hot heat source to the cold heat sink. On a gradient too shallow to cause convection cells to emerge, the movement of material or energy is by unstructured, three-dimensional diffusion. The convection cells that emerge on a steeper gradient do not rely on simple diffusion but move fluid, material, and energy in a directed fashion. The coherent flux of material in the convection cell moves heat away from the heat source much faster than mere diffusion. Thus the convection cell dissipates the gradient on which it forms. The manufacturers of double-glazed windows are very careful to space the panes so that convection cells, called Benard cells, do not form. Should Benard cells emerge, the window pumps heat outside with great efficiency, defeating the purpose of double glazing. Convection cells persist only if there is a continuous input of energy into the hot spot at the top of the gradient. Far-from-equilibrium structures therefore are also known as dissipative structures.

The thermodynamic emergence of particular structures depends on exquisite details at the outset, the initial conditions at the beginning of the emergence. Emergence can be considered a process of amplification of the details of the slight imperfection that seeded the growth of the emergent structure. Emergent structures therefore are unique every time they emerge. Each snowflake is unique, as is each species and each product of meiosis. We assert that one class of explanation of life, and the unique forms it presents, is the existence of gradients that generate structure. The unique nature of emergent structures is responsible for the richness of life, as each exemplar—each organism and each species—takes on its particular identifying characteristics.

Schneider and Kay (1994) note that emergent structures are a means of resisting the move away from equilibrium that comes with strong inputs of energy or material. For instance, summer heat on the ground would keep increasing, except that storms dissipate the energy gradient, taking heat up to the cool end of the gradient high in the atmosphere. Because of their strong throughputs, thunderheads have the effect of resisting forces that would otherwise push the system up the heat gradient, farther from equilibrium. Conversely, the formation of snowflakes resists the lowering of the air temperature in winter.

Not limited to physical systems, gradients at the edges of complex social systems stimulate energy dissipation as well. Roman coins in gold and silver are found as far north as Scandinavia, as far south as the central Sahara, and as far east as southern India and Sri Lanka. They were known in Han China (Harl 1996:290–314). The structures that emerge at such gradients have played central roles in history and have, indeed, dissipated energy quite efficiently. The peoples on Rome's northern frontier formed first into raiding parties, then into tribal confederations, and finally into nations. Some of them raided the empire, and others of their number were hired to defend it. The order manifested at the frontier blurred, and ultimately the gradient could no longer be maintained. Similarly, the Abbasid surplus in silver flowed out from the Middle East through Viking trade routes and into northern Europe (chapter 3). Soon a political structure emerged in northern Europe (the Carolingian Empire) that claimed status as an equal. As the Abbasid Caliphate collapsed and this gradient grew shallower, the

European effort at political unity proved short-lived (Hodges and Whitehouse 1983).

Energy gradients stimulate political change today just as surely as in antiquity. Manufacturing plants along the U.S.–Mexico border represent an attempt to control the flow of resources southward, while northward migration blurs the gradient uncontrollably. As Spanish becomes the dominant language of Los Angeles, earnings flow south to families still in Latin America. The recent civil war in Somalia had much to do with who controlled (and thereby profited from) the flow of foreign aid into Mogadishu (Besteman 2002). The end of the Cold War made such aid increasingly scarce and valuable. The energy gradient of foreign aid stimulated civil war (by being worth fighting over) while also making it possible. Correspondingly, Somali political unity could not be maintained without the levels of foreign aid to which the Cold War had accustomed it.

Despite the unique characteristics of each emergent structure, a gradient of a given type spawns similar structures. Although each structure is unique because of initial conditions, its similarity to other structures arises because the particular gradient directs the positive feedbacks toward a particular type of constraint. Nature uses only what possibilities are available as it resists being pushed away from equilibrium. Unique as they are, every snowflake is hexagonal because freezing of water is grossly limited by molecular possibilities. A head of water tends to generate whirlpools, but hot summer air generates thunderheads, again because of the nature of the gradient and the material on which it acts. In summary, the potentiality of constraint is realized only when there is a strong enough gradient. The details of the initial conditions amplify only in a suitable thermodynamic environment, and what can emerge is limited by the nature of the material involved.

THE THERMODYNAMICS OF ECOSYSTEMS

Life itself is a dissipative emergent structure (Schneider and Kay 1994). It dissipates the temperature gradient between the surface of our planet and outer space. The emergence of dissipative structures in the full history of the development of life has left behind dissipative structures at all levels of biological organization. For example, locking together photosynthesis and respiration, where one generates and the other consumes oxygen, is an example of the emergence of an energy-dissipative structure. Respiration is far more efficient, that is to say more energy dissipative, than oxidation processes such as glycolysis, which preceded respiration in the macroevolution of living systems. The elaboration of life appears to amount to an increasing capacity for dissipating the temperature gradient by removing heat from the planet's surface. Note that living systems are emphatically open systems with strong throughput, as one would expect for energy-dissipative structures. The high degree of order in life is thermodynamically far from equilibrium. Remove the input of free energy, and life shuffles to a halt as it starves or suffocates. Life has all the characteristics of an energy-dissipative, far-from-equilibrium structure.

All of life is set in the context of the temperature gradient from the ground to the upper atmosphere, and life emerges by elaboration of chemical patterns, principally carbon chemistry. Each manifestation of life is different in exactly the way that all snowflakes are different. On the other hand, much as all snowflakes are similar in having six sides and showing radial symmetry, there are general constraints that entrain life so that all of it shares general properties. Some characters are shared through homology and descent; one suspects that adenosine triphosphate was the first dominant carrier of chemical energy, and that is why it is universal across all of life. However, in ecosystem science we are less interested in the constraints and universalities that arise through homology under the evaluative function and are more interested in those that are imposed from the universals of the generative function. These general constraints on life arise from the limits imposed by a mixture of organic molecules and organismal forms emerging on a temperature gradient at a planetary scale.

All predictions are that the constraints on a system will remain in place and hold system behavior within limits. System behavior changes radically when constraints change. Rosen (1989) refers to changes of constraint in his description of the phase change of gases turning to liquids in an analogy to biological systems. In social systems such changes could be a societal collapse or the emergence of a new resource that reorders the society. Radical change in system behavior is an indication that there has been a phase change of the matrix of the system. With the phase change from water to ice, the context of each molecule of water becomes a much stronger positional constraint; ice is crystalline, not fluid. When there is a phase change, the matrix that offered the previous constraint is no longer there. Many of the predictions we can make about living systems come from the constraints associated with water on the gradient from warm planet to cold outer space.

When one makes mayonnaise, fat is held in the context of a water matrix, and it is that matrix that turns a liquid mixture of water and fat into a thick, rubbery emulsion. If one adds the olive oil too fast, then the colloid loses its integrity. The fat escapes the matrix, and the mayonnaise suddenly curdles. The mayonnaise is transformed from a colloid into a simple liquid mixture, with neither the water nor the fat phase in control. Thus disorganized, the curdled mayonnaise loses all capacity to constrain further additions of olive oil into a thicker emulsion. The order of the emulsion is lost. The phase that constrained the system, the water, no longer does so. Phase changes tell the scientist the time and place to look for the critical new constraints. All of life is colloidal, like mayonnaise, and when an organism dies the constraints offered by the water matrix of life are lost. When life is lost, in a sense the organism curdles. The carbon chemistry remains remarkably constant in the corpse, but the water starts leaking out almost immediately (Allen, Havlicek, and Norman 2001).

Living systems are all colloidal, and water is the constraining matrix wherein all life functions. Unfortunately, most biological discussion turns on issues of carbon chemistry, such as photosynthesis, and the water is taken for granted. Consider, then, the amount of water that is in your head as you think about these ideas. Your thoughts are held in a brain that is more than 80 percent water. Might it not be foolish to take that water for granted? Water is the medium in which life is constrained. The controls on this planet that Lovelock (1979) calls Gaia work through water as the medium of operation. There is no life on Mars because there is little or no water to get organized. Living system sustainability might well be characterized as sustaining the pathways of water. When we take water seriously as the matrix of life, living systems are an emergent property not of carbon and its chemistry but of planetary water. Despite the importance of carbon chemistry in life, the dissipation of the gradient is only marginally a matter of carbon energy and depends to a much greater extent on changes of state associated with water. In terrestrial ecosystems the principal engine for doing work is the latent heat of vaporization of water.

Photosynthesis captures about 2 percent of sunlight, but that amount is far too small to service all work that plants need to do. Plants use sugar to fuel animals to move information in terms of pollen and seeds and to build the solid phase of the plant so that its colloidal phase can be supported. This leaves the heavy lifting and movement of materials, the day-to-day dynamics within the plant, to depend on some other source of energy. In aquatic systems, such as streams and oceans, the principal energy used to do work is water movement. The sun lifts the water to the headwaters or creates waves and ocean currents (Mario Giampietro, March 2002, personal communication). That source of energy to do work on land is heat from the sun to evaporate water. The latent heat of evaporation of water is very large, so that when the plant has used the heat of the sun to drive its transpiration stream, there is little heat left to warm the plant. The more the plant uses the heat of the sun to lift water and minerals from the roots to the top of the canopy, the cooler will be the upper surface of the canopy. Most of the energy in a plant is used to feed the latent heat of vaporization of water and has little to do with the energy of carbon chemistry.

Life as an energy-dissipative structure gives patterns that we would not have thought to notice without explicit acknowledgment of the generative function. However, once we do look in the light of the generative function, the patterns are startling. If the gradient is one of temperature from the planet's surface, and emergent structures dissipate that gradient, then the upper surface of biological systems should be unexpectedly cool (they have dissipated the heat) and all the more so when life is more fully functional.

The cooling of vegetation is an indication of ecosystem functionality. In that light, Akbari (1995) showed that the capacity for cooling can be seen in the surface temperature of remotely sensed agronomic vegetation, as a function of the disturbance regime. Lawns generally are very warm, mowed fields are cooler, unmown pasture is cooler again, and finally old fields are the coolest of all the agronomic herbaceous vegetation. Furthermore, the act of mowing the pasture once makes the vegetation temporarily warmer. Work in the laboratory of Steven Murphy, at the University of Guelph, shows that increasing nitrogen fertilization of crops produces cooler vegetation (Akbari 1995). This increase in cooling capacity occurs with diminishing returns on nitrogen investment, until the point acknowledged by agronomists, using entirely different criteria, as the maximum useful nitrogen application. Luvall and Holbo (1991) showed that the degree of forest maturity and its history of natural regeneration also can be seen as a capacity for cooling. Mature, uneven-aged Douglas fir stands emerge as cooler than 15-year-old plantations. At a global scale, images from satellites indicate that tropical rain forest biomes are the coolest of all biomes in terms of low-wave reradiation back into space (Sellers, Mintz, Sud, and Dalcher 1986). Shulka, Nobre, and Sellers (1989) used a calibrated model and substituted grassland and forest for the same location as they discussed Amazonian deforestation and climate. They too found that the model predicts that vegetation is cooler on its upper radiative surface when it is modeled as having tropical forest. In real-time observation of tropical forest, one is looking at the tops of clouds at 35,000 feet, but the vegetation put them there, so the cool temperature reflects the behavior of the full system, not just the clouds. At many scales looking at many different sorts of vegetation, the larger, more elaborate, and more functional is the vegetation, the cooler it is on its upper radiative surface (Schneider and Kay 1994).

Note that in a conception of life that focuses on energy in ecosystems, the carbon lattice configuration, which would be viewed as central in evolution, takes a back seat to the energetics surrounding water. In terrestrial ecosystems, energy dissipation is largely evaporative. Certainly there is carbon energy in life, and indeed most contemporary biology addresses energy in studies of photosynthesis and respiration biochemistry. In ecology this focus on carbon manifests itself in food chains, productivity studies, and the like. Even so, carbon energy is only the small amount of energy that is used to maintain the elaborate structures that dissipate the main energy flux, the evaporation of water in terrestrial systems. The carbon energy embodied in a tree builds over centuries. Over that same period, the total energy involved in a tree transpiring through its life is much greater.

One might expect a certain resistance to a view of life as an emergent energy-dissipative structure, for it relegates whole sectors of biology to just a supporting role. Genetics becomes a set of blueprints for making a wick. Biochemistry becomes making a wick. From the perspective of the generative function, the central function of terrestrial life is to act as a wick that pumps water from the soil into the air. Trees allow water in the soil to interact with that in the air over an interface that is meters deep. All of botany collapses into an elaborate explanation of transpiration. Given this wholesale relegation to subservient roles of many sectors of biology in a thermodynamic view, one might ask what one gets in return. Taking a thermodynamic view of ecosystems, particularly one that emphasizes emergence under the second law, takes ecosystem science into new sectors of ecological prediction that are particularly pertinent for sustainability.

EXPERIMENTS ON THE GENERATIVE FUNCTION

The field data cited in Schneider and Kay (1994) indicate that surface temperature might be a useful measurement as a guide to management to achieve sustainability. If cool upper surfaces of vegetation indicate increased vegetation function, then managing for cool upper surfaces may well be part of achieving sustainable ecosystems. Impressive as the field data may be, experimentation under controlled conditions would give a clearer view of ecosystem and vegetation function. Experiments would give more precision in predictions for planning management toward sustainability. Allen, Havlicek, and Norman (2001) investigated the temperature of soybean vegetation in wind tunnels so that the emergence of vegetational and ecosystem structure could be calibrated properly (fig. 6.3). Their results do indeed indicate that management for the whole vegetation function in terms of vegetation temperature gives more robust cropping systems, engendering at least agricultural sustainability.

Allen, Havlicek, and Norman (2001) raised soybean plants in either understory or canopy wind speeds. They tested the two treatment

ECOSYSTEM IN A CAN

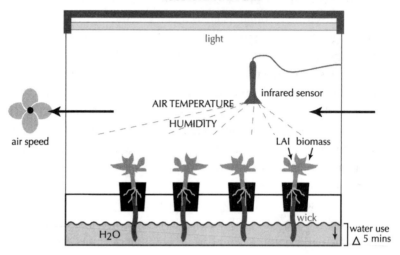

FIGURE 6.3

The wind tunnel used by Allen, Havlicek, and Norman (2001) to investigate the cooling of vegetation under various wind speeds. Water use was measured by the level of water in tanks below the pots. Humidity, air temperature, leaf area index (LAI), and biomass were all measured. Light was held at a constant of about one-third of full sun. The radiative thermal temperature was measured over an area of canopy of the order of 0.25 square meters. Fans could vary air flow between tests to wind speeds of 0.3, 0.5, 1.8, 2.5, and 3.0 meters per second.

vegetations for surface temperature under wind conditions from the understory and under those at the canopy in a strong wind. The infrared thermal sensor showed that the vegetation was coolest in the respective conditions in which the plants were grown (fig. 6.4). This means that when plants are less stressed, they are cooler. The assumption is that plants less stressed are using more heat from radiant input and the air. This does the work of moving water and minerals.

If rain forests are cooler than other biomes, then we might assume that their richness of species form and function plays a role in their particularly high vegetational functionality. Therefore we might expect a mixture of plants from the two experimental treatments to elaborate

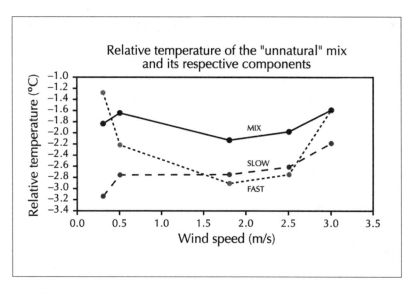

FIGURE 6.4

The temperature of the upper radiative surface of the experimental vegetations expressed relative to air temperature. The abscissa shows the various wind speeds of the test. The ordinate shows vegetation temperature. The three lines are for plants grown in the slow wind treatment, plants grown in fast wind treatment, and a mixture of plants from the two wind treatments. Note that the one-treatment vegetations are coolest at the wind speed in which they were grown.

the vegetation and increase its functionality. On the face of it, this should create a cooler vegetation. However, because the heterogeneity in the mixture created an aerodynamically rougher profile, the air flow was more turbulent, so the warm air could warm the plants. The mixture was warmer than the vegetations of which it was composed (fig. 6.4). This result pressed the experimenters to realize that there must be a distinction between mere complicatedness, which is at best independent of functionality, and the functional complexity in vegetation (as described later in this chapter; see "Evolution, Emergence, and Diminishing Returns"). If plants are mixed together with no function to the juxtaposition, this only complicates the vegetation. Beyond structural elaboration, complexity entails an elaboration of organization. Organization is not just the addition of more parts with symmetric relationships to each other but the addition of upper-level components that bear an asymmetric relationship to lower-level components of the whole (fig. 6.5).

In the attempt to achieve more complexity, the experimenters had only made the mix more complicated. The plants grown in slow wind, characteristically experienced by understory plants, grew faster and so

Complication

solves problems and moves on to the next problem, leaving structure behind. Complication makes structure that is difficult to control, predict, or mend. It causes horizontal differentiation.

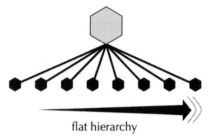

flat hierarchy

Emergence

creates energy dissipative far from equilibrium structures. It causes COMPLEX structure through self-simplification. Behavior becomes simple but energetic cost is high. Emergence causes vertical differentiation.

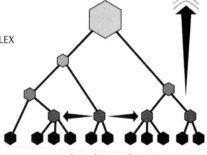

deep hierarchy

FIGURE 6.5

The distinction between systems that are complicated and those that are complex. Complexity invokes asymmetric relationships that reflect organization and constraint. Complicatedness increases the number of parts but does not necessarily increase system organization. If the new parts are at the lowest level under consideration, then there is only an increase in complicatedness for the system so considered. If the new parts are at middle or higher levels, then complication of that level increases constraint and amount of organization.

were taller. Therefore, in the mixed vegetation, plants grown in slow wind were up in the overstory of the mixture, up in the fast wind. The taller plants emerging above the mixed vegetation experienced winds higher than those of their growing conditions. The converse applies to the shorter plants raised in high wind but now down in slow wind. The mixed vegetation was more elaborate, but it had an unnatural sort of elaboration. The elaboration was only of the number of parts, not an elaboration of organization within the whole. Lacking the functional organization of rain forest, the mixed experimental vegetation was warmer, not cooler, than the homogeneous vegetation in which its constituents were grown.

In the mixed vegetation that manifested warming, not cooling, the plants are all of the same age, be they short or tall. To achieve a vegetation in which the components were elaborate as well as organized, plants grown in high wind were planted earlier. At the time of the test of temperature of a new, more complex mixture, the plants grown in high wind were tall, but the younger plants raised in slow wind were short. Now the uneven-aged mixed vegetation had tall and short plants, but this time the tall plants had been grown in high wind. Therefore, the new mixed uneven-aged vegetation had the same degree of complicatedness as the even-aged mixed vegetation from the previous experiments. There were two types of plant, and the difference in height of the components led to a canopy surface that was exactly as rough as the mixed vegetation in the previous even-aged experiments. The more complex uneven-aged vegetation mixture was more natural in that the taller plants had indeed been raised in overstory wind speeds, whereas the understory had been raised in understory wind. With the plants positioned in the vegetation so that they were in a favorable position for each type, the vegetation was more organized. The vegetation was organized into overstory and understory, as opposed to the merely complicated vegetation mixture, in which plants happened to be of different sizes. Compared with even-aged mixed vegetation of equivalent roughness, the uneven-aged mixed vegetation was cooler (fig. 6.6). Artificial vegetation is cooler when plants actually in the overstory are adapted to overstory wind speeds. At this point we have two conclusions. Less stressed vegetation is cooler, and complex vegetation is cooler than its merely complicated

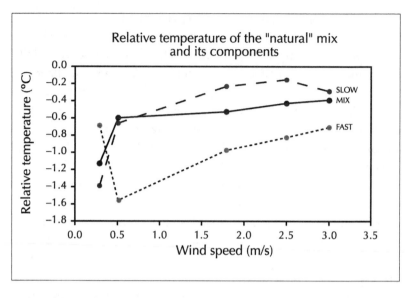

FIGURE 6.6

The labeling here is the same as figure 6.4, but the plants from the two treatments are of different ages. Note a similar result in figure 6.4 for the relative temperatures of the simple vegetations, consisting of plants from only one treatment. Again the plants were cooler at the wind regimes in which they were respectively grown. The difference here is that the mixed vegetation had plants from the faster wind treatment in the overstory. The mixture was cooler relative to its respective fast and slow components than was the mixture of the even-aged plants in figure 6.4 relative to its respective fast and slow components.

counterpart. Our complex vegetation was just as it would be had it grown naturally, with overstory plants being older.

The results were then compared at a deeper level. The temperatures of the two sorts of mixture, the natural mixture and the unnatural mixture, were expressed as deviations from the temperatures of their respective component vegetations. These deviations were then compared with their counterparts in the other experiment, that is, the respective other type of mixture of plants. For the plants raised in fast wind, these comparisons showed values close to zero (fig. 6.7). That means that the plants raised in fast wind respond identically whether in complex or complicated vegetation, whether they are in the overstory or the understory of

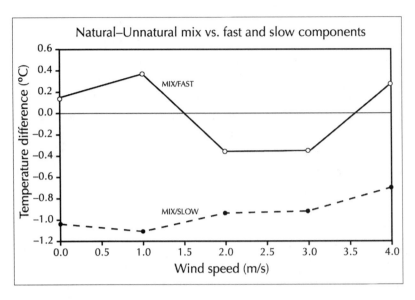

FIGURE 6.7

Subtractions of normalized responses. The data in figures 6.4 and 6.6 were transformed against the respective mixed vegetation data so that the plant temperatures of the plants grown in fast and slow wind were expressed as deviations from the mixed vegetation to which they contributed. When the normalized data from the even-aged and uneven-aged results were subtracted from each other, the only signal left is the differences between mixing plants of even age as opposed to uneven age. The values close to zero degrees difference in the plants raised in fast wind suggest that plants raised in these conditions behave the same in overstory or understory. However, the slow-wind plants were a degree cooler if they were in the slow-wind environment of the understory of the mixed vegetation.

the vegetation. The more significant result was that the plants raised in slow wind were a full degree cooler if they were in the understory of the mixture. That is to say, slow-wind plants were cooler if they were in properly complex vegetation that puts the components of the vegetation in the places for which they have been prepared.

This deeper level of comparison leads to an interpretation of how complexity works. Complexity comes at a price (chapters 2 and 3), and in vegetation that price is the slow growth of the overstory as it pays for its accommodation to the wind. If benefit may be considered as increased functionality, the benefit is derived from the protected under-story manifesting coolness, which itself indicates functionality. The benefit of being a homeotherm comes from keeping the overall metabolism of the animal in a constant temperature regime. The cost is high heat loss. In a moderate environment with lots of food, complexity manifested in homeothermy works well because the high metabolic rate pays for itself in increased resource capture. In a cool environment with little food, it is better to be a snake because the cost of warm-blooded complexity is too high.

If we are to achieve ecological sustainability, it appears from this analysis of the generative function that cost–benefit analysis of complexity is a key. If elaborations of organization generate commensurate benefits and increased functionality, then managing for complex organization is worthwhile. On the other hand, if the cost of organizing and managing for a complex system is great and the benefits appear minimal, then the resources saved by being simple can be deployed elsewhere. Here the analyses of complexity in social and biophysical systems clearly converge (chapters 2 and 3). Moreover, it does appear that we can use temperature of the upper radiative surface of vegetation as an indicator of the sustainability of vegetational systems. The wind tunnel work of Allen, Havlicek, and Norman (2001) suggests that managing for the complexity of whole vegetational structure and function is worthwhile.

In crop science, at present the management is for superior individual varieties of plant and for making simple homogeneous vegetation of the crop. The assumption is that having 100 percent of a superior plant gives superior vegetation. However, a whole new direction of plant breeding is indicated by the experiments in the wind tunnel. By breeding different varieties of the same crop species, intercropping within one species could treat agricultural fields as complex organized vegetation instead of as simple aggregates of a single type of plant. Coming from vague notions of the benefits of diversity as opposed to specific understanding through complexity theory, there have been some remarkable reductions in rice fungus infection of more than 90 percent by mixing varieties of rice. Instead of seeking a single strain of super-resistance, researchers mixed varieties, only one of which is resistant to the fungus. Zhu et al. (2000) achieved huge reductions in infection in this way. Usually the benefits of using a resistant variety are short run because resistance is quickly overcome. However, if a mixture of strains is deployed, the benefits of resistance are used to damp down endemic levels of infestation, but the resistance in the resistant strain is sustainable. This resilience is achieved by lowering selection pressure on the fungus. Complexity theory offers a sophisticated and particularly generalizable interpretation of these results. Diversity is not in fact the basis of this success, for it is organization of the parts of the vegetation, not the number of parts, that offers the advantage. By intercropping different varieties, the rice vegetation is made more complex, not just more diverse and complicated. Traditional rice agriculture did indeed use mixed-strain planting, not that we are recommending slavish return to traditional unscientific methods. Sustainability is achieved very differently in a complex vegetation than in a monoculture with a single genome.

OBSERVATIONS ON ECOSYSTEMS AND SUSTAINABILITY

Given that thermodynamics and complexity are fundamental to understanding and managing ecosystems, we may put in perspective some of our current environmental disputes. These disputes often turn on individual species or use the vulnerability of such species to advance a broader agenda. In a thermodynamic conception, individual species may have less ecological significance than the public commonly supposes. Lose a species and another will use the energy suddenly available, much as coyotes now capture some of the energy formerly used by wolves. Reducing species diversity does not necessarily mean that an ecosystem has lost very much functional complexity. Ecosystems therefore are most intelligently managed not just for their constituent parts but more fundamentally for energy flows, stores, and biogeochemical cycling.

Of course there are often emotional and legal reasons to be concerned about individual species, and sometimes there are ecological reasons as well. It is a serious matter to remove the last large predator species from an ecosystem, for example, or to simplify a system too much. We acknowledge these points but want managers to understand that such concerns increase the cost of management and reduce its effectiveness. Too often, this is management for structure rather than for function. Supply-side, cost-effective management of ecosystems would concern itself more with thermodynamics and as much as possible let the ecosystem take care of individual species.

EVOLUTION, EMERGENCE, AND DIMINISHING RETURNS

In chapters 2 and 3 we discussed how complexity arises from adaptive problem solving and how this process can make societies vulnerable to collapse. New organization (i.e., complexity) can also arise after the appearance of new energy gradients. This route to complexity provides further insight into our possible future, for as with Renaissance and modern Europe, new energy is a primary route to sustainability. In continuing our discussion of complexity and sustainability, we distinguish here between three types of elaboration. The first is complexification internal to a system, arising from adaptive problem solving (as in the social systems we have discussed). The second is complexification external to a bounded system, in which a new energy gradient, amplified by positive feedback, generates emergent structure and organization that constrain the system in new ways. The third is complication, involving only horizontal differentiation with no increase in organization. These distinctions will lead to new understanding of the relationship of energy to complexity and sustainability, and of our potential energy future.

Living systems, including human societies, are energy dissipators. They will elaborate new structure and organization in the face of rich, new sources of energy. In human societies that elaboration ramifies throughout the system and changes it thoroughly, as any fundamental economic transformation must. Within a short time the society has new structures and organization, a new configuration that we capture by such short-hand terms as "Iron Age," "industrial age," "the era of fossil fuels," and the like. The emergence of such "ages" typically involves external complexification based on new resources.

Biological evolution is a problem-solving process much like societal problem solving. Biotic systems find themselves not just in a physical environment, which is predictable and offers the same challenges over again, but must also deal with a biotic environment. The biotic environment of a given species is composed of populations and species that have a capacity for evolution, making them unpredictable. Evolution of the biotic context itself causes it to offer ever greater challenges to the species in question. Much as societies in a changing cultural and climatic context face ever greater challenges, species undergoing evolution also face successive challenges. Of course, there are sentient humans sometimes planning cultural change, and that is different from biological evolution, where there is no planning. However, human planning often is political in nature, with the plan being perhaps only marginal to the material issue at hand (March and Simon 1958; Simon 1997). As a result, much change in human society is survival of action patterns and ideas that turn out to be suitable by happenstance. The selection is by humans in hindsight, much as the environment in natural selection acts to perpetuate what already has survival characteristics coming into the evaluation.

An obvious characteristic of living systems is that they are often elaborate. In biological systems, this elaborate structure comes from a process somewhat similar to the elaboration of social systems. Societies solve problems, often leaving the resulting infrastructure in place, and then move on to new problems. Biotic systems also continuously solve problems, and they too accumulate ever more elaborate infrastructure. The baroque structure in organisms is accumulated infrastructure. Some elaborations still serve the purpose for which they were evolved. Many elaborations no longer serve their original purpose but are the structural context for more recent solutions. Often, apparently adaptive elaborations only look adaptive because they were the structural context in which more recent, still pertinent problems were solved. The old context often appears as if it emerged to deal with a contemporary problem even though it did not.

In both societies and biotic systems some of that accumulation of elaborate infrastructure can be described as a very flat hierarchy. There is a large number of components, all at the same low level, giving a high degree of horizontal differentiation. We call such systems, or subsystems, complicated. Complicated systems have a large number of different but generally equivalent parts, such as plants of the same age and of the same species. A monoculture, or an even-aged stand, is a complicated system. Similarly, a long computer program with insufficient hierarchical structure of subroutines is complicated. Any programmer knows that happenstance generates unrecognized, unplanned relationships between parts of a program. In a complicated program with little grouping of parts of the code by design, the unforeseen connections get out of hand and come to dominate behavior. Change any part and there can be a large number of apparently unrelated effects. Complicated, flat hierarchies are hard to manage because they lack organization.

Flat hierarchies may come to differentiate vertically. Faced with the difficulties of managing the flat hierarchy of a complicated system, systems with enough free energy available often segment into subsystems. These might be subroutines in a computer program, organs in an organism, plant species in vegetation, or administrative units in a society. Segmentation amounts to a process of vertical differentiation, whereby the flat hierarchy differentiates new layers that constrain and coordinate the subsystems. Pattee called this a process of self-simplification, where complex hierarchical structure, namely organization, is achieved to facilitate more coherent and simple behavior of the whole. Flat complication arises out of accumulating clutter, whereas vertical differentiation arises from creating organization.

Viewed from the context, an aspect of a society may appear as a local complication amounting to a horizontal differentiation, but there is a twist. If the societal system is highly bounded locally, what was local becomes global when viewed from within. That local complication is now everything and as such may qualify as a vertical differentiation that amounts to an increase in organization. So as we proceed in the present discussion, we do not suggest that horizontally differentiated complications as distinct from vertical complexifications of organization are in any way absolute. As with so much, it all depends on level of analysis. Some of the muddle in the literature as to the metrics for complexity and its nature comes from particular definitions of complexity giving precedence to one mode of hierarchical description over another. A definition of complexity that emphasizes only horizontal or vertical differentiation is incomplete. Definitions that use them both together sometimes are muddled. The solution is to use both horizontal and vertical differentiation, but separately. Accounts of complexity in social change should not give precedence to either function or type of differentiation. Rather, they should require switching between the effects of horizontal differentiation, internally driven vertical differentiation, and externally driven vertical differentiation of the system with each new question.

Systems become successively less manageable as they differentiate horizontally. A fix in some part of the system becomes a problem somewhere else. In modern society it might manifest itself in transfer between different types of pollution. Keeping a pollutant out of the air means that it accumulates in solid or liquid form. Avoid a spill from an oil tanker, and the oil ends up in the air, as we intend. The fiscal choices for politicians generally become harder as the downside of spending in one sector emerges as a problem in some other sector of a society. For instance, the state of Wisconsin now spends more on prisons than on higher education, although a more educated society might need less incarceration. Politicians cannot afford to appear to be "soft on crime" and so knowingly spend money on the penal system ineffectively because they must. The system is complicated, not organizationally coherent, and so becomes unmanageable.

Despite solid research data and sometimes grim prospects, contemporary decision makers act rationally only in the short term. In historical examples (chapters 2 and 3), the process of stressing a society toward its collapse appears to be guided by rational behavior at all stages. The decisions of the historical elite to complicate infrastructure and complexify organization, the acquiescence of the rank and file to shoulder the increased burden, and sometimes the abandonment of the enterprise were all reasonable at their respective times (Tainter 1988). In contemporary society, there is almost always a rationale for behavior that appears in hindsight to have been unfortunate. For instance, jobs are created as mines are built, and associated regulations appear to protect the environment. But then mines shut down intermittently as globally determined prices make them uneconomical. Therefore mining counties organize around potential rather than actual jobs and so have lower incomes and less developed economies, despite political rhetoric to the contrary. Environmental regulations are relaxed as closure is threatened (Freudenberg 1992; Freudenberg, Gramling, and Schurman 1998). The situation in contemporary society is different from historical societies only in our unique capacity to anticipate the whole process of complication, complexification, and collapse. As no people in any other age, we can foresee unsustainability and act to preempt unmanageable burdens and potential collapse (Tainter 1995). With our computer-modeled projections and the examples of past civilizations better known to us than ever before, the ultimate rational decision is for our society to confront the larger processes that make short-term rational behavior nonsustainable.

The cost of maintaining an ever more elaborate infrastructure or organization continuously increases as successively harder problems are solved (chapter 2). Elaboration of structure and organization that arises in response to local problems may be independent of the size of the resource base to support that internal complexification. Thus decoupled, problem solving keeps on going, even when it exceeds the resource base. When the cost of structure and organization overshoots the resource base, the society strains for a while in deficit until costs associated with further elaboration alter the society's defining organization, often resulting in collapse.

The benefits of internal complexification of this sort are ephemeral. Internal complexification is a process wherein the system never gets ahead for long, for it is a continuing struggle to deal with the issues arising in business as usual. Furthermore, the benefits of internal complexification in solving problems are at first great, but they diminish from there. The diminishing returns are those to which we made reference earlier (chapters 2 and 3), where society is losing on both ends of the cost–benefit ratio (fig. 3.19). Societies become vulnerable to collapse when, eventually, the benefits of further elaborating structure and internal organization become small. New benefits become not worth achieving, and eventually even the structure and organization achieved are not worth their maintenance.

Self-simplification through vertical elaboration (that is, external complexification) is one way to rescue an overly complicated system from an inability to adapt in the face of new challenges. Self-simplification invokes emergence and makes the system more complexly organized as new levels appear. With emergence, the change in system structure is so great as to warrant calling the aftermath of emergence a new structure. As changes occur to supersede or replace the highest levels in society, it can be argued that a new system has emerged, despite clear continuity in some respects. Examples here could be the new society that emerged in the Industrial Revolution in Britain or in the change from the Roman Republic into an imperial system. Vertical differentiation is the best way to deal with excesses of horizontal differentiation. As Rome, for example, acquired province after province, and new wealth poured into Italy, the republican system of governance (senatorial oligarchy, elected officials, and plebeian right to vote) was strained beyond its organizational capacity. After a period of civil wars a new hierarchical level was established, that of a permanent, full-time emperor and his administration, who constrained all others: senate, officials, populace, and provinces.

The cost of emergence of vertical differentiation is great. There is an energy or resource cost in creating and maintaining a new far-from-equilibrium dissipative system. Where there is a gradient to dissipate (that is, a new energy source, such as Rome's new provinces), that is what pays the cost. Some societies do not capture the energy to meet that cost, and so they collapse under the burden of internal elaboration. The collapse may occur because human effort needed to run the system is seen as not worth the benefit. If there is successful emergence, after what is often a painful transition, the new energetic costs of vertical differentiation are not noticed. This is because, in external complexification through emergence, the energy to meet the cost of the new configuration is a prerequisite for the elaboration of organization in the first place.

The new levels emerge generally for reasons other than sentient human planning. Often a new level develops in response to the unexpected and unplanned availability of new energy inputs. For instance, the Industrial Revolution in Britain (chapter 2) was an elaboration that occurred in the aftermath of the development of steam engine technology and pumps to mine below the water table. New energy inputs created a steeper gradient, which spontaneously generated vertical differentiation in the form of new, upper-level organizations. The populace was not eager to use coal because one had to pay for it instead of cutting trees. The first phase of the change to coal was viewed as a distinct and unwelcome cost. But as the new energy dissipation came on line, resistance was swept aside by positive feedbacks in coal consumption. The fact that the changes in society were costing supplies of coal was not considered because there was so much of it, and air pollution from it was at first local. However, those same energy demands later became a real cost in twentieth century industrial disputes and pollution, but that was long after the emergence of the new industrial level, when it had greatly elaborated internally. Following the emergence, there is a period of internal differentiation at the new level of complexity. As the benefits derived from new energy inputs begin to decline or are taken for granted, the increased energy that supported the original emergence becomes burdensome.

The difficulty in interpreting the tension associated with horizontal and vertical differentiation comes from the fact that one is usually looking at a particular end result that invites a narrative of rational or adaptive behavior. We are therefore dealing with the tension between prospective and retrospective views of the system. The new energy source releases constraints, generating elaboration, but we see the effect only when the consequences have been played out. Although imperialism looks to be planned when seen after the fact, it is as often something that the victors find themselves facing rather than the outcome of a plan. There is little thought of administrative costs and vertical differentiation as a war is waged, either for defense or for plunder. The energy captured in agriculture, imperialism, irrigation, or industrialization causes vertical differentiation and energy dissipation. A new energy source inevitably causes vertical differentiation. Without planning or forethought, Native Americans in California, for example, found themselves in such a productive but unpredictable system that they developed social hierarchies, something hunters and gatherers normally do not do (chapter 4). Note that their burning habits caused something akin to agricultural field mosaics to emerge, but the similarity cannot have been by design, for they used the mosaics largely for hunting. By making the environment more productive, Native Californians produced a steeper energy gradient than would have existed otherwise, and higher population densities and greater social complexity followed inevitably.

The emergence of the dissipative structure of a given society occurs without noticing the full cost of energy dissipation. After the initial costly start-up phase, the ride through the positive feedbacks in emergence is essentially free. Positive feedback amplifies the utility of the gradient, as steam engines, canals, and railways did in Britain. But after a long period of internal elaboration, the cost of the original emergence makes itself felt. Emergence is unpredictable and even unrecognizable as it is happening, and so sentient calculation and planning in those early phases are neither urgent nor really possible. At the other end of the phase, where positive feedback has been replaced by diminishing returns, planning and active management are indeed urgent, but the planning is not any easier to achieve.

From the example of the rejuvenation of the Byzantine Empire, it appears that clever planning in downsizing is a possible scenario, although there was much less of a plan in this case than one might imagine. Deliberate fiscal simplification ramified through the society unpredictably, until even literacy and numeracy were nearly gone. In the process of emergence the actors are swept along unknowing, but at the end it is possible, albeit very difficult, for the actors to see the situation as a new very large problem that needs solving in a conscious process of invention. The modern era involving information, including historical information, gives its actors a better chance than most people had in the past to plan and execute a solution to diminishing returns. Inventing complexity is easy, and humanity does it all the time. Inventing simplicity is hard.

Implications for the Contemporary Period

The problems we face turn significantly on issues surrounding energy. Often the gradient responsible for emergence in social systems is directly an energy gradient, as in physical systems. Sometimes in social systems, the gradient is more easily understood as some energy surrogate, such as wealth in a social system. As noted in chapter 3, the riches of the Abbasid Caliphate in Mesopotamia were such that an elaborate set of structures emerged as their silver flowed across Eurasia. At one point the King of Mercia minted a gold coin to mimic the respected design of Abbasid coins. It is the only coin ever struck in Britain that had inscribed on it, "There is no God but Allah and

Mohammed is his prophet," if only the Saxons could have read it. When the Abbasid silver was gone, the gradient disappeared, and towns as far away as Southampton went into decay (Hodges and Whitehouse 1983). The Roman gradient of riches and artifacts is also understandable in thermodynamic terms. It caused many barbarians to converge on the Danube and the Rhine, despite the larger hinterland available for them to use. In time they dissipated the gradient, blurring the distinction between the empire and themselves.

When societies experience very large changes, such as the development of imperialism or industrialism, the change may be an external complexification brought about by a set of positive feedbacks. Such change can be beneficial. The benefit of complexification through positive feedback is escape from the high costs and low benefits at the end of the previous system of problem solving. For instance, the move to dependency on agriculture obviated the problems of moving large distances to the next hunted or gathered resource. The marginal cost of search and pursuit, which is very important to foragers (Winterhalder 1981), becomes moot in sedentary agriculture.

In the case of internal complexification, the cost is met by burdening the existing system's resource base. For the Romans, this permanent burden proved fatal. In the concept of external complexification that we elaborate here, the burden on extant infrastructure may be only temporary. The development of the coal-based economy in Britain is again a case in point. There was clearly a significant burden on the populace when Britain switched from wood, which could be cut in the environs, to coal, which had to be bought. But as coal came on line, technological developments enabled such benefits as industrial production and transportation. This case illustrates the point that in acquiring a new, abundant, concentrated resource, positive feedbacks facilitate paying for the change. In the case of coal and industrialism, there were positive feedbacks wherein coal itself facilitated the adoption of coal. The need for coal necessitated pumping of mines with steam engines. Pumping capacity made more coal available in a positive feedback. With increased availability of coal, iron smelting necessitated moving either the coal or the iron ore into the proximity of the other. The pump engines were modified and put on wheels to make steam railways that ran on coal and moved on iron rails, thereby placing a demand on fuel and metal. The availability of iron led to better machines for extracting coal, thus closing a second positive feedback.

The cost of externally initiated complexification that pertains to our particular issue has two phases. First there is the cost of setting up the positive feedbacks for increased resource exploitation. This can be politically forced, much as a problem-solving response, and the cost may be borne by straining the extant system. In industrialization, this was the early phase, when people had to shift from cutting wood gratis to buying coal. Once the feedbacks were in place, however, the system paid for itself by sequestering ever more of the new resource. The cost of further movement toward the new way of doing things is then paid by dissipation of the new energy or resource gradient. In complexification associated with positive feedback, lowering the gradient of the new resource leads to further exploitation of that new resource. By tapping into the new resource, the complexification pays for itself. We suspect that there is typically positive feedback underpinning this sort of external complexification.

The costs and benefits of internal and external complexification are largely but not entirely separate. A new resource gradient, such as coal, will complexify a society from without, or the society may elaborate internally under duress, based on existing resources, to solve a problem. If in the case of duress there is positive feedback that elaborates society and exploitation of resources, that may produce positive feedback and increasing returns. If there is no positive feedback, however, and little extra resource is extracted, the situation produces diminishing returns. The long-term benefit of external complexification is entrapment in the cost–benefit relationship of a new resource base. Beyond the disruption of old, comfortable ways of doing things, one cost of external complexification is that a gradient becomes dissipated. Early on, dissipation of the gradient is of no importance because resources are abundant. The cost of internal complexity produces the diminishing returns that attend the problem solving in the new system. As the gradient is dissipated, that cost becomes harder to bear.

A biological example of resource use is most illuminating for the present and future use of energy resources in the industrial nations. There is a cluster of related genera of ants called attoid ants, including

361

the genus *Atta*, that farm fungi on various organic resources. One species uses caterpillar droppings to grow their food. Droppings are a high-quality resource, rich in nutrients and energy. The problem with droppings is that they are a highly processed resource, several steps away from the source of energy: the sun, leaves, fruit, caterpillar, caterpillar droppings. High quality as droppings may be and useful as they are for growing fungi, they are in short supply. This limits the size and degree of organization of the colonies of that ant species. Other fungus-farming ants in the genus *Atta* use leaves instead of caterpillar droppings. Leaf-cutting ant societies are far more organized than their excrement-collecting counterparts. Transporting leaves to where they will be used involves long trails with underpasses and overpasses through the forest. The limits on those systems appear to involve efficient organization of waiting time in lines, as described by queuing theory (Burd 1996). The enormous quantity of leaves needed to produce the same energy as a smaller amount of caterpillar droppings imposes a need for higher organization.

It is no accident that the first entry into fungus farming by attoid ants was through the high-quality resource embodied in droppings. It is far too unlikely that any sequence of events would lead to using leaves de novo. Caterpillar droppings, on the other hand, are high quality. In a reasonable but not essential scenario, droppings could have been brought into the nest attached to collections of "wild" fungi. Caterpillar droppings accumulated by happenstance on the floor of the nest could then have become the basis of "domesticating" fungi inside the colony. It is clearly easiest to start a process of exploitation with high-quality material, such as droppings. Such material gives the highest return on effort.

Among the fungus-farming ants, there appear to have been at least two complexifications: first, the emergence of fungus farming itself, and second, the shift from droppings to leaves. The first complexification depended on the exceptionally high quality of the resource. That quality imposed order on the excrement-harvesting ants. The second complexification depended on the leaves as a resource but was less driven by the quality of the resource. Leaves are a low-quality resource, not far removed from the sun. Accordingly, it is not the quality of the resource that imposes organization. Rather, leaf-cutter society is driven by the ants becoming more organized to deal with the massive amount of resource that must be transported and processed.

Human history displays a pattern similar to that of the fungus-farming ants. It took a high-quality resource, caterpillar droppings, to let the ants into a fundamentally new mode of operation, farming. Subsequently, a second ant species was driven to develop larger and more elaborate systems to harvest a low-quality resource, leaves, of which the ants must obtain much more. There is a similar pattern in human use of resources, exemplified any time people shift from resources that give higher returns on effort to those that give less. The problem of first plucking the lowest fruit is that only higher fruit then remains. Hunter–gatherers, as discussed in chapters 2 and 4, often concentrate on abundant, high-quality foods, moving to where they may be found. Hunters and gatherers set the scene for agriculture, much as fungus-farming ants using caterpillar droppings set the scene for leaf-cutter ants. Agriculturalists move closer to the sun by raising tracts of homogeneous food, which they are able to do because their hunter–gatherer forebears found and manipulated plants that were potentially productive and amenable to cultivation. Similarly, leaf-harvesting ants operate closer to solar energy because the apparently earlier species set the stage by initiating cultivation on droppings, which are high in energy but dispersed, rare, and hard to find.

Agricultural societies are limited ultimately by solar energy, and when they encounter problems with local resources one solution is to capture more territory where solar energy falls. Conquest speeds up energy capture temporarily by appropriating other peoples' accumulated surpluses. Yet once these surpluses have been looted, a conqueror (such as the Romans) must thereafter garrison, administer, and defend a province. These responsibilities are paid from yearly agricultural production (chapter 3). As with caterpillar droppings, looted surpluses are valuable but in short supply. Surpluses are a transitory resource. There is always a need to shift to a new phase of low-quality resources, taxes. Taxes are hard to collect and demand a large infrastructure, but in an agrarian landscape there are a lot of people to tax. Taxes are a low-quality resource because one can extract only small quantities from each person. As with highly organized leaf-cutters, peasant taxation systems tend to be elaborate and costly. In the later Roman Empire, every person was considered a taxable entity, and each agricultural field across the vast empire was evaluated and taxed. The preceding looting takes a high-quality resource, namely accumulated riches. Whereas the end of a period of high-quality resources comes when the scarce resource becomes insufficient, low-quality resource cycles end when there is too much demand put on the large resource base. Imperial taxation produces a large, highly organized system, but the cost of increased taxation becomes unbearable. Long frontiers and an increasingly disgruntled populace erode political stability. Local people reassert autonomy. Imperial systems seem typically to end with too much burden placed on a taxation system.

Analogously to the Romans, the state of Alaska funds itself today largely by taxing a high-quality resource, oil. Oil is an accumulation of past solar energy. Not only does oil provide sufficient revenue, but there is a surplus that is distributed each year to Alaska's citizens. Eventually the high-quality oil will decline. When that time comes, if Alaska is to continue the same suite of government services, it will have to increase taxation of a lower-quality resource, its citizens. Citizens are a lower-quality resource than oil because they are dispersed, and their per capita yearly production is low. The taxable yearly production of citizens is a shallower gradient than the accumulated solar energy captured in oil, just as the taxable production of Roman peasants was of lower quality than the accumulated surpluses of defeated peoples.

When Renaissance Britain found itself without sufficient wood for everyday needs, living standards declined, and the next best alternative was sought. In retrospect, the move to coal was a very good idea, for coal is an energy source of high quality, but at the time it was resisted. The move to coal meant in time that increased productivity did not depend on human muscle, powered by food. Fewer people could achieve more. There have been shifts to other fossil fuels and carbon fuel technologies, and these are reflected in the Schumpeter (1950) curves, well understood in economics. The questions we face today are, "When will the industrial world come to the end of the high-quality period that depends on carbon, and what will we then do?"

One scenario for our future is change to an economy centered more around information than around matter, powered to a large degree by hydrogen. This scenario is worth discussing, for it illustrates some consequences of different energy paths. Pure hydrogen itself is a high-quality energy source akin to fossil fuels, but it must be extracted rather than merely mined. Hydrogen for use in fuel cells can be generated by consuming fossil fuels, which works well as long as fossil fuels remain affordable. In a future less dependent on carbon energy, however, the resources that will be used to extract hydrogen may be renewable, low-quality sources such a wind, wave, and solar. Hydrogen would entail a shift to low-quality resources.

Unlike the industrial era, which solved problems with larger tools using more energy, the information era depends on miniaturization. When industrialism began people still needed to eat, and so the previous agrarian system remained necessary although it became a lesser route to prosperity. Industrialism rode atop agriculture. So also in an information economy, there will be an industrial base that is taken for granted. In industry material things are moved through a distribution system, and people move to and from places of work. An information economy, in principle, minimizes moving material things and increases efficiency by moving information instead of people. The flux of information allows the human system to be distributed, a characteristic of low-quality resource capture. It is no accident that, while industry with its high-quality energy is concentrated, agriculture is distributed. The energy captured in agriculture is low quality, and must be distributed to supply 6 billion people. The critical features of renewable resources are that they are low quality and dispersed but abundant in quantity.

Renewable energy gains political support from the environmentally sensitive, but there is great irony in this support. Although environmentalists criticize industry and fossil fuels, the environmental damage done to the world is minimally from industrial sources. That is because the energy used by modern industry is principally of high quality and so works in a focused fashion with small side effects. There is measurable environmental degradation coming from industry, but it is not the main villain. The principal cause of loss of species and environmental degradation is agriculture. The distributed nature of agriculture means that habitat is removed and landscapes are grossly altered. Increased flooding, soil loss, and nonpoint sources of pollution are principally caused by agriculture. Furthermore, although environmental advocates pillory agribusiness with its agro-chemicals, Third World peasants are equally destructive. Environmental degradation is greater when the resource is of low quality and distributed. The environmental impact of ants that use droppings is minimal compared with the impact of those that strip trees of leaves. Such is their consumption of leaves that some species of *Atta* are agricultural pests. Thus a switch to renewable energy sources, ironically, would be much worse environmentally than anything done by industry using fossil fuel. In a further irony, although industry has not been anxious to embrace renewable energy sources, enormous profits would be made from building the infrastructure to capture and concentrate renewable resources. Politicians would be less in the pockets of road builders and more influenced by businesses that recreate whole coastlines for wave capture and cover huge tracts of land with solar collectors.

Based on the differences between high- and low-quality resources, we discuss two potential paths to a low-quality energy economy. Much as leaf-cutter ants could not have evolved without the preliminary step of the species that uses high-quality droppings, a hydrogen economy powered by renewables could never emerge without a high-quality carbon energy source as a precursor. In the first scenario, we make the transition to a renewable future by first burning our way down the mountains of sulfurous coal that dwarf the coal consumed heretofore. Coal would provide the financial capital needed to build vast systems to capture renewable resources. The environmental pollution coming from dirty coal over the coming centuries would be much worse than that coming from the cleaner-burning oil and gas we use at the moment. Under this scenario, the present energy distribution system would remain intact, while the new, huge, decentralized system built to capture renewable energy (transformed coasts and deserts) would dwarf the infrastructure seen heretofore. In time this infrastructure, powered by renewables, would enable continuation of systems of production, distribution, consumption, and settlement recognizably like those of today. Although there would be much political conflict over the environmental damage we can foresee in this scenario, people would find the transition gradual, without wrenching change, and thus probably favor it.

In the second scenario, industrial societies decentralize and much existing infrastructure becomes unnecessary. The diffuse information system would keep human activity integrated while enabling decentralization. This is possible because information technology allows, for example, international businesses to be run from remote locations. Urban decay would accompany rural gentrification. Energy would be captured directly by small individual units scattered across the landscape. For many the transformation would be catastrophic as the system decentralizes and infrastructure workers become redundant. At the same time, new opportunities would emerge in producing and repairing small, dispersed sources of energy production. The infrastructure to capture renewable resources would develop slowly because it would have to be built in a gradual process of accumulation by individual households or small communities. Useful parallels might be the millennia of capital accumulation following the first agricultural systems, and the early English textile industry consisting of small, dispersed entrepreneurs commissioning piecework. Hydrogen might be generated as part of local, distributed energy capture systems, so that at least some high-quality energy is available for special situations that need it. Once enough capital has been accumulated and expanding local energy production coalesces, the highly organized, massive infrastructure needed to concentrate low-quality energy might emerge. Hydrogen could be very important in this fully developed system. The end point would be similar to the scenario of burning sulfurous coal, but the cost would be disruption of social systems and change in living standards more than environmental pollution. Some sort of intermediate scenario that uses coal and involves decentralization is also a possibility. Many people would prefer such a decentralized existence, but others might find it wrenching. It would require capital investment by each family or community. These investments would be largely redundant and would not initially enjoy economies of scale. Living standards, as currently defined, would thus decline. Presented with the option of such a future, many people would demur.

Engineers are impressive in their ingenuity. They are able to invent many alternatives within fossil-fuel technology that can stretch the supply of high-quality energy to give a surprisingly long glide-down of the carbon-based energy system. The *immediate* problems with fossil fuels may be atmospheric pollution, international balances of payments, and political conflict, not running out of fossil energy sources. In the short term that may be true, but it would be imprudent to ignore the inevitable decline of carbon-based energy. Recall that the information economy is built on the industrial, as the industrial was built on the agricultural. It is noteworthy that whereas the price of modern information technology goes down, the price of a car or rotor tiller moves up with inflation. We should not be lulled into overconfidence by successes in information systems because the costs of technology involving moving parts appear real and pressing. Although the energy for electronics is large at the societal level, the greater problem is energy for electric motors. The industrial underpinning of an information society is an essential weakness of the modern human situation.

It will not be possible to finesse our way past the coming obstacles, whenever they come, engineering brilliance notwithstanding. Surviving the inevitable energy crisis will of necessity entail radical social reorganization. We will pass through a complexification if and when renewable resources become the primary source of energy. Such transition points are critical, and societies are not guaranteed to make it across them. The practice of less than full-cost accounting can last only so long before proving disastrous, as the Romans found to their dismay. Carbon-based energy can continue to support society only at higher cost and diminishing returns. At the end of a phase of diminishing returns, the whole system is supported by a sort of shell game, again exemplified by the late Roman Empire. The carbon-based energy system is already generating higher and higher costs to capture energy. Extraction costs will increase as we deploy more complicated industrial machinery to drill for oil under deep water and pollute more waters and the air. By moving to hydrogen and low-quality, renewable energy, we could escape from under the pressing constraint of diminishing supplies of high-quality carbon resources. Moving to hydrogen, extracted by renewables, would get us away from the complications of drilling in the Antarctic and over the edges of the continental shelves. The change to a complexification on hydrogen would have its immediate downside, but hydrogen technology would be simpler than the difficulties of a continuing carbon energy crisis. The renewable nature of low-quality energy would generate positive feedback, generating expanded sequestration of those resources, just as coal stimulated technical development that enabled the production of yet more coal.

We have already used the fossil carbon energy of which nonindustrial people could avail themselves. The Industrial Revolution, as we experienced it, cannot be repeated, for we have used the high-quality resources that made it possible. Thus it appears that we only have one shot at an economy based on renewables, and this is it. Our version of caterpillar droppings inevitably will become too costly, and we will need to transition to something equivalent to leaf cutting. Much as coal and industrialism initially were not inviting, neither is the initial prospect of a hydrogen economy based on low-quality renewables. It will take great political skill to ease the transition, for it must happen through the support of the mass of people on whose backs the costs will fall. Many people are going to have to change their ways of life as fundamentally as did the displaced yeoman of Britain when coal mining became the best work available.

It is encouraging that once external complexifications get under way, they simplify systems and may pay for themselves. We need only to begin to tap the sources of renewable energy at a sufficient scale for the system to move by itself. Renewable energy at a workable cost will encourage further movement in that direction. Engineers will be essential, but a fundamental problem is one of bringing the mass of stakeholders along with us. Great transitions do not turn only on bright ideas, as many entrepreneurs have learned. The origins of agriculture cannot be explained by genius theories. The important transition was accepting agriculture as worthwhile or inevitable, just as the shift from wood to coal was inevitable if initially undesirable. The hard part of the transition to renewable resources turns on willingness, values, and faith in the system. The two scenarios we offered are interesting as intellectual abstractions, but people will choose one path or another based not on abstractions but on the pressures and prospects of their daily lives. Engineers and ecologists will work in this transition alongside social scientists, politicians, business leaders, educators, pollsters, and social and environmental advocates.

369

SUPPLY-SIDE SUSTAINABILITY AND
RESOURCE MANAGEMENT SCALE

The message from the past for modern management of ecological systems is to avoid using a complicated plan that shifts its focus between the separate renewable resources. When a system becomes complicated, the parts have gained too much influence and autonomy. Management for renewable resources directly gives too much influence to the separate resource plans. In the Roman Empire, problems were exacerbated by the way in which the central administration worked not on a contextual level (the support system of land and peasants) but on the particular problems of specific revolts or invasions and the economic crises they produced. The vision of the future used by Roman administrators was from a retrospective posture, using forecast, not prediction (fig. 6.8). They cast the future as an extension of the details that pressed on the system in their present. In the end, the Romans had so much payload from the accumulation of local responses that adaptive courses of action were prohibited. Much as the Romans would have been well advised to manage their whole system, in ecosystem management today it is important to manage the whole system and avoid focusing exclusively on the local issues of resource management one resource at a time. All resources, actual and potential, are linked together by the material ecosystem.

Managing for the function of the entire ecosystem is the only way to manage for multiple resources simultaneously. The future must be cast through prediction coming from the general constraints that apply to the whole ecosystem. It is a mistake to manage for the resources directly. The future seen in terms of individual resources uses a forecast that depends on the detailed narratives that explain the present state of each resource. Managing for each resource separately assumes a complicated flat hierarchy with a multiplicity of separate parts, the resources. Attempts to manage for disparate resources separately can be expected to fail in a gridlock of conflicting management actions, as occurs in all complicated flat hierarchies.

There are clear cases of failure in the exploitation of renewable resources. We see the central problem as being the management of the

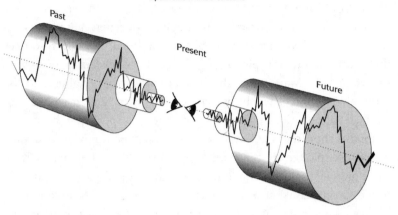

Prediction
Explain Future in Terms of Constraints

Past

Present

Future

FIGURE 6.8

Constraints forwards and backwards. Although a commonsensical approach would indicate that constraints determine what happens, constraints do not determine the details of outcomes. Rather, they appear to indicate what will happen by drawing a contrast to states outside the limits of the constraint. It is from that contrast against a much larger set of possibilities, namely the behaviors that are excluded, that constraints give an impression of determining what happens. This clarifies the distinction between the often questionable notion of environmental determinism and the more often valid notion of environmental "constraintism." At the level of analysis that determines how constraints work, constraints do not determine what happens; they just indicate what will not happen. As time passes, the particular behavior of the system is likely to move beyond the range of immediately past variation. Much as recent behavior is more limited than more temporarily distant behavior, predictions are forced to become more general as they are cast further into the future. Think of past and future behavior as a series limited by tubes of constraint that are narrower closer to the present in both the past and the future.

lower-level parts, the renewable resources, to the neglect of the entire ecosystem. Underlying the availability of renewable resources is a material production system that is the source of renewal. It is that contextual renewal system that should be managed, not the resources themselves and their exploitation. The entire ecosystem is the engine of production. Manage for the functioning of that engine, and the renewable resources will be available. True, management will be constrained by the limits imposed by that context, so perhaps jobs may be lost in the short term, and resources that might be sorely needed must be left unexploited for the time being. However, in the final analysis, there always was a material limit imposed by the capacity of the system to produce, like it or not. Therefore managing for jobs and temporary resource demand, to the detriment of the functioning of the production engine, is self-defeating. Managing for the functionality of the entire contextual engine gives steady employment at sustainable levels and resources that will be available to meet the pinch in supply that occurs next time.

Poor management uses a middle number system specification and so cannot predict the long-term outcome of management action. Managing for the resources directly casts the system in middle number terms. The renewable resource as a resource is subject to pressures of local market and political considerations, the long-term significance of which can be seen only in hindsight. Effective management translates the inviting middle number version of the system into some large or small number system specification that acknowledges ultimate limitations of renewal. The translation away from a middle number specification lets the planner see what prospective management action will do, so good choices can be made. Managing for the functionality of the entire ecosystem invokes a small number system specification because of the small number of actions that make any sense. Small number system management is achieved by playing the role of the constraining context of the ecosystem through measured and long-term action. By contrast, a middle number specification leads quickly to an unworkable number of actions and side effects, some of which have unexpected, undesirable effects. A large or small number system specification will lead to a small number of choices about how to keep the ecosystem that underlies production in a functional condition. True, there will be local loss of optimality, but pursuing local optimality leads to successive bouts of overexploitation that eventually destroy the capacity of the resource to be renewed.

Modern use of renewable resources clearly has some problems, with stocks of fish being driven below harvestable levels and forest capital being removed faster than the regeneration rate. The Gulf of Maine fishery, as we discussed in chapter 4, is an illustrative example of failure of resource exploitation. The model of exploitation heretofore has been one of managing for the resource. The logic of it has been to take fish at the size immediately above that of the maximum growth period. Thus the fish catch favors survival of population members up to the size that can capture resources from the ecosystem with greatest efficiency. Explicitly the old large fish classes are fished out. The record of the successive overfishing to near extinction of the most desirable species shows that the present model of resource exploitation is inappropriate. If the resource is managed at maximum efficiency, the resource is pressed to the limit. The resource has patterns of aggregation, so it is possible to continue to catch fish in those aggregations beyond long-term sustainable levels until it is too late.

The alternative is to manage the entire ecosystem for long-term sustainability. This is the reason for the title of this book; *supply-side* refers to management for the components of the systems that generate natural resources. In managing for the function of the whole ecological system that generates the supply, immediately the heterogeneity of the resource ceases to be a problem because there is no more overfishing in the aggregations. In managing the entire ecosystem, we take into account the places between the aggregations that have fallen below adequate production. To implement management for the entire system, we must work at the level of the constraints that operate on the resource, not at the level of the behavior of the resource itself. Prediction, as opposed to forecasting, comes from a prospective view of the system that emphasizes system constraints. We have recommended that the constraint be sought in the oldest fish, the ones that are explicitly targeted for extinction in the present management regime. At any given instant, older fish, having shown their durability, have a longer life expectancy than younger fish. They therefore integrate the changes in the physical environment more completely. Also, there are indications that big fish constrain the process of regeneration of the young. Their eggs are of higher quality, and their hatchlings disproportionately survive the holocaust of destruction in the early years when the negative exponential of loss between age classes is steepest.

The manner in which older fish represent the entire ecosystem function is less than transparent. But this is to be expected, for entire ecosystem function is necessarily large, complex, and intangible. By contrast, the manner in which the middle age classes capture the most resources from the ecosystem is fairly clear; the process is easily measurable, and the model is entirely intuitive. However, that does not make fish in the middle age classes the best indicators of how to manage the system. Retreat to a depauperate population model, just because it captures an identifiable detail of the system, is, as we have noted, akin to the drunk looking for his cufflinks under the lamp post because he can see there, even though he lost them down the street. Just because the facts of the population model that favors middle age classes appear valid in detail, it does not mean that those details are an appropriate criterion for management. Managing with a narrowly focused population model is managing by forecast, not prediction. In forecasting, the details of the present that are used to see the future are true, but despite that verity they do not remain relevant for long. This is because the outcome is directed by the larger context and other unseen details that emerge as significant after the model is cast. In the case of the failure of present fishery management, the unseen but crucial detail is the heterogeneity of the fishing effort in response to a heterogeneous positioning of the resource. This causes a lag in the manifestation of disastrous overfishing. The trick is to find effective management based on prediction not forecast. One finds the constraints of the system, or surrogates for the constraints, and then uses them to identify the pertinent limits of the outcome.

Identifying constraints is by no means easy, but there may be some guidelines that can help. System theory indicates that behaviors are likely to be unreliable, whereas persistent structures should be better indicators of the future. Resource capture is a behavior and therefore is a less reliable indicator of the functionality of the ecosystem engine than is presence of persistent structures such as the individuals of the oldest class in the system. The effects of structures are likely to be still present after the passage of time. If one is to use behaviors, then behaviors that are apparently slower are likely to be contextual and so offer reliable constraints. For example, changes in oxygen in the atmosphere are sufficiently slow that an oxygen-rich environment is extremely likely for the foreseeable future. Planting trees to replenish oxygen is an irrelevant ecofad, although there may be other reasons for reforestation. Carbon dioxide levels are changing much faster, but even so the biologically pertinent effects of carbon dioxide usually are slow enough for the level at any time to be taken as a constant under which our calculations for the future can be made with some reliability. On the other hand, levels of ozone near the ground vary wildly, and we are reduced to forecasting the damaging conditions. We deal with ozone excesses near the ground by inefficient emergency measures such as keeping susceptible people indoors and planting species that can take the abuse.

The capacity of ecosystems to perform human services and supply their own needs is remarkable. The material ecosystem performs all the integration that is needed, and so our attempts to manage from the level of individual functions closely related to short-term resource needs is a waste of effort. Continuing to manage piecemeal from local market-based decisions about a resource, not the engine that produces the resource, is likely to bloat infrastructure that societies use to exploit renewable resources. Furthermore, energetic fine-scale activity has a history, as in the Gulf of Maine, of being at odds with the functioning of the whole ecosystem. The whole ecosystem is the unit of genuinely renewable resources. Because of the integrated nature of ecosystems, management at the level of the whole ecosystem should remain lean and yet deliver abundantly for a society that plans on what is sustainable in the long term. We give examples in chapter 7.

If management of whole ecosystems can be achieved, it will change the relationship of modern society to its energy source. Byzantium cut back on the complexity of its exploitation of the photosynthetic energy shining on the empire. In doing so, Byzantium also increased net energy that was captured in a useful form by the agricultural system; no longer were lands abandoned as frequently as before because taxes on them could not be paid. While Byzantium prospered, overexploitation of agriculture caused the population of neighboring

Mesopotamia to fall to its lowest level in five millennia. As it was for Byzantium, so can it be for modern civilization if we can align our decisions with the long-term health of our resource base at a global scale.

Given the emergence of global international corporations, a system that relies on government control of policy at national levels is in effect a laissez-faire system at the global scale. In modern democracies there is a direct vote for those who manage societal affairs at a national level; we hope and expect populism to continue and expand to fill the mid-level in the present international trend toward democratization. However, government control of policy at national levels is unable to generate positive feedback systems at the global scale. At a global level, people exercise constraint by "voting" as customers of corporations. Whereas governments can be expected to regulate at a national level, global-level agreements have always been difficult for governments. Already commerce is reaching upscale in the arena of quality resource management to fill the regulatory void left by weak international government institutions. Some important regulatory functions are being taken over by standards generated by international business. It is telling that international standards of quality management (ISO 9000) and environmental management (ISO 14000) emerged independently of government regulation and appear to be a sign of spontaneous emergence of a new level of complexity in society.

The renewable resource crisis resides largely at the global level, and so it appears to be crucial to achieve a context for management that sets long-term goals. We can expect management in such a setting to moderate short-term resource extraction at a certain cost in immediate productivity. However, this cost would be minimal compared with the accumulation of costs deferred by short-term maximization. The adage "Think globally but act locally," to which we referred in chapter 5, is embodied in our proposal. The difference is that this time *globally* can be used literally, not figuratively. Only in a postmodern world of cyberspace is it possible to communicate globally fast enough to preempt, for instance, shady deals by logging operations with government officials in Third World nations. Both the logging operations and the corrupt parts of government can be constrained to act sustainably. This is not a call for national autonomy to be overridden but rather a suggestion that empowering local peoples will force better policy for the long term. Local peoples informed by global information systems as to their best interest will act appropriately locally, defying the corrupt part of their government. The same information systems that inform locals can also inform the global customers of the logging companies, imposing a constraint above even international corporations. The global level of organization that stands the best chance of constraining international commerce probably is a global customer base.

Failure to achieve a global management context invites an alternative system in which one local resource is depleted after another, until all is lost. Under local, nepotistic regimes, eventually we should expect that the national level would come to be viewed by the populace at large as not worth the effort, as is now sometimes happening in the Third World. The speed with which the political institutions of Indonesia have changed is a case in point, even to the bright side of an opportunity of self-determination for East Timor. The general populace appears to have developed apathy toward major civilizations before our own (Tainter 1988). We cannot anticipate the final stress that will be viewed as not worth overcoming: Global epidemic is one candidate. Certainly a general populace with no ownership of the problem in a world of gross inequity would walk away faster than one in a more equitable system. A version of an ownership crisis was identified by Gates (1998), who wants to rework capitalism so that it is better at generating capitalists. He pointed out that the present system concentrates ownership in the hands of a few. Gates (1998) suggested some very specific ways of investing a large number of the regular citizenry, such as workers in the firm. With regard to sustainability, local workers invested in the firm live in the environment of the factory and have incentives to demand viable local action while supporting their company in its global plans.

We recommend adopting decision-making strategies that use long-term goals and planning. Such a change in human capture of renewable resources would be a human system that represents a new level of organization, in the technical sense of an emerging complex system. Humans have never managed anything at a global scale, or for the long term, and doing so will demand a change in level of conscious planning. Having never attempted a leap so far upscale, we need to employ the best new theory of emergence and resource management available. The crisis is at a global scale. It can be met only by renewable resource management at a level commensurate with the problem. We see constrained, responsible commerce as a major part of that resource management, in part because it is the only institution that operates effectively at the global level.

CONCLUSION

The process orientation of the ecosystem criterion is unique. Of course, all the other criteria do have processes associated with them, but none has processes so central to the conception of the criterion as do ecosystems. This focus on process has led us to consider the side of biology that starts with process and eventually generates structure. The generative function often is the missing half of biological discussions.

The generative function draws attention to a central issue in management: how to achieve prediction as to the outcome of management action. The difficulty of prediction in biology and ecology is one of changing constraints. Change in constraints often can be seen as a new higher-level structure that emerges with its new constraints coming apparently out of nowhere. The new constraints come from the emergent structure that appears far from equilibrium in response to the steepness of some gradient. There is not likely to be much equivalence between the old and new higher-level constraints, except in the unusual condition in which the system is fractal. We used the example of a fractal system in turbulence, but for the most part, upper-level structures in biology bring with them constraints that are so different as to demand ascribing a new type of structure for observation. It is essentially impossible to predict particulars across radical changes in constraints, and one must settle for general classes of organizing principles to get any sort of reliability in telling the future.

The process orientation of the ecosystem leads to an aspect of the ecosystem criterion beyond the generative function. Processes are responsible for linking the parts of an ecosystem. Accordingly, ecosystems are viewed as being particularly strongly integrated. In management issues this leads us to management of the integrated whole. Management actions often involve working on a mix of different materials that are not easily unified, except in an ecosystem conception. One might be harvesting an organism, a tree, but the soil and air flow, both ecosystem considerations, are also influenced in the act of taking the harvest. The renewal of renewable resources is very often an ecosystem process such as primary production. Even if a population is generating the resource, the material flows are the feature of interest, and that is an ecosystem consideration. Thus ecosystem management, in the vernacular sense and in the policy arena, is very close to management of ecosystems in the technical sense of the term *ecosystem*. Ecological sustainability is inescapably an ecosystem issue.

7

Retrospect and Prospects

We have ranged widely in this book, from societies to ecosystems, both sustainable and not, and from the past to the future. In chapter 1 we introduced five lessons to emerge from the discussion, which are the principles of supply-side sustainability:

* Manage for productive systems rather than for their outputs.
* Manage systems by managing their contexts.
* Identify what dysfunctional systems lack and supply only that.
* Deploy ecological processes to subsidize management efforts, rather than conversely.
* Understand diminishing returns to problem solving.

In this chapter we offer concluding observations on these principles, including where and how to apply them.

In synthesizing these points, we find ourselves ending much as we began: with some principles of supply-side sustainability that are best illustrated by social examples and others that are more clearly seen through ecology. This division arises primarily from our respective backgrounds and the subject matter with which we are each familiar, rather than from how we perceive our topic. We actually see the topic as indivisible: Sustainability is a matter of socio-ecosystems, and although we can divide that topic analytically, we should not do so in practice.

Dividing analytically, we begin this chapter by examining the implications of our discussion of problem solving. The discussion is primarily social but with essential ecological elements. We progress then to topics of ecological management, with important social examples.

SUSTAINABILITY AND PROBLEM SOLVING

Commonly we think of sustainability as maintaining an accustomed way of life. Sustainability in this conception comes from ensuring the continuity of the resources on which a way of life depends. It is short-sighted, though, to think of sustainability as the achievement of stasis. The human environment is of course dynamic; new challenges continuously emerge. Moreover, as we have discussed throughout, deciding what to sustain is a matter of the values of the moment. Human values are variable, mutable, and transient, and our conception of what we want to sustain ranges accordingly. A few years ago a Russian colleague related to Tainter a joke then making its way around Moscow. The scene is ancient Rome, and slaves are demonstrating in the forum. They march back and forth, carrying placards that read "Feudalism: the progressive future of humanity." The point for sustainability efforts is that our expectations for the future can never be anything but a function of today's perspective. Yet our sustainability goals, and the environments in which they are possible, change endlessly. Sustainability arises not from societies or environments but in the context of their dynamic coupling.

Human sustainability does not refer to an absolute condition of the biophysical world. Challenges to sustainability emerge from changing human values, dynamic environments, and the interaction between them. Sustainability is, and will always be, a moving target. Sustainability is in part a matter of continually adjusting to changing circumstances, which we do through our problem-solving institutions. Problem solving, as discussed in chapters 2 and 3, is an economic process in which costs are assumed for benefits to be gained. Yet problem-solving systems, over time, tend to grow in complexity and costliness and to suffer from diminishing returns. As the net benefits to complexity in problem solving decline, complexity itself becomes less suitable for solving problems. Problem-solving institutions that develop in this way eventually will be terminated, collapse, or need large subsidies. Thus, *a primary characteristic of a sustainable society is that it will have sustainable institutions of problem solving.* These will be institutions in which problem-solving efforts produce constant or increasing returns to expenditures through positive feedbacks, or diminishing returns that can be financed through subsidies (such as energy) of assured cost, quality, and supply. We suggest that this condition of positive feedback and sustainable problem solving can be achieved in part by applying the first four of the principles of supply-side sustainability: Manage for systems, not outputs; manage systems by their contexts; supply dysfunctional systems only what they lack; and harness ecological processes to subsidize management.

Among our problem-solving efforts today, not all meet this sustainability criterion. We illustrate these points in application through three examples, of which the first exemplifies unsustainable management. Yet like the Byzantines (chapter 3), we also engage in problem-solving efforts that do appear sustainable, and these suggest approaches to a sustainable future. We present these positive examples shortly and then turn to a matter raised in chapter 2: whether we should hope for a technological *deus ex machina* to descend from the clouds and establish sustainability.

Unsustainable Problem Solving in Natural Resource Management

In southeast Alaska today, the Tlingit Indians try to sustain their cultural identity by continuing to hunt, fish, and gather the traditional foods used by their ancestors. These foods are essential to ceremonial life, and the hunting or gathering of them is used to socialize younger Tlingits to many facets of their culture, which is fundamental to sustaining it.

Sitka black-tailed deer (*Odocoileus hemionus sitkensis*) is a primary game animal for Tlingit hunters. In addition to cultural value, deer meat is economically important in an area with few opportunities for employment. A prime determinant of deer population levels is the amount of forage and shelter in severe winters. During periods of snow deer rely on shelter and forage in high-density, old-growth forests, usually at lower elevations. Deer health and population levels depend on this old-growth forest. Unfortunately, the best and most easily harvested timber is also found in the same old-growth stands. Thus the lands most highly prized for timber production are also those most valuable for deer populations and subsistence hunting. In recent years Tlingit hunters have not been able to obtain the deer they need.

Southeast Alaska offers few ways to earn a living. Because timber workers have few employment alternatives, subsistence conflict tends to be a zero-sum game. Neither side can win without the other losing. The forests seemingly can produce economically viable levels of timber, or they can produce satisfactory numbers of deer by Native hunting techniques, but they cannot produce both. Favoring timber production reduces old-growth forests, local deer population levels, and opportunities for subsistence harvesting. Conversely, managing for subsistence would cause timber workers to lose their jobs. Political leaders are always concerned with maintaining high levels of employment and personal income, which jobs in the timber industry provide. Contented workers vote for incumbents. Thus forest managers are under intense political pressure to produce timber.

Subsistence conflict is a classic example of the problems that arise from single-resource management. It is also a textbook example of unsustainable management. There are conflicting state and federal jurisdictions and much litigation. For Native Alaskans to hunt, each community must demonstrate that it has traditionally harvested a species. Approval of a hunt takes nearly a year. Hunters must determine whether the place they intend to hunt is federal, state, or private land and whether federal or state law applies (which changes with court decisions). It is a byzantine system of many obstacles and seems certain to elicit high levels of noncompliance.

The conflict between timber harvesting and subsistence hunting is producing increasing complexity and costliness in land use and management. In the past, a Tlingit wanting to hunt or fish could do so at will or with the consent of the elders in another clan's territory. By

1996, 6 federal agencies had 74 staff members working on subsistence, with an annual budget of $11 million. In the same year, 10 Federal Regional Advisory Councils had 84 members and 5 coordinators. Subsistence regulators hold 50 to 100 public meetings each year and annually produce 50 evaluations and findings. This new bureaucracy constrains subsistence hunters, sport hunters, and indeed much of Alaska's citizenry and political system. As a result of the decision in a recent lawsuit, the U.S. federal government has a new responsibility for subsistence fishing. The initial estimate of the cost to implement this mandate was an additional 250 federal employees and $52 million per year. In 1994 the median economic value of the subsistence fishing harvest was about $103 million. If the initial estimate of the cost of implementing the legal decision, $52 million per year, should prove correct, this one judicial decision would reduce the net economic value of subsistence fishing by 50 percent.

As the complexity and costliness of subsistence management grow, the net value of subsistence harvesting declines. Even for the Tlingit, who count the value of traditional foods in cultural rather than economic terms, the personal cost of hunting and gathering has grown. Subsistence production and management are encumbered in a problem-solving system that is experiencing diminishing returns (Tainter 1997).

Alaskan subsistence exemplifies the course taken generally by natural resource management, in which there is much conflict. In industrialized nations, where public opinion is diverse, there are many opportunities to appeal and litigate land use decisions. Therefore a land management agency achieves much less actual accomplishment per unit expenditure today than was the case a generation ago. As noted in chapter 3, public participation in land management generates a competitive spiral. The more the public disputes decisions, the more land management agencies must invest in justifying decisions, in training their employees in public relations, and in litigation. As agencies improve their public performance, those intent on challenging decisions must invest more resources in their cause and devote more effort to securing those resources. As this competition ratchets upward it becomes more complex and costly both to promulgate decisions and to contest them. Conflict in public policy has the same complexity-amplifying effect as an arms race (Tainter 1992).

It may be that the decisions made by this tortuous process are better than they would be otherwise. Yet in a case such as Alaskan subsistence we may wonder whether today's decisions are *better enough* to warrant the extra complexity and cost. Regardless of the answer one matter is clear: Even in wealthy nations, land management systems that are forced down a path of rapidly diminishing returns to problem solving cannot be sustained indefinitely. The Alaskan stalemate continues only because it is subsidized by the wealth of Alaskan oil and the federal government, but that subsidy depends on shifting political whims.

Sustainable Problem Solving: Managing Systems

We have noted throughout that environmental management becomes ineffective when we manage for the outputs of productive systems rather than for the systems themselves. We have traditionally managed resources to achieve steady and predictable, or increasing, outputs—of maize, rice, soybeans, meat, timber, and the like. Such production systems experience endless variations in output, so that managing for their outputs entails attending to myriad details. This approach was described as like sticking one's finger in the dike. Leaks spring here and there, and the problem causing the leaks is never addressed. In such a management system there is little opportunity to make the effort more cost-effective. Note that Alaskan conflict is over outputs: deer versus trees.

A primary function of natural resource management is to replace a missing ecological context. As we suggest, management should be focused not on the system of interest, the outputs, but on the context that regulates the outputs. Management identifies what dysfunctional systems lack and supplies it (Allen and Hoekstra 1992:275–276). Efficient management uses ecological processes to subsidize management rather than conversely. In the case of the Rio Grande gallery forest discussed in chapter 1, regenerating the forest by hand planting would be highly inefficient. In Tainter's home community of Corrales, New Mexico, a village official once suggested that the forest detritus, which now accumulates but does not decay (thus posing a threat of fire), could be cleared by hand. This task would be impossibly laborious. Yet both decomposition and reforestation can be accomplished inexpensively if we allow spring flooding (now controlled by upstream dams)

385

to resume (Crawford et al. 1999). In this strategy the management objective is subsidized by processes that are free and available whether we use them or not: winter precipitation and gravity.

Two examples illustrate the supply-side approach. The first, Rishi Valley, shows how this approach can be implemented with little more than commitment and a small amount of scientific knowledge. The second, fishery management, exemplifies a fusion of social and ecological science.

Rishi Valley of India was once a deforested landscape converted to secondary growth, in which farmers managed barely a marginal existence. In the 1930s, philosopher Jiddu Krishnamurti and his followers established a school in this barren place. From this small beginning a program of environmental restoration emerged that is locally based and self-supporting. Missing contexts have been identified and are initially provided. As these contexts assume their functions, the system restores itself.

To restore the forest it is necessary to restore its context, the landscape. Drought-resistant trees and shrubs anchor soil and provide mulch; 20,000 are planted each year, year after year. Check dams slow drainage and catch nutrient-rich silt, the silt in turn being transported manually to areas where it is needed most. Stone terraces channel runoff toward large percolation tanks, from which the water table is replenished. Cash crops support self-sufficiency, and villagers are educated in the principles of sustainable land use. Neither irrigation nor artificial fertilizers are used. The results are that the water table has risen from 12 to as high as 3 meters, biomass has increased by 300 percent, and 150 migratory bird species have returned (Kaplan 1996:354–368).

The key to restoration of Rishi Valley has been to use not external funds (which are used so heavily in most restoration efforts) but the subsidies of solar energy (producing rainfall), gravity, and dedicated people. These subsidies are available whether we use them or not. The ecosystem both subsidizes the restoration effort and is enhanced by it. The subsidies minimize restoration costs so that the effort and its benefits are both localized and of appropriate scale. It appears to be a sustainable exercise in restoration.

We recommended in chapter 6 that on the ecosystem criterion, managers should concentrate on energy flows rather than structures. In their management of water and silt, this is what the people of Rishi Valley are practicing. The structure, in the form of biomass and returned birds, emerges on its own without costly human management.

In addition to general principles, Rishi Valley may offer specific lessons for urban residents in industrialized nations. Albuquerque, New Mexico, is a desert city, rapidly depleting its water table in an area that receives about 20 centimeters (8 inches) of precipitation annually. Yet the problem may not be an imbalance between people and water but a failure of soil management. As in other urban areas, much land is either paved or stripped of vegetation so that precipitation runs off and is lost. The ground surface is the context of the aquifer, and part of the solution to Albuquerque's sustainability could be to capture precipitation now lost to runoff or evaporation. Little precipitation now reaches the aquifer, but greater amounts could be captured by such practices as clustered rather than dispersed housing, surfaces that are covered with porous materials, and open spaces that are biologically diverse, not artificial monocultures (Horst 2001). Failure to manage the landscape in this way will in time produce a city that exceeds its water supply and then stagnates economically.

Acheson, Wilson, and Steneck (1998) described a scientific approach in what they called parametric management of fisheries. Parametric management seems to have much in common with supply-side sustainability. These scientists advocated managing the ecological variables that limit population levels. Noting the failure of common stock and recruitment models to predict population levels, they advocate instead using rules governing fishing locations, fishing area, and techniques. In this approach, one manages fish populations by managing the human influence on critical population control factors such as breeding grounds, migration routes, nursery areas, and spawning. In short, one manages not for the outputs, as in stock and recruitment models, but for the context that affects critical processes. This context is human use. Acheson et al. (1998:396–399) observed that this is the approach taken in many tribal and peasant societies. In these societies, religious sanctions and community pressure regulate such factors as time and location of fishing, technology that may be used, and the stage of a fish's life cycle during which it may be kept. None of the soci-

A HIERARCHICAL APPROACH TO ECOLOGICAL SUSTAINABILITY

eties Acheson et al. (1998) surveyed attempts to manage for outputs by establishing quotas. All manage the human context that affects life processes, allowing population, community, and ecosystem to work largely on their own. Such social controls of the human influence on reproduction can work even in industrial societies. In the Maine lobster industry, for example, a conservation ethic emerging in recent decades has prevented overharvesting so that today the lobster catch is high but apparently sustainable. The conservation measures include catching only midsized lobsters and notching the tails of egg-bearing females (which notifies others that the animal should be released if caught again). The practices are largely voluntary, but even so, today there are estimated to be millions of notched lobsters in the Gulf of Maine (Acheson et al. 1998:399–405).

Supply-side sustainability was in effect the approach used by the Byzantines when they shifted the bulk of the state's employees down the political food chain, from being consumers to producers of the empire's wealth. When the government could no longer squeeze enough outputs (i.e., taxes) from peasants to support the army, it transformed the army to being a part of the productive system that was largely self-supporting. Soldiers provided most of their own salaries, plus surplus food and sons who in turn would become soldiers. Combined with overall simplification of the society and economy, this transformation not only sustained the Byzantine state but also nourished one of history's most remarkable recoveries.

Technological Optimism and Sustainability

As noted in chapter 2, conventional economists and technological optimists believe that in free-market societies, resources do not matter. The financial rewards to ingenuity, in this view, guarantee that new resources will always be developed. This is known in economics as the principle of infinite substitutability, and it is exemplified in the conversion from wood to coal in post-Medieval England (Barnett and Morse 1963:139; Gordon 1981:109; Sato and Suzawa 1983:81). Technological optimists tend not to favor conservation. Their message is, "Don't worry; technical innovation will always descend like a *deus ex machina* to sort all problems and establish a sustainable future."

One difficulty with this view is that it ignores a dilemma we have raised: diminishing returns to problem solving. Technological optimists assume that creativity is the only human endeavor not subject to diminishing returns. When we consider the factors of increasing complexity and declining benefit–cost ratios in knowledge production, we realize that technological optimism must be tempered by knowledge of history. Human creativity in problem solving often is constrained by the factors of complexity and costliness. The late Roman Empire did not lack creativity or flexibility; it could not deploy them given its circumstances. Technological optimism seems to be most warranted where knowledge allows us to develop new energy subsidies of great potential, as with Watt and the steam engine, or to do existing work more efficiently, as with today's developments in microprocessor technology. In such a case the cost of problem solving becomes less significant, for it can be subsidized by new, abundant energy or by more efficient ways of working.

A limitation to the view of conventional economists is that many problems do not lend themselves to technological solutions. The conflict between subsistence hunting and timber production in southeast Alaska, for example, has economic, social, cultural, and political dimensions but no technological solution. This will be the case in much of environmental management, where goals reflect intangible values. In such a sphere problem solving is truly at risk of diminishing returns and ineffectiveness.

In short, although we would be pleased to see a technical miracle that guarantees abundant resources, solves problems, and adjudicates contending values, a more prudent approach to the future is to *prepare* for it. We offer the principles of supply-side sustainability as a route to the future that is based on rational planning rather than on faith.

MODELS OF SUSTAINABLE AND UNSUSTAINABLE FUTURES

The major historical case studies of chapters 2 and 3 describe potential outcomes to long-term trends in problem solving and provide bases from which to foresee options for our future.

The Roman model. Continued attempts to resolve challenges drive increasing complexity and costliness in problem solving that cannot be subsidized by external resources. Problem solving reaches the realm of diminishing returns. New challenges must be addressed, so problem solving continues by extracting more and more resources from the productive system. In time this trajectory leads to fiscal weakness and disaffection. If the problem-solving institution stands in isolation it will in time collapse, and if the region is environmentally fragile, as in Mesopotamia, collapse may have ecological and demographic consequences that last centuries. If the institution is part of a larger system, such as an unproductive government agency or branch of a firm, it is likely to be terminated. Natural resource management seems in danger of such an unproductive dilemma.

The Byzantine model. An institution, faced with challenges and not having the resources to complexify further, voluntarily simplifies. Overt simplification of the Byzantine type is not easily accomplished, for every problem-solving system has people in powerful positions who benefit from complexity and high intake of resources. If the transition can be accomplished, costs are greatly reduced and the productive system may rebound. This is like the strategy used by many American businesses over the past 20 years, where simplification of management and elimination of costs contributed to improved competitiveness and recovery. It is a strategy that is costly in the short term to individuals but may enhance both institutional and individual prospects in the long term.

The European model. Uncontrolled competition or other problems lead to ever-increasing complexity. Problem solving drives consumption of resources regardless of cost, for the alternative may be failure or even extinction. Competition-induced ingenuity leads to continuing technological development, but luck plays an important role in sustainability: New resources are found, and the technology is developed to use them productively. The European success in this strategy established the empirical basis for the technological optimists' view of our future. They predict that the future can continue always as an extrapolation of what has passed before—that nothing changes. It is important to remember the component of luck in the European experience, which the technological optimists overlook. Over the horizon they found new worlds to exploit, and then new and abundant sources of energy. Had European luck proved otherwise, the world today would be a very different place.

These models, then, suggest three potential futures: complexification, diminishing returns, and unsustainability; resilience based on simplification; or growing complexity based on further subsidies. There may be a fourth scenario, which emerges not only from comparison to the past but also from emerging conditions that we create. This is an approach to sustainability that relies not on the brute force of the industrial era but on the knowledge-producing abilities of the information age. Knowledge is the basis of supply-side sustainability. By this we mean not the mechanical knowledge on which technological optimism relies but knowledge of how to use ecosystem processes so that they support us, rather than conversely. This approach is meant to minimize the costliness of problem solving. Within this approach, we use higher levels of information and less energy in management. We use ecological processes whenever possible. We let the ecosystem subsidize management (as in the Rio Grande gallery forest and Rishi Valley) rather than conversely. The alternative is to continue traditional management, in which we manage for the outputs of productive systems, endlessly plugging leaks and trying to keep outputs at an even level and never understanding why the task grows more challenging. The remainder of the chapter explores this lesson.

Central to supply-side sustainability is to understand problem solving itself, its processes, costs, benefits, consequences, and historical development. Problem-solving systems follow a trajectory that may take decades, generations, or centuries to complete. Decision makers need to know where any problem-solving institution is in this historical process. Is problem solving producing stable, increasing, or diminishing returns? If problem solving seems ineffectual in a specific sphere, is this an old or a recent development? Are current problem-solving efforts sustainable, or should new approaches be tried? If sustainability is a function of successful problem solving, then we need to develop a science of problem solving that will address such questions. It will be substantially a historical science (Tainter 1995, 2000) in which the development of problem solving comes for the first time to be understood. The first step toward countering diminishing returns to problem solving is to understand the process by which it happens. The great advantage we have today is that we are the first era that has sufficient historical knowledge to begin the task. The only thing that is certain about the future is that it will present challenges. We can gamble that our problem-solving institutions will suffice to meet those challenges and accept the consequences if they don't. Or we can increase our chances of being sustainable by understanding problem solving itself, the trends by which it develops, and the factors that make it successful or not.

MANAGEMENT AND BASIC RESEARCH

Much as Oscar Wilde characterized the British and Americans as divided by a common language, managers and applied ecological scientists are divided from each other by a set of common concerns. The British and the Americans are astonished at each other and unforgiving because, speaking English as they do, respectively, "They should know better." At the same time, both Anglophone groups might forgive almost any other foreigners because they do not speak English as a first language. Similarly managers and ecological scientists might forgive economists their misguided ecological pronouncements because economists are unfamiliar with material natural resources in the field. Nevertheless, managers and scientists expect altogether far too much of each other because each group thinks the other should know better. Ecological scientists complain that managers slavishly follow scientific models in situations where they do not apply, and managers complain that scientists appear unable to tell them what will happen; both accusations are somewhat unfair.

Managers have different concerns from those of ecological scientists, even from scientists studying in areas that are applied. Whereas the scientist might be prepared to wait for the next insight, managers have to act in the present, such that even no action is an action in itself. It is not surprising that scientists and managers have separate cultures and have views of each other that do not correspond to the respective self-perceptions. A more profitable pair of reciprocal views of each other might be along the lines of demarcation between physics and engineering. Even within ecology as a science, theorists and experimentalists exist without the healthy exchanges that occur in physics between theorists and experimentalists. Ecological theory is arcane, and ecological empiricism often is less than fully systematic. Resource managers can validate ecological ideas derived from both theory and basic experimentation (chapter 1). In the presence of insufficient systematic testing of theory by experimentalists in ecology, managers can perform many real-time tests of ecological ideas and should be viewed as sometimes testing basic theory. Of course, at other times they can be seen as ecological engineers, working strictly as consumers of ecological science for application. Managers will be able to play this second role better in the presence of more effective ecological theory that has been tested by managers in their theory-testing mode.

The unhappy relationship between managers and ecological scientists comes from a no-man's-land that lies between management and science. An attempt to bridge that gap occurred at the meeting, under the auspices of the U.S. Forest Service, of a group of managers and scientists that gave rise to the special edition of *Ecological Applications* (Parsons, Swetnam, and Christensen 1999). That meeting offers an encouraging counterexample, in which managers, applied scientists, and basic scientists engaged in spirited exchange. Nevertheless, for the most part there is an area of discourse that lies between the scientist and the manager in which neither is well equipped to engage. Although *technology transfer* is not the fashionable word it once was, something like technology transfer is the specialist field between science and management. Just translating the findings of science into practical terms is insufficient because there must be an active accommodation on both ends of the conduit. In the middle ground, the scientist needs to address scientifically what is special, as a class of scientific activity, about the science that faces management. On the other side of the middle ground, managers would do well to understand how their demands must be tailored to what the scientists can do, as opposed to what they would like them to be able to do. Give the scientists a better target, and applied ecology will aim differently. Offer managers insights with more general utility, and they will have a model for what to request of ecological science.

Part of management must operate in a world between theory and practice, for there must always be some predictive model that underpins action. However, given the material constraints and the concrete outcomes, the manager cannot afford to be driven most of the time by esoteric, theoretical issues. Effecting action in a tangible place is the goal of the manager. Management action ideally is informed by ecological science, but often it cannot wait for the scientist to deliver a measured assessment of the facts of a given situation. However, if they can learn to be guided by the tests of managers, ecological scientists can develop general principles, which managers can apply, if they in turn would develop the skills to do so.

Scientists who do occupy that middle ground are restoration ecologists. The danger in restoration ecology is that it becomes gardening in the wild. There are too many places for gardening to be the solution to a world in a configuration that is undesirable. In a small garden, the enthusiast can weed, mulch, and transplant to achieve the desired configuration. On large estates, small armies of gardeners and undergardeners, at huge expense, can groom an area of an agricultural field such that every plant is where it should be in the design (fig. 7.1). Even so, the areas involved are so small as to be mere details at a scale that makes any ecological sense. Putting each part where one wants it is not an option in an ecological situation. Restoration of any import must use unseen ecological processes to reconfigure the system. Adjusting the system to be restored by physically putting the physical entities in the desired place is futile. Fortunately, the best of restoration ecology is indeed ecologically scientific.

Many definitions of restoration have been proposed. Some of them do translate to gardening in the wild, but the best are not only more profound but also more practical. Restoration sometimes is a matter of meeting some goal of mitigation. Mitigation often is proposed in a legal setting. Perhaps a highway department is required to mitigate impacts to a stream over which a bridge has been constructed. Unfor-

FIGURE 7.1

(A) The garden of novelist Norman Lewis, who is an enthusiastic collector and gardener. (B–D) Bodnant Gardens in North Wales takes a small army of gardeners to maintain the grounds in a contrived wild condition.

Photographs T. Allen.

tunately, the law is a blunt instrument in such matters because it is facile only at prescribing the state of system, preferably using the positions in space where certain tangible things should be found. As regulation and mitigation are written, the weaker definitions of restoration apply. There is therefore a distinction to be made between meeting a mitigation criterion and performing a proper restoration. A mitigation refers to achieving a state. A restoration restores the underlying processes so that the state of the system can be maintained with minimal inputs.

In her restoration of a small, besieged wetland in San Diego, Joy Zedler (2001) used a definition of restoration and made it distinct from meeting a mitigation requirement (fig. 7.2). In her definition of restoration, Zedler requires that there be a capacity for self-sustainability in the restored system. True, some management action may be needed to maintain the system within a desired range, but it should not be the work of moving components of the system in large numbers to make up for unremitting loss of material. That is what gardeners do with their compulsive tending of flower beds as they hold them in a bizarre configuration that could never occur without heavy human intrusion. Acidified patches of soil for rhododendrons next to more basic soils for other flowers is a monstrous condition that is of no interest beyond a good show through the seasons. Joy Zedler's preferred definition of a restoration requires self-sustaining processes in place, to which the occasional overt management action might be applied. It matches our conception of a supply-side approach to restoration.

Without fire management, which could involve actively torching vegetation, the Curtis Prairie in the University of Wisconsin Arboretum would become overgrown. However, those fires would occur naturally if surrounding vegetation were allowed to burn. The arboretum in which the Curtis Prairie exists is surrounded by a golf course, a six-lane highway, and suburbs. None of these lands is allowed to burn, so

FIGURE 7.2 (*opposite page*)

The small, besieged wetland in which Joy Zedler made the distinction between restoration and mitigation. The freeway was delayed because of the mitigation requirements.

Photographs by Joy Zedler.

the Curtis Prairie needs to have fires set in it. In the absence of an adequate context, even restorations of Joy Zedler's preferred type need a nudge in the right direction. However, the Curtis Prairie, and any other adequate restoration, functions internally in a self-sustaining fashion. The only thing missing in a proper restoration are the services of the context over the midterm.

Under the principles of supply-side sustainability, one approach to management is to compensate for a removed context. The difference between trivially meeting mitigation requirements and performing a proper restoration is a matter of scale and therefore a matter of degree. When Joy Zedler managed to meet the mitigation requirements imposed on the San Diego highway project, the mitigation lacked sustainability and so was not a proper restoration. The essential reason it was inadequate was that the patch of wetland was so small that it would need constant attention like a flower garden if it were to persist in its mitigated configuration. The context of even small patches of the wetland had been so completely removed that it would need attention to the same degree as would a flower bed. One might then respond that the Curtis Prairie, for the same reasons, needs the attention of a regular burning, and that is a fair comment. However, the issue is one of scale. Zedler's wetland mitigation is so small that only a very few of the processes of a wild system have room to exist in it. The Curtis Prairie, by contrast, is sufficiently large that most of the processes that sustain local populations are in place, as are most of the processes for mineral recycling. Missing from the prairie are bison and an exogenous source of fire. But apart from that, most of the pathways and processes one expects of a prairie are in place. By contrast, the population of mitigated fish in Zedler's wetland is so small that random fluctuations in the short term would cause local extinction. The narrow genetic base not only can cause inbreeding depression but it also narrows the response potential to environmental challenges. The mitigation in Zedler's wetland is so small that it is inadequate, whereas the Curtis Prairie is large enough to be substantially more self-sustaining. But it is a matter of degree, with the threshold for adequacy being set by the judgment of ecologists, managers, and society.

The strength of Joy Zedler's approach is found in how the work transcends the normal boundaries of restoration ecologists. She does develop scientific insights into how to achieve particular mitigations and restorations, but that is not the central issue. Rather, her work is a guide to those who would write mitigation plans and regulations. That distinction might at first appear small, but it is crucial and addresses in particular what is missing in most of restoration ecology. There is a big difference between the arena of the science of restoration and the implementation of its findings in regulation.

To understand how distinctive is Joy Zedler's approach, look at the boundary between science and society. Biological and ecological scientists have important contributions to make in areas of central human concern, but they are neither trained nor skilled in how to make the judgments associated with implementation in society. In the arena of societal regulations that surround human reproduction and death, biologists can tell society when the heart starts beating in an embryo and what happens when, in an adult, it stops for particular lengths of time. However, biologists are not qualified to say when human life starts or stops, for those are matters of judgment and morality in the context of societal values. Of course, biologists in a democracy have the right to an opinion like everyone else, and that opinion might be informed by their science. If there are experts who can help society with such issues, they are not scientists but are more likely to be lawyers. That is what lawyers do, and good lawyers are as expert in that matter as are good biologists in finding out what is likely to happen in such-and-such a biological situation.

Similarly, in matters of writing regulations with regard to mitigation of compromised ecological situations, the science of restoration can certainly help, but it does not allow the regulations for mitigation to write themselves. There are very few scientists who work at the edge of the science to offer the regulators the critical models on which to base their wordsmithing. It is straightforward to define a desired state of affairs specified by numbers of tangible entities. However, it is asking a lot of those who craft regulations to include in a meaningful way the unseen processes that allow an ecological configuration to be sustainable. Regulations and mitigations do not achieve significant results until they cause action that lines up the unseen forces that will do the work of sustaining the system in some desirable condition. Mitigations that resemble a plan for a garden in the wild cannot be sustained because it is too expensive to keep putting the bits back in place. It is therefore incumbent on the scientists to lay out in the clearest fashion the differences between three conditions. The first are states of a local system that mimic some prior state. The second are situations that are desirable and sustainable without elaborate and constant manipulation. The last are situations that are desirable but unsustainable without constant gardening.

It is frustrating for scientists to deal with the language of regulation when it is cast in general terms. The particular size of a population in a given area is more the language of science, and science can make measurements to see whether indeed the numbers are achieved. By contrast, more general policy and regulation statements, such as the 1978 Great Lakes Water Quality Agreement between Canada and the United States, are more troublesome (International Joint Commission, Great Lakes Regional Office 1978). It is hard to know what some statements mean, and it is equally difficult to measure the system to see whether standards have been met. To manage for "the integrity of the waters" is a less-than-fully informative statement for a scientist. First, what is meant by *integrity*? Second, how would you measure it to see whether the parties had achieved the stated goal?

Beginning law students are first taught how to speak and write the language of legal documents. Legal documents are at once elaborate in their use of words and open to interpretation. The word *integrity* is evocative, but it requires scientists to work around it to identify a suitable set of meanings. Integrity is a general intent of good and proper functioning, but how to deal with it can be described only in lengthy scholarly papers (one of the best is Kay et al. 1999). Thus the notion of restoration can take on several meanings. Joy Zedler's work points to the tension between merely achieving a restoration to meet a standard of mitigation and achieving a restoration that rests at the bottom of a basin of attraction, a system that is self-sustaining.

ENERGY SUBSIDIES

With all else equal, better criteria for management and restoration necessitate less energy subsidy. In the preceding section we joined Joy

Zedler in her requirement that restoration and other management activities be self-sustaining. The more management is self-sustaining, the less energy subsidy is needed for the management. Managed systems that are self-sustaining align with critical underlying processes of the system unmanaged, with all else equal. Management that aligns itself with processes that emerge as critical in the managed situation can entrain those processes in the cause of the desired outcome. Much as a catalyst does not have to provide the energy to make a process work, the most effective management only reconfigures the elements so that the productive process flows by itself. A catalyzed reaction depends on minimal energy to get it going and drive it. The catalyst merely arranges a critical detail that reduces the activation energy. In the same way, the most effective management does not actually do the work itself but canalizes system processes, which do perform the work using energy and material gradients already present. Thus California Indians set fires to initiate cascades of beneficial ecological processes. Such management is akin to the process of external complexification that uses a neglected gradient, such as coal in preindustrial Britain, to effect desirable change.

In a trivial example, planting trees seems to be a low-cost, high return activity, given that such an act sets in motion a solar-driven process of maturation that produces a massive accumulation of desirable material. As we see in Rishi Valley, planting a tree can affect drainage, soil protection, and mineral recycling (fig. 7.3). But even planting trees takes a lot of effort of a focused sort, and so we might seek to go further toward minimal energy inputs to management. Better than planting trees would be to create an environmental setting that plants trees of desired species by natural reseeding.

Exploiting natural resources will cost less to both humans and the ecological system that is exploited if human activity goes with the flow of the processes in the exploited system. The cost will be less for humans because we will not have to overcome natural tendencies in the system as we capture the resources. It will cost less because human activity will encroach less on the natural functioning of the ecological system. If we work against the natural tendencies of the ecological system, the cost can become prohibitive. Far more energy input is needed to achieve the desired outcome if the nature of the managed system is

FIGURE 7.3

The Scottish highlands are being reforested, with mineral recycling, water drainage, and forest products as benefits.

Photograph T. Allen.

compromised. That is why temperate farming methods that put mineral nutrients in the soil do not work in the tropics, because the soil is not the place in which tropical ecosystems store their nutrients. Rather, tropical mineral nutrients are held in the biota. At great expense one could farm the tropics like an Iowan growing corn, but the costs would overwhelm the enterprise. So we must align management systems to natural process so as to use minimal energy. That is what happens when we set up conditions for natural forest regeneration rather than forcing the issue.

But this raises the question of what is natural. The reason we need to manage ecological systems is that we need services from them. With that as a given, we want to manage efficiently and sustainably. The very presence of our needs will mean that more energy almost always will be channeled through ecological systems, so that the ecological system can meet our needs with the increased productivity we engineer. What is indeed natural depends on the consequences of the increased energy flux. The nature with which we must deal is that which emerges with the energy flowing through the system at the rate that human exploitation imposes. As more energy is applied to a system, processes of emergence will redefine what is natural. Withdraw water from a tub fast enough and the vertical air–water interfaces that are the walls of the whirlpool will be the natural state of affairs (fig. 7.4). At that point, flowing with nature means riding the wall of water in the direction of the gyre. Over certain ranges of energy increment, the natural ecological processes that dominate the behavior of the material system will change as the applied energy causes emergence. That emergence defines what is natural. These qualitative changes generate different qualities of system productivity, some of which become available as human resources. Because there are more than 6 billion people on the planet, we have no choice but to apply significant extra energy to force production from pieces of landscape so that there is increased productivity. Whatever emerges when that energy is applied is the nature with which we must deal and with which our production processes must align.

On the face of it, it might appear contradictory to be putting more energy into system management while simultaneously attempting to minimize energy consumption by managing in concert with natural

FIGURE 7.4

Tornado in a bottle: a toy made from two soda bottles that forms whirlpools when the system receives random perturbations.

Photograph T. Allen.

fluxes. However, there are two levels of analysis here; one level is the need for increased production, and that increases energy inputs, but taking that as a given, there is still a need to minimize energy commitments within those constraints of increased demand for output. Thus it is not contradictory to increase the amount of energy put into the natural resource system while minimizing energy needed for management by aligning management action with natural processes. The processes are natural, like the gyration in a whirlpool, but they are natural processes that are associated with a system that has greater demands for output imposed on it. The decision to put more energy into management is considered and is viewed as a strategic move that is on balance worth the effort. Failed management is not so much caused by the strategic decisions to put more energy into the system as it is a failure to flow with the dominant processes that emerge in the new management arena. Consider one of the most energy consumptive ecological managements: agriculture. The immediate failure of agricultural systems is not in the decision to raise the level of intensity of cultivation. Failure is more immediately a matter of ignoring the dominant ecological processes that have come into play, such that management finds itself fighting upstream. When the whirlpool emerges, there is a right and a wrong direction to ride the vortex.

Turning to a specific agricultural example, the failure of Sumerian agriculture during the Third Dynasty of Ur (chapter 2) was not intrinsically irrigation itself. Indeed, the existence of Sumerian society and political power depended on irrigation. Without irrigation, there would have been no Sumerian society for us to discover, and there would have been no failure for us to critique. Because irrigation can be conducted in moderation without soil salting, it was not the irrigation in principle that was the problem. The demise of Sumerian agriculture occurred because the irrigation was conducted ignoring the underlying potential for soil salting, which need not but certainly can determine system behavior. People who turn to irrigation are not necessarily doomed to failure because of it. Irrigation represents the strategic phase of increasing energy input. The Sumerian failure was the tactical mismanagement of irrigation, such that it was overcome by allowing salt to accumulate. Irrigation can be conducted without salting, but it takes the proper management of local ecology of the agroecosystem, bearing salinization in mind.

Salting of soil to the detriment of agriculture does not become an issue until irrigation is imposed. Dry agriculture positions itself on the landscape at sites where natural salting is constrained by the local environs. Such agriculture performed wisely does not increase soil salt, for it leaves the local constraints at the site in place. These constraints had kept the site salt-free up to the time the dry agriculture was imposed, and they keep it that way as the agricultural system works. However, dry farming does not give returns of the same order as irrigation, so there is an incentive to intensify in a move to irrigation. Although there are sites without salt in dry climates, there is always potential for soil salinization in such climates. The actuation of those potentials is manifested in the salt deserts of the North American Great Basin. In some places the process of soil salting occurs intensely with or without human manipulation. In those places, the thing to do is to mine the salt as efficiently as possible. The wise manager of irrigation systems notes natural salting as a phenomenon and remembers that there is a potential for disaster along those lines. In irrigated regions, the process of salt accumulation is there in the wings, always available to become the principal actor. Irrigation need not increase soil salt, but there is great potential for it to do so. The strategy of irrigation is sound but dangerous. Only in the tactics of its implementation can the strategy of irrigation be used long term without disastrous soil salting.

The reason that irrigation is dangerous is that it involves many of the processes that create salt flats in the absence of human manipulation. Salt pans form naturally when water stands and evaporates (fig. 7.5). Irrigation involves an intensification of management and causes the emergence of situations that certainly can cause large-scale soil salinization for the same reason natural salt flats occur. Although hydraulic agriculture may even tempt salinization, the soil can remain sweet under irrigation. For example, drip irrigation, with water delivered underground by tubing, produces the same remarkable increases in production over dry farming that occurs in more conventional irrigation, but drip irrigation appears to be long-term viable. Salting occurs because of evaporation of water as it is delivered and stands in the field. Conversely, applying sweet water to the soil reverses the

FIGURE 7.5

Salt flats on the Utah–Nevada border in the Great Basin. (A) They periodically flood and then dry out to leave the salt. (B) Opinion is divided as to whether this sculpture is art or a violation of the great 20-mile-wide flats in western Utah, near Bonneville.

Photographs T. Allen.

process of salinization. Thus there are two sets of processes with potential to dominate the outcome: one that increases soil salt and the other that decreases soil salt. If irrigation is to avoid salting the soil, the management of the system must align itself with processes that keep the soil sweet, and it must divert processes that cause salting.

A general principle is illustrated in the preceding discussion of irrigation. The input of more energy sometimes yields more output in exactly the same terms as before, and then the ecological system is not more complexly organized, it is merely more productive under the same constraints that prevailed before. But often an input of more energy does indeed lead to qualitatively different outputs, invoking qualitatively different constellations of processes. That more complex system then presents the human actors with a need to increase the complexity with which they manage the system. Thus growth of output necessitates not only a capacity to deal with more material but also an elaboration of the model that is used to address the system.

There is a principle in system analysis called requisite variety (Ashby 1956) It states that to control a material system that is complex, there is a minimal level of variety of model components and relationships without which control is impossible. As more energy is applied, production systems emerge as more complex, and that raises the stakes on requisite variety. Increased production eventually will require an elaboration of the management system. In the case of irrigation, this elaboration of the management system might take the form of laying tubing underground and controlling the flow of water with a servomechanism. By the standards of just letting water into the field and flooding it, drip irrigation is an elaboration of control. But the complexification is worth the effort, for such a system amplifies the freshwater input while minimizing the evaporation that is the crucial step in salinization. Water is used more economically, and the processes that cause salinization are blunted.

In the industrial era, the idea of progress was associated with overcoming problems by scaling up to a bigger engine. In the information age the reverse scaling operation applies. Progress today is achieved by making things smaller, so that faster computation can be brought to bear on many details. Progress now is growth by intussusception, not by increase in size. The elaboration of management that must accompany the emergence of new levels in the material system is a matter of growth by intussusception to deal with the consequences of growth in size. More elaborate management is in line with modern conceptions of progress. More elaborate design or control is commonly associated with increases of size or input of power, but it is the increase in complexity, not the increase in size, that is the hallmark of the age.

The legacy of the industrial age is the move to apply more energy to the production system. In the industrial age the increased application of energy generally was not met with an elaboration of the control system beyond what was obvious. Certainly, the notion of elaboration of control is less impressive than the increase in system size, although that did bring in new controls. An example of increased control was universal time as railroads linked distant places by a schedule. Management that is not more elaborate will suffice only as long as the increase in energy input does not cause emergence of new levels in the system to be managed. When new levels emerge, the system is likely to fail. The increase in energy input and size of the management action is a change of scale; when there are large changes of scale, insignificant processes exhibit positive feedback and bring the system down. The linear increase in size and energy input causes nonlinearities to come to dominate the system. The failure of Victorian technology often was a failure to elaborate control that was commensurate with an increase in the size of the system and the exertion of greater power. Elaboration of control or design is needed to overcome the feedbacks or nonlinearities, or at least turn them to work with the new, more elaborate design. When bridges fall down it is because there has been an increase in size using a design that was created for a much smaller span (Petroski 1993). Wind and wave do not matter much for small road bridges going over creeks, but great iron girder bridges across estuaries can fail catastrophically because of fluid mechanics. The original design for smaller bridges turned on gravitational forces calculated into the girder construction, not amplifying oscillations in wave mechanics. In this century suspension bridges have been torn down by the span moving like a lashing bullwhip under emergent effects of the wind. Elaboration of the design to deal with these emergent effects was understood as necessary in the case of the Tacoma Narrows Bridge, but it was not implemented because of fiscal constraints.

The increase in the world population, as well as demands for a higher standard of living, create demands for greater production of natural resources. The industrial solution is simply to move to larger means of production. A failure to recognize the likely emergence of new levels as more energy is applied to larger systems can be expected to lead to catastrophic failure of production of natural resources. Merely increasing the energy applied to manipulation of ecosystems will give increased production of natural resources at first, but it is an outdated Industrial Revolution approach that will fail. Historically, the demise of irrigation in Mesopotamia, from Sumer to the Abbasid Caliphate, has come from pressure on those who farm the land to be more productive to meet increased taxation. The response of the producer is to irrigate more to meet the demand, unaware that salting of the soil and a catastrophic collapse become inevitable. The problem is that we now have the capacity to move upscale in our solutions as never before. Although the destruction associated with the Industrial Revolution should not be underestimated, the Victorians did not have the capacity for increases in energy input that are available now. The general impression of improvement with only minor setbacks in the last century occurred because increases in energy application were important but not so large as to cause frequent, radical changes in the manipulated systems through emergence.

The contemporary world, with its larger demands for scaling upwards, offers a new class of problems. First, the application of industrial levels of energy to ecosystemic production has been going on longer now, and accumulated effects should be expected. Second, we have a capacity for applying enormously more energy than heretofore, and at ever-larger spatial scales. The temptation is to use an industrial rather than a postindustrial model and only apply more energy, without elaborating control and design systems to deal with the commensurate increases in complexity. Applying more energy to natural resource systems is not particularly difficult in the modern world. The new information age takes production systems of the industrial age for granted, as well it might. The real challenge is in growth by manipulating information in an arena of growing complexity. Applying more energy is easy; the hard part is elaborating our control and design system to deal with the new processes that gain ascendancy under the new energy regime. The real challenge is to learn to line up with the emergent natural processes. Failure to align with the emergent processes in the end costs so much that failure is inevitable.

To ride the vortex in the wrong direction takes a lot of effort, and of course it destroys the emergent production structure. Management that is insufficiently elaborate to ride the emergent elaborations in the material flows often pays for conflicting inputs to the system. It wastes resources while destroying beneficial flows. More effort applied to an inappropriate management action, like the Sumerians increasing irrigation, is always counterproductive. The energy one puts into the system goes directly to feeding undesirable positive feedbacks—salting in the case of Mesopotamia. Doing nothing might be better. This would amount to the Sumerians accepting that they could not maintain their original levels of production. As it was, they brought on total system collapse in short order and spent a lot of effort doing it. The case of Alaska, discussed earlier, is a counterproductive system that seems only to generate more conflict.

In turn, all the arguments in this chapter can be translated into the language of diminishing returns that was developed in the early chapters of this book. Faced with a problem, a complex society such as ours complexifies. In the beginning this gives great returns and is the equivalent of the first energy infusion that causes the beneficial emergent structure. The later diminishing returns come from further energy subsidies, or energy diversions. This energy has been introduced to deal with undesirable aspects of the emergent structure or cope with further problems. Our recommendation to deal with the whole system, instead of the small emergencies that arise, is a way of saying that one should go with the natural emergent flows rather than try to divert them to a degree that diminishes their long-term stability. Diminishing returns can arise when the emergent system elaborates more than does the attendant management system. Undesirable feedbacks not only are permitted to form but are fed by the increased effort applied in the cause of meeting new emergencies. By treating the symptoms, ad hoc management exacerbates the cause.

In this information age we have the opportunity, if we will only seize it, to elaborate the management system in a way that is commensurate with the emergent elaboration of the material system. For example, by investing in software engineers, steel mills can meet demands for increased production while saving heat input to the furnaces through more sophisticated design and practice. Less heat applied to the furnaces means a saving in energy costs as well as a diminished set of side effects. The energy saved can be applied elsewhere, and there are diminished pollution costs. Some of this information-based saving of energy inputs is emerging in agriculture, but the savings often are reinvested in further increases in production. For example, satellite positioning technology used to control the amount of fertilizer applied to a piece of land, on the face of it, should decrease fertilizer inputs. Preliminary data indicate that there is increased production, but the amount of fertilizer applied by a producer stays about the same. Technology does appear to offer benefits, but it is the soft technology of information rather than the hard technology of bigger industrial engines.

SOCIETAL DEMANDS

Energy subsidies enter into issues of management and ecological sustainability at two levels. First is the level of energy subsidies for management action itself, as we have just discussed. Second is the level of energy subsidy that is required to maintain the human population and societal infrastructure so that there are social resources available to be put toward ecological management. Therefore, an extension of energy-efficient management is work on societal values.

Mark Stevens (1996) suggested that we can reduce the nature of problems in general down to the difference between the state of the system and the desired state. After Chen (1973), Stevens identified how three types of problem arise. First, the material system has moved to a state that is different from the original desired state. Second, the material system has not altered its state significantly, but the desires have changed. Third, both the material system and the desires have changed.

We started our discussion of energy subsidy by noting that planting trees is a small input that harnesses solar power for centuries. Less demanding again is to use natural reseeding, but that does not always produce the desired species. In the absence of natural reseeding producing the desired species, another strategy is to change what is desired. By altering the context, by altering the demand, desires can be adjusted so that natural reseeding works to give the desired effect. In the case of reseeding, the material system is in change after a timber harvest, but the desires are also in flux. First the material system is undergoing change, a change from a standing crop to a stand after harvest. A further change is in the direction of movement of stand composition in the aftermath of harvest. All of these changes are of Chen's (1973) type 1 system.

Our recommendation is to be flexible as to the desired crop, by working on the consumer end of the process, so that a workable supply-side configuration is created. Given the material changes and the changes in desires, the problem appears as a Chen type 3 problem. The material system is changing, the desires are in flux, and the solution lies in accommodating the changes in demand to match the changes that will occur in the material system if minimal energy is applied to it. The problem is one of tailoring needs by changing social values so that desires that are achievable are also sustainable. While social justice is a part of sustainability—and democratic systems appear to be the best way to achieve that—it is irresponsible to allow populist demands to destroy the means of production. There will always have to be some regulation, but much can be achieved by educating an informed public. With its ability to manipulate demand through advertising, commerce again has a role in sustainability.

Changes in the demand side are hard to achieve. In the material ecological system, we are prepared to accept a certain amount of cost of doing business. If some trees must die in a managed fire, then that is acceptable. Accordingly, there is a certain amount of room for maneuver. However, in a social system, what is desirable or acceptable has much tighter constraints imposed on it than the constraints on material manipulations of ecosystems. Ecosystems can't talk back. In a social setting, setbacks for any individual person or, worse, some class of people are hard to defend in public, even though the consequences of doing nothing, or anything else, may be more harmful to more people. We have already given the example of Freundenberg's work on mining communities (Freundenberg 1992; Freudenberg, Gramling, and Schurman 1998). The data are counterintuitive in that they say unequivocally that, with all else equal, counties with mines in them are less economically developed. When the mine is put in place with good environmental regulation, only a few people see a stake in it. However, by the time the mine threatens to close unless environmental regulations are softened, the number of workers in the mine significantly compromises the political will to maintain adequate environmental regulation. And at the time the workers have a point, although they are probably unaware that the mine is going to close soon anyway. The tangibility of the downside of management in human terms adds a layer of difficulty that does not apply to the effects of management on the material ecological system.

People live their lives and develop myths for their functioning at a scale that is tangibly human. By contrast, the effect of societal values on sustainability is altogether larger scale. The ecological effect of values is a collective effect, inside which individuals act with little tangible effect. Garrett Hardin's lifeboat ethics may not be an appropriate extension of his model for the tragedy of the commons. But when that model does work, it illustrates how individuals can act in a reasonable fashion without much effect individually but with disastrous consequences in aggregate. By each person acting in his own short-term interest, without sinister motives or flagrant disregard for the common good, the whole system breaks down and the commons is nonsustainable. At the upper level of societal effects, innocent action at the lower level of individual behavior can have large and deleterious effects. It is a failure to think globally.

Yet notwithstanding Hardin, sustainability is much more difficult than thinking good thoughts and doing good deeds. It embodies challenging contradictions. More than 200 years ago, Adam Smith established modern economics by showing that when individuals seek their own good, an "invisible hand" maximizes good for the society. The contradiction may only be surficial. Commerce, as we have noted, is context for both social and ecological sustainability. Smith's "invisible hand" will provide both short-term well-being and long-term sustainability when short- and long-term goals coincide. A primary route to achieve this is through knowledgeable citizens who understand and incorporate in their daily lives the principles that we call supply-side sustainability and who think globally as they act locally. We are under no illusion that this would be easy to accomplish, and recognize that there will always be individuals who cheat. We raise the contradiction between Hardin and Smith to emphasize that sustainability depends on knowledge and will be achieved when it has the moral force of societywide, even universal, values. Recall the Malian villagers in chapter 1 who ascribed their unsustainability to Allah. Sustainability requires that we situate the system context properly, recognize that context is the true invisible hand, and respect the contexts of sustainable systems. Raising sustainability goals to the level of morality provides the situation in which short-term, individual well-being and long-term sustainability coincide. The Maine lobster industry, discussed earlier, is a case in point. Dan Bromley (1992) provided many examples of common access to resources where Hardin's model of tragedy patently does not apply. Those are often situations in which local moral imperatives have been achieved exactly by a society engineering a coincidence of individual well-being and sustainability of the common ecosystem.

Even rational approaches by actors at the societal level are blunted by the movement upscale of local events. Every action at a societal level has some local victims. There is much to be gained by searching for win–win scenarios, in which all parties apparently gain from cooperation, but even in win–win scenarios there are victims. If there is gain by cooperation, then those who are not part of the discourse are in competition with the cooperative unit. In biology, mutualistic success costs competitors of the mutualists. If ferns could think and talk, they would probably take a dim view of pollination. In the communication age, the victims of government programs make better copy for news media than does an analysis of the long-term economic gain or ecological conservation. Victims paraded in the spotlight make impotent the actors at the level of government (Wolf and Allen 1995). Signals from the local level amplify upscale and emerge as strong random disturbances in the policy arena.

If in the policy arena measured action is blunted, how then do societal values change, as surely they do? For change one must look for positive feedbacks and for ratchet mechanisms. Intelligent planning and attempts at regulation are derailed at random by the amplification of incidental local events such as the faces of redundant workers in news

415

segments about job losses under timber-cutting moratoria. Therefore, some other force must be responsible for directional change at higher levels. If ideas and actions are derailed at random, then the ideas and actions that arise in larger numbers will survive the process of attrition and have an aggregate effect. One should look for positive feedbacks where successful action generates more action of the same sort. The tendency for self-replication of actions of a certain sort will overcome the random effects in aggregate. The aggregate effect will appear as an upper-level constraint, and social values will be constrained by positive feedback effects. In the economic arena, Herbert Simon (1997) and his colleagues (March and Simon 1958) showed that rational behavior is not an effective model to describe human activity, and so it is certainly not reasonable to expect societal values to be guided by informed rationality.

Given random disruption of plans for the common good, the plans that stand a better chance of enactment are involved in a move upscale to a more complex society. For example, once military spending has increased, recipients of military pay or armament workers become victims when there are attempts to cut the military budget. An increased budget is defended once it is in place. George Bernard Shaw once noted that every profession is a conspiracy against the lay public. Systems with a propensity to self-perpetuation survive, and so professions elaborate, protected by professional societies and their organs. It is not that the arguments of vested interests are invalid or unreasoned, for many of them are entirely defensible on grounds of rational argument. Given the random disruption of reasoned argument at upper levels, the survival of policies or recommendations does not turn on rational argument one way or another. Rather, the direction of change is a product of feedbacks that link change in a certain direction to more change in the same direction, creating the emergence of constraints and higher complexity. We have indicated at several points in this book that complexification usually is a one-way process until the complex society simplifies its burdensome infrastructure.

Given the positive feedbacks associated with commerce, it is no accident that mercantile ethics have now displaced almost all other ethical systems in the industrial world. Until remarkably recently there were farming ethics, scholarly ethics, and various craftsmanship ethics, all of which barely survive today. For example, given success in grantsmanship on one hand and shrinking university budgets on the other hand, any academic who functions other than as an entrepreneur is marginalized. And the same applies to farmers, clerics, doctors, and many other professionals. Activity that operates without positive feedback will fall under the constraint of activity that does develop positive feedbacks to create new, higher levels.

Government regulation has a static quality that means that it often functions without leading to positive feedback. As a result, institutions such as the Endangered Species Act (ESA) are at first triumphs but can only be eroded. On the other hand, successful commercial activity leads as a rule to positive feedback. Thus, for good or bad, the North American Free Trade Agreement (NAFTA) appears headed for greater expansion. Both of these institutions, the ESA and NAFTA, have significant ecological implications for sustainability, but only one appears likely to grow in influence. The value of the respective outcomes of each piece of government action is not at all clear. If (and we emphasize the proviso here) over the long run trade agreements support stable human populations by raising living standards, then they will do more to preserve species than all legislation that is more tangibly directed toward that end. Although the particular outcomes seem uncertain, it does appear certain that there will be no stasis; always the institutions that are expanding under positive feedback will win out. It is therefore crucial that issues of sustainability be cast in terms of the prevailing positive feedbacks, be they in trade or whatever.

Although lessons from history might indicate that certain courses of action are unsustainable in the long term, at the time the decisions were made, most of the unfortunate actions seemed reasonable enough to the actors and were directed toward sustaining the system, at least in the short run. When our society faces modern dilemmas it is largely without hindsight, just as Roman leaders had to make their decisions without it. If we are to manage actively in the modern world, we should probably accept the emergence of new levels of organization as given. Each emergence will be unique, and that means that in principle, the quality of emergence cannot be predicted. At each new level, the rules will change. Therefore flexibility is the first rule of effective management for sustainability. At the time actors make their decisions, reasonable values are unlikely to be important in the long run and may easily lead to action that is in the long run deleterious or even patently immoral. Lacking foresight and recognizing the limited utility of present values in the future to which they lead, planning must be guided by more neutral values, values that may span the barriers of time, space, and culture. If values are ephemeral in their application, then planning should work with transcending values that can be useful over the long run. Although we have found wisdom in the adage to think globally but act locally, we suggest also that planning over the long run must create meta-institutions that override the tendency to adapt too closely to locally prevailing situations. Notwithstanding innumerable exhortations, most people don't think globally and perhaps never will. Long-term sustainability, which we take as the ultimate positive social value, might well involve plans and action that appear in the short term to be at best value neutral or even disadvantageous.

ECOSYSTEMS, COMPLEX SOCIETIES, AND SELF-REFLECTIVE SCIENCE

The ultimate guiding principle for working toward sustainable production of natural resources is to deal with the entire unit of biogeophysical production and manage for its long-term functioning. This does not mean leaving the system alone to do what it will, for a world with as many people in it as this one must demand more of ecosystems than they will produce without human manipulation. As more energy is put into manipulating ecological systems, they will change their character. That character often will involve more production of natural resources. Because we are prepared to countenance manipulating the system, with the ultimate goal of producing more natural resources, it might appear that we feel that present management strategies are adequate. We are not suggesting that at all, for standard management is for the resource directly. As each significant change in the market, political arena, or technology comes into play, an ecosystem managed directly for the natural resource will be buffeted this way and that. Eventually that ecosystem will be compromised. Managing for the resource directly will burden the ecosystem and make it nonsustainable. Rather, the focus of the management must be on the entire ecosystem, not just on the part that produces the resource of interest.

As more energy is applied to the ecological system, perhaps in the form of fertilizer in an agroecosystem, many aspects of the system will change. Soil microbes will change, as will plant species and the primary productivity of plants. As the form of the vegetation changes, so will its relationship to the meteorological, microclimatic factors. Not only will the system produce more resources in terms of fiber or human food, but a large number of other changes will arise. There probably will be greater production in several compartments of the system, including those that are of no interest with regard to the natural resource that is taken. These changes in production must be taken as part of the equation if management action is to be predictive. Management must be for full functioning of the entire ecological system, not for the natural resource itself.

Taking a more integrated approach to management is not a new concept. Even the name *integrated pest management* indicates a trend in that direction. So what is new about what we are recommending? The answer is in our recognition of the integrated nature of the response of the system to human management practices. The material ecological system is already doing the integration for us, so integration per se is not the central issue. Manage for the entire material ecological system and there will be integrated management whether we aim for it or not. Whereas the integration is done by the material ecosystem itself, managing for the sustainability of the entire system is very much a matter of skilled and thoughtful intervention. When ecosystems fail under management, it is because the material ecological system gives an integrated response that leads to failure.

As more management effort is applied, the ecological system will respond by complexifying. New levels will emerge that will be the context for parts of the system that existed before the management effort. Changes can be expected to be gradual and continuous at first but then radical and discontinuous as more effort is applied. The new system order will depend on the continued existence of the old system parts now held together in a new form. The new form of the system parts is brought about by their context being changed. For example, fencing will change the patterns of activity of wildlife species so that they function as if they were a different species. Knowledge of the system before the application of the management effort will be helpful but inadequate because the old system parts will operate in new ways under intensive management constraints. Too much planning and direction will be applied to the system if the management is for particular levels of particular resources. An overdirected system will demand much more effort to manage because the upper-level constraints will be engineered rather than emerging through natural processes. Management directed at the long-term sustainability of the material ecological engine will achieve system self-repair and will demand only the energy to achieve the naturally emergent behavior. Let the emergence occur as it will and then deal with the resources to be taken in the context of the natural emergence. The limits on the resources to be taken are defined by the emergent behavior of the system, not by the demands of consumers.

Twinned with the emergent ecological engines will be emergent social institutions. In much the same way as the material ecological system should be managed as a whole entity, the society must be treated as a whole unit with its own emergent properties. Like the intensely managed ecosystem, society will show its own emergent properties, and these should be taken as part of emergent nature. The emergent nature of society is half of the equation in the management of ecological systems for renewable resources. The collapse of complex societies comes from the emergence of infrastructure as the system becomes more complicated and complex. Other complex societies have developed this burden of infrastructure as they have dealt with their particular class of problems. For some societies these problems that cause societal elaboration arise from military challenges or conquest; for others the problems come from irrigation systems, their management, and their declining production. The challenge for the industrial world appears to be sustainable management of ecological systems for renewable resources. Our barbarians, our plague, our climate change appears to be ecological in nature. Our problems are things such as striving to increase production of natural resources on a rising tide of pesticides and in the context of anthropogenic global warming. In response to these issues, modern industrial societies are becoming more elaborate in both structure and organization.

In this societal elaboration, we need to return to the distinction between a complex and a complicated situation. External complexification involves the emergence of new overarching levels, the emergence of new constraint structures. By contrast, in a complicated situation, the constraint structure stays roughly the same, and the elaboration is at a level that already exists. In chapter 6 we identified the subtleties of viewing different levels at which complexity resides. Increasing complicatedness is a quantitative matter of increasing the number of units residing at a given level. Increasing complexity is qualitative in that new levels emerge de novo and qualitatively change the functioning of levels below.

External complexification involves the emergence of new constraints over the whole structure and is the means of sustaining an enlarging social system. Complexity has the effect of simplifying system behavior because the parts are all pressed up against a single limit. Remember Howard Pattee's (1972) term *self-simplification*. He noted that complex systems, like organisms, possess upper-level structure that simplifies the behavior of the material of which they are made by imposing a small set of limits. Complexification (that is, self-simplification) improves predictability. All management depends on adequate predictions of the effects of management action, so predictability is critical for success. Complex systems are manageable, whereas equivalent complicated systems are not. For example, an organism can be managed, whereas an equivalent mass of chemical material cannot.

In contrast to complex systems, complicated systems become less predictable. Complicated structure amounts to burdensome infrastructure. Complicated structures emerge piecemeal as local exigencies generate more parts and infrastructure. Each problem increases the elaboration of the system without the creation of new, overarching, emergent constraints. A social system becomes more complicated if the response to local problems is to treat them as emergencies and manage them from the highest level in society. Each emergency increases the clutter of remedial machinery, creating each piece as a response to some local exigency. Putting out each brush fire seems the sensible thing to do at the time, but in the long term the system becomes burdened with the material consequences of its history. If there is only multiplication of parts at a single level, the cost of management goes up, and predictive power used to guide management goes down.

It is to avoid the situation becoming more complicated that we recommend managing for the entire suite of ecological production, both useful and otherwise. In our management strategy, management emerges as an external complexification, a set of unified, overarching constraints. The more usual and less effective management strategy is to deal with the resources one at a time. Managing for resources directly produces a surfeit of administrative units and duplication of effort in the society. So managing for an integrated ecosystem engine is part of dealing with society in an integrated fashion. Managing for whole ecosystems and social contexts has two effects. On one hand it invokes new, emergent complex systems that make the situation simpler to handle. On the other hand such integrated management diminishes the degree of complication and the amount of societal clutter that stands in the way of effective problem solving.

Of course, in a complex system there are quasiautonomous parts that need their own levels of management. Inside the integrated management for whole ecosystems and whole societies are subsystems that require local management action. For example, in executing an integrated management plan, the lumberjack should be left alone to make decisions about how best to fell a particular tree within the context of its particular neighbors. Problems of diminishing returns occur if local decisions are made at too high a level. Local decisions and action must be planned and executed locally, leaving the upper levels to perform the task of managing the whole. When we referred to local victims thwarting national-level decisions, the pathology is the intrusion of local-level considerations into upper-level decision making. Local victims are best compensated and accommodated locally. Trying to patch in accommodations to each local issue at the level of national policy produces an expensive and ineffective national-level instrument. When Forest Service timber sales are planned in the White House, national government cannot be effective.

There is nothing intrinsically good or bad about societies becoming more global. The standard of living and the quality of life commonly improve when societies become smaller and less complex (Tainter

1999). When they become less complex, societies often become less complicated, and that is a good thing. It then becomes a fair question to ask why we discuss increasing the complexity of society as it manages resources for a burgeoning world population. Our answer is that we are stuck with a prevailing situation that has enormous potential for human suffering and ecological destruction. For all the inequities in the world, the First World provides antibiotics and anesthetics, without which simple appendicitis would again be not only fatal but a gruesome way to die. The recent break-ups of the Soviet Union and Yugoslavia indicate that sometimes contracting systems can be a brutal experience, and we fear that a collapse of the First World would be more so. For the moment, we find ourselves unable to dispense with complexity, and managing global ecosystems may demand more of it.

Sometimes the currency of a society changes as the old way of doing things begins to suffer diminishing returns. The move to an information-based society, away from a simply industrial society, appears to be a case in point. Management by information is complex and has its own costs, as every user of a computer knows. However, these costs are lower than those of managing by brute force, as in the industrial approach. The state of science and technology in the arena of management, ecology, and cognition is clearly primitive yet, so science and technology will have to move much further if the western standard of living is to be maintained, or extended, as Third World people wish.

We are still shaking off the ethics of the industrial age, for some celebrated scientists still see interplanetary travel as an important part of the human future. It is a holdover from the age that foresaw bigger engines projecting industrial society through its problems. It is significant that the late Carl Sagan, a man of vision, was inspired to be a scientist by the 1939 World's Fair, one of the great hurrahs of the industrial vision. His vision is not ours. It is significant that the space shuttle uses computers from the 1970s that do not have enough power to run modern video games for children. There are too many moving parts in an industrial solution to meet modern challenges. Why does the space shuttle often not fly? Consider the problem if it has, say, 1,000,000 parts, each with a 1 in 1,000,000 probability of failure. Although the price of electronics continues to decline, even small engines as appear in garden appliances hold their price and keep up with inflation because they have moving parts.

Most of us are bound by the myths and values that are part of the modern world, and so we probably do consider sustainability to be the continuance of western and industrial civilization. The future will assign its own judgment to those values. Even so the authors here surrender to a feeling that there is an urgent need to begin the transition to a sustainable future. True, we are equipped only with what comes with membership of modern society. But we are hopeful that there appears to be more self-knowledge in modern society than there was in the major civilizations that have come and gone before. For all our reservations about the extent of the transition to a postindustrial society, the times are changing. We now have computational power so that we can take the calculations in most simulations for granted. The irony is that increased computation is having its critical effect outside computation per se. Computation is gradually liberating scientists to think more about what should be simulated and why. Computers are particularly poor at semantic and lateral, syntactic problem solving. Whereas some people are trying to overcome those limitations, we are accepting those limits and are moving to study the consequence of having a sentient scientist in the process of discovery.

The ideas in this book reflect a new science that is emerging in the information revolution. The cutting edge of science is adopting a postmodern posture. If we cannot escape the consciousness and subjectivity of scientific investigation, then that itself becomes a matter to investigate self-consciously. The discourse in this book has been moving away from an investigation of the material system to be managed and toward the general nature of management and the role it plays in modern society. This emerging area of discourse in science appears under the rubric of "post-normal science" for Silvio Funtowicz and Jerry Ravetz (1994; Ravetz 1999). James Kay is studying ecological management as a self-conscious process using hierarchical notions (Kay et al. 1999). Historian William Cronon (1992) is asking hard questions about the role of the narrator in historical ecology. Never before has there been such a self-referential discussion among practicing scientists in environmental studies (Allen, Tainter, Pires, and Hoekstra 2001). The confidence of the naive realist scientist begins to look like the self-assurance with which Soviet leaders applied scientific socialism and insisted on Lamarckian genetics. Industrial might has shown us its limits, and computational power someday will do so also. Active self-reflection will increasingly be a fruitful way forward, as Funtowicz, Kay, and Cronon have shown. Perhaps it is not possible for complex societies to avoid collapse forever, but if there has ever been a time when the problems of complexification could be managed to avoid collapse, the time is now. Traditional approaches to finding the mechanics of ecological systems will increasingly be of less and less use, even if they succeed by their own standards. What appears to be needed is a flexible, strongly self-referential study of society and its exploitation of natural resources.

REFERENCES

Acheson, J. M., J. A. Wilson, and R. S. Steneck. 1998. Managing chaotic fisheries. Pp. 390–413 in F. Berkes and C. Folke (eds.), *Linking Social and Ecological Systems: Management Practices and Social Mechanisms for Building Resilience*. Cambridge, England: Cambridge University Press.

Adams, R. E. W. 1973. The collapse of Maya civilization: A review of previous theories. Pp. 21–34 in T. P. Culbert (ed.), *The Classic Maya Collapse*. Albuquerque: University of New Mexico Press.

———. 1980. Swamps, canals, and the locations of ancient Mayan cities. *Antiquity* 54:206–214.

———. 1981. Settlement patterns of the central Yucatan and southern Campeche regions. Pp. 211–257 in W. Ashmore (ed.), *Lowland Maya Settlement Patterns*. Albuquerque: University of New Mexico Press.

———. 1983. Ancient land use and culture history in the Pasion River region. Pp. 319–335 in E. Z. Vogt and R. M. Leventhal (eds.), *Prehistoric Settlement Patterns: Essays in Honor of Gordon R. Willey*. Albuquerque and Cambridge, MA: University of New Mexico Press and Peabody Museum of Archaeology and Ethnology, Harvard University.

Adams, R. M. 1983. *Decadent Societies*. San Francisco: North Point.

Adams, R. McC. 1978. Strategies of maximization, stability, and resilience in Mesopotamian society, settlement, and agriculture. *Proceedings of the American Philosophical Society* 122:329–335.

———. 1981. *Heartland of Cities*. Chicago: Aldine.

Ahl, V. A. 1993. Cognitive development in infants prenatally exposed to cocaine. Unpublished Ph.D. diss., University of California, Berkeley.

Ahl, V. and T. F. H. Allen. 1996. *Hierarchy Theory: A Vision, Vocabulary, and Epistemology*. New York: Columbia University Press.

Akbari, M. H. 1995. Energy-based indicators of ecosystem health. Unpublished M.S. thesis, Department of Crop Sciences, University of Guelph.

References

Alcock, S. E. 1993. *Graecia Capta: The Landscapes of Roman Greece*. Cambridge, England: Cambridge University Press.

Allen, T. F. H. 1993. *The Ecosystem Approach: Theory and Ecosystem Integrity*. Report to the Great Lakes Science Advisory Board. Washington, DC: International Joint Commission.

———. 1998a. Community ecology. Pp. 315–383 in S. I. Dodson, T. F. H. Allen, S. R. Carpenter, A. R. Ives, R. L. Jeanne, J. F. Kitchell, N. E. Langston, and M. J. Turner (eds.), *Readings in Ecology*. New York: Oxford University Press.

———. 1998b. The landscape level is dead: Persuading the family to take it off the respirator. Pp. 35–54 in D. Peterson and V. T. Parker (eds.), *Ecological Scale: Theory and Application*. New York: Columbia University Press.

Allen, T. F. H., B. L. Bandurski, and A. W. King. 1993. The ecosystem approach: Theory and ecosystem integrity. Initial report to the Great Lakes Advisory Board. Washington, DC, and Windsor, Ontario: International Joint Commission.

Allen, T. F. H., T. Havlicek, and J. Norman. 2001. Wind tunnel experiments to measure vegetation temperature to indicate complexity and functionality. Pp. 135–145 in S. Ulgiati, M. T. Brown, M. Giampietro, R. A. Herendeen, and K. Mayumi (eds.), *Advances in Energy Studies, Second International Workshop: Exploring Supplies, Constraints, and Strategies*. Padua: Servizi Grafici Editoriali.

Allen, T. F. H. and T. W. Hoekstra. 1992. *Toward a Unified Ecology*. New York: Columbia University Press.

———. 1994. Toward a definition of sustainability. Pp. 98–107 in W. W. Covington and L. B. DeBano (tech. coordinators), *Sustainable Ecological Systems: Implementing an Ecological Approach to Land Management*. General Technical Report RM-GTR-247. Fort Collins, CO: USDA Forest Service, Rocky Mountain Forest and Range Experiment Station.

Allen, T. F. H., R. V. O'Neill, and T. W. Hoekstra. 1984. Interlevel relations in ecological research and management: Some working principles from hierarchy theory. *General Technical Report* RM 110, Fort Collins, CO: USDA Forest Service. Republished 1987 in *Journal of Applied Systems Analysis* 14:63–79.

Allen, T. F. H. and T. B. Starr. 1982. *Hierarchy: Perspectives for Ecological Complexity*. Chicago: University of Chicago Press.

Allen, T. F. H., J. A. Tainter, and T. W. Hoekstra. 1999. Supply-side sustainability. *Systems Research and Behavioral Science* 16:403–427.

———. 2001. Complexity, energy transformations, and post-normal science. Pp. 293–304 in S. Ulgiati, M. T. Brown, M. Giampietro, R. A. Herendeen, and K. Mayumi (eds.), *Advances in Energy Studies, Second International Workshop: Exploring Supplies, Constraints, and Strategies*. Padua: Servizi Grafici Editoriali.

Allen, T. F. H., J. A. Tainter, J. C. Pires, and T. W. Hoekstra. 2001. Dragnet ecology—"Just the facts, ma'am": The privilege of science in a postmodern world. *BioScience* 51:475–485.

Alley, R. B., D. A. Meese, C. A. Shuman, A. J. Gow, K. C. Taylor, P. M. Grootes, J. W. C. White, M. Ram, E. D. Waddington, P. A. Mayewski, and G. A. Zielinski. 1993. Abrupt increase in Greenland snow accumulation at the end of the Younger Dryas event. *Nature* 362:527–529.

Anderson, K. 1993. Native Californians as ancient and contemporary cultivators. Pp. 151–174 in T. C. Blackburn and K. Anderson (eds.), *Before the Wilderness: Environmental Management by Native Californians*. Ballena Press Anthropological Papers 40. Menlo Park, CA: Ballena Press.

Anderson, R. H., D. E. Fleming, R. W. Reese, and E. Kinghorn. 1986. Relationships between sexual activity, plasma testosterone, and the volume of the sexually dimorphic nucleus of the preoptic area in prenatally stressed and non-stressed rats. *Brain Research* 370:1–10.

Andrewartha, H. G. and L. C. Birch. 1954. *The Distribution and Abundance of Animals*. Chicago: University of Chicago Press.

Apollonio, S. 2002. *Hierarchical Perspectives on Marine Complexities: Searching for Systems in the Gulf of Maine*. New York: Columbia University Press.

Ashby, W. R. 1956. *An Introduction to Cybernetics*. London: Chapman Hall.

Bailey, G. N., C. Ioakim, G. P. C. King, C. Turner, M.-F. Sanchez-Goni, D. Sturdy, N. P. Winder, and S. E. van der Leeuw. 1998. Northwestern Epirus in the Paleolithic. Pp. 71–112 in S. E. van der Leeuw (ed.), *The Archaeomedes Project: Understanding the Natural and Anthropogenic Causes of Land Degradation and Desertification in the Mediterranean Basin*. Luxembourg: Office for Official Publications of the European Communities.

Bailiff, I., J. Bell, P. V. Castro, E. Colomer, M.-A. Courty, L. Dever, T. Escoriza, N. Fedoroff, M. Fernández-Miranda, M. D. Fernandez Posse, A. Garcia, S. Gili, M.-C. Girard, P. González Marcén, M. K. Jones, G. P. C. King, J. L. López Castro, V. Lull, C. Marlin, C. Martín, J. McGlade, M. Menasanch, R. Micó, S. Montón, L. Olmo, C. Rihuete, R. Risch, M. Ruiz, M. A. Sanahuja Yll, M. Tenas, and S. E. van der Leeuw. 1998. Environmental dynamics in the Vera Basin. Pp. 115–172 in S. E. van der Leeuw (ed.), *The Archaeomedes Project: Understanding the Natural and Anthropogenic Causes of Land Degradation and Desertification in the Mediterranean Basin*. Luxembourg: Office for Official Publications of the European Communities.

Barker, E. 1924. Italy and the West, 410–476. Pp. 392–431 in H. M. Gwatkin and J. P. Whitney (eds.), *The Cambridge Medieval History*, Volume 1, *The Christian Roman Empire and the Foundation of the Teutonic Order* (second edition). Cambridge, England: Cambridge University Press.

References

Barnett, H. J. and C. Morse. 1963. *Scarcity and Growth: The Economics of Natural Resource Availability*. Baltimore, MD: Johns Hopkins Press.

Bart, J. 1995. Amount of suitable habitat and viability of northern spotted owls. *Conservation Biology* 9:943–946.

Bates, D. V. and R. Sizto. 1987. Air pollution and hospital admissions in southern Ontario: The acid summer haze effect. *Environmental Research* 43:317–331.

Bean, L. J. 1974. Social organization in native California. Pp. 11–34 in L. J. Bean and T. F. King (eds.), *?Antap: California Indian Political and Economic Organization*. Ballena Press Anthropological Papers 2. Ramona, CA: Ballena Press.

Bean, L. J. and H. W. Lawton. 1973. Some explanations for the rise of cultural complexity in native California with comments on proto-agriculture and agriculture. Pp. v–xlvii in H. T. Lewis, *Patterns of Indian Burning in California: Ecology and Ethnohistory*. Ballena Press Anthropological Papers 1. Ramona, CA: Ballena Press.

Beer, S. 1979. *Heart of Enterprise*. Chichester, England: Wiley.

——. 1981. *Brain of the Firm* (second edition). Chichester, England: Wiley.

——. 1985. *Diagnosing the System for Organizations*. Chichester, England: Wiley.

Bent, A. C. 1932. Life histories of North American gallinaceous birds. *Smithsonian Institution, United States Natural History Museum Bulletin* 162. Washington, DC: U.S. Natural History Museum.

Besly, E. and R. Bland. 1983. *The Cunetio Treasure: Roman Coinage of the Third Century* A.D. London: British Museum.

Besteman, C. 2002. The Cold War, clans, and chaos in Somalia: A view from the ground. Pp. 285–299 in R. B. Ferguson (ed.), *The State, Identity, and Violence: Political Disintegration in the Post-Cold War World*. London: Routledge.

Blackburn, T. 1974. Ceremonial integration and social interaction in aboriginal California. Pp. 93–110 in L. J. Bean and T. F. King (eds.), *?Antap: California Indian Political and Economic Organization*. Ballena Press Anthropological Papers 2. Ramona, CA: Ballena Press.

Blackburn, T. and K. Anderson. 1993. Introduction: Managing the domesticated environment. Pp. 15–25 in T. C. Blackburn and K. Anderson (eds.), *Before the Wilderness: Environmental Management by Native Californians*. Ballena Press Anthropological Papers 40. Menlo Park, CA: Ballena Press.

Boak, A. E. R. 1955. *Manpower Shortage and the Fall of the Roman Empire in the West*. Ann Arbor: University of Michigan Press.

Bolin, S. 1958. *State and Currency in the Roman Empire to 300* A.D. Stockholm: Almquist and Wiksell.

Boserup, E. 1965. *The Conditions of Agricultural Growth: The Economics of Agrarian Change Under Population Pressure*. Chicago: Aldine.

Bromley, D. 1992. The commons, common property, and environmental policy. *Environmental and Resource Economics* 2:1–17.

Brown, A. K. 1967. *The Aboriginal Population of the Santa Barbara Channel*. University of California Archaeological Survey Report 69. Berkeley: University of California.

Burd, M. 1996. Server system and queuing models of leaf harvesting by leaf cutting ants. *American Naturalist* 148:613–629.

Bureth, P. 1964. *Les titulares imperiales dans les papyrus, les ostraca et les inscriptions d'Égypte (30 A.C.–284 P.C.)*. Bruxelles: Fondation Égyptologique Reine Élisabeth.

Canham, C. D. 1978. Catastrophic windthrow in the hemlock-hardwood forest of Wisconsin. Unpublished Ph.D. diss., University of Wisconsin, Madison.

Carey, A. B., S. P. Horton, and B. L. Biswell. 1992. Northern spotted owls: Influence of prey base and landscape character. *Ecological Monographs* 62:223–225.

Carneiro, R. L. 1970. A theory of the origin of the state. *Science* 169:733–738.

Carpenter, S. R., W. A. Brock, and P. C. Hanson. 1999. Ecological and social dynamics in simple models of ecosystem management. *Conservation Ecology* 3(2):4. [online] URL: http://www.consecol.org/vol3/iss2/art4

Carpenter, S. R., D. Ludwig, and W. A. Brock. 1999. Management of eutrophication for lakes subject to potentially irreversible change. *Ecological Applications* 9:751–771.

Caughley, G. 1994. Directions in conservation biology. *Journal of Animal Ecology* 63:215–244.

Chayanov, A. V. 1966. *The Theory of Peasant Economy* (trans. by R. E. F. Smith and C. Lane). Homewood, IL: Richard D. Irwin for the American Economic Association.

Chen, G. K. C. 1973. An anatomy of problem solving systems. Pp. 17–29 in M. D. Rubin (ed.), *Systems in Society*. Washington, DC: Society for General Systems Research.

Clark, C. and M. Haswell. 1966. *The Economics of Subsistence Agriculture*. London: Macmillan.

Cohen, M. N. 1977. *The Food Crisis in Prehistory: Overpopulation and the Origins of Agriculture*. New Haven, CT: Yale University Press.

Colinvaux, P. 1979. *Why Big Fierce Animals Are Rare*. Princeton, NJ: Princeton University Press.

References

Conklin, H. C. 1957. *Hanunóo Agriculture: A Report of an Integrated System of Shifting Cultivation in the Philippines*. FAO Forestry Development Paper 12. Rome: Food and Agriculture Organization of the United Nations.

Cope, L. H. 1969. The nadir of the imperial antoninianus in the reign of Claudius II Gothicus, A.D. 268–270. *Numismatic Chronicle* Series 7, Volume 9:145–161.

———. 1974. The metallurgical development of the Roman imperial coinage during the first five centuries A.D. Ph.D. diss., Liverpool Polytechnic, Liverpool, UK.

Coren, S. and L. M. Ward. 1989. *Sensation and Perception* (third edition). New York: Harcourt Brace Jovanovich.

Crawford, C. S., L. M. Ellis, D. Shaw, and N. E. Umbreit. 1999. Restoration and monitoring in the middle Rio Grande bosque: Current status of flood pulse related efforts. Pp. 158-163 in *Rio Grande Ecosystems: Linking Land, Water, and People*. Proceedings RMRS-P-7. Fort Collins, CO: USDA Forest Service, Rocky Mountain Research Station.

Creveld, M. van. 1989. *Technology and War, from 2000 B.C. to the Present*. New York: Free Press.

Cronon, W. 1992. A place for stories: Nature, history and narrative. *Journal of American History* 78:1347–1376.

Crumley, C. L. 2000. From garden to globe: Linking time and space with meaning and memory. Pp. 193–208 in R. J. McIntosh, J. A. Tainter, and S. K. McIntosh (eds.), *The Way the Wind Blows: Climate, History, and Human Action*. New York: Columbia University Press.

Culbert, T. P. 1973. Introduction: A prologue to Classic Maya culture and the problem of its collapse. Pp. 3–19 in T. P. Culbert (ed.), *The Classic Maya Collapse*. Albuquerque: University of New Mexico Press.

———. 1974. *The Lost Civilization: The Story of the Classic Maya*. New York: Harper & Row.

———. 1988. The collapse of Classic Maya civilization. Pp. 69–101 in N. Yoffee and G. L. Cowgill (eds.), *The Collapse of Ancient States and Civilizations*. Tucson: University of Arizona Press.

Cumont, F. 1936. The frontier provinces of the East. Pp. 606–648 in S. A. Cook, F. E. Adcock, and M. P. Charlesworth (eds.), *The Cambridge Ancient History*, Volume 11, *The Imperial Peace, A.D. 70–192*. Cambridge, England: Cambridge University Press.

Curtis, J. T. 1959. *The Vegetation of Wisconsin*. Madison: University of Wisconsin Press.

Darwin, C. 1964 [1859]. *On the Origin of Species* (facsimile of the first edition). Cambridge, MA: Harvard University Press.

Dill, Samuel. 1899. *Roman Society in the Last Century of the Western Empire* (second edition). London: Macmillan.

Drake, J. A. 1992. Community-assembly mechanics and the structure of an experimental species ensemble. *American Naturalist* 137:1–26.

Duncan-Jones, R. 1974. *The Economy of the Roman Empire: Quantitative Studies*. Cambridge, England: Cambridge University Press.

———. 1990. *Structure and Scale in the Roman Economy*. Cambridge, England: Cambridge University Press.

Eldredge, N. and S. J. Gould. 1972. Punctuated equilibria: An alternative to phyletic gradualism. Pp. 82–115 in T. J. M. Schopf (ed.), *Models in Paleobiology*. San Francisco: Cooper and Co.

Erickson, E. E. 1973. The life cycle of life styles: Projecting the course of local evolutionary sequences. *Behavior Science Notes* 8:135–160.

———. 1975. Growth functions and culture history: A perspective on Classic Maya cultural development. *Behavior Science Research* 10:37–61.

Etnier, D. A. and W. C. Starnes. 1993. *The Fishes of Tennessee*. Knoxville: University of Tennessee Press.

Evans-Pritchard, E. E. 1940. *The Nuer*. Oxford, England: Clarendon Press.

Evenson, R. E. 1984. International invention: Implications for technology market analysis. Pp. 89–123 in Z. Griliches (ed.), *R & D, Patents, and Productivity*. Chicago: University of Chicago Press.

Fairbanks, R. G. 1993. Flip-flop end to last ice age. *Nature* 362:495.

Ferrill, A. 1986. *The Fall of the Roman Empire: The Military Explanation*. London: Thames & Hudson.

Feyerabend, P. F. 1962. Explanation, reduction, and empiricism. Pp. 28–97 in H. Feigl and G. Maxwell (eds.), *Minnesota Studies in the Philosophy of Science*, Volume 3. Minneapolis: University of Minnesota Press.

Finley, M. I. 1968. *Aspects of Antiquity: Discoveries and Controversies*. New York: Viking Press.

———. 1973. *The Ancient Economy*. Berkeley: University of California Press.

Fox, G. A. and D. V. Weseloh. 1987. Colonial waterbirds as bio-indicators of environmental contamination in the Great Lakes. Pp. 209–216 in A. W. Diamond and F. L. Filion (eds.), *The Value of Birds*. Cambridge, England: ICBP Technical Publication No. 6.

Frank, T. 1940. *An Economic Survey of Ancient Rome*, Volume 5, *Rome and Italy of the Empire*. Baltimore, MD: Johns Hopkins University Press.

Frere, S. S. and J. K. S. St. Joseph. 1983. *Roman Britain from the Air*. Cambridge, England: Cambridge University Press.

Freundenberg, W. 1992. Addictive economies: Extractive industries and vulnerable localities in a changing world economy. *Rural Sociology* 57:305–332.

References

Freundenberg, W., R. Gramling, and R. Schurman. 1998. Natural resource extraction and rural economic prospects: A closer look. *Western Planner* 19(8):6–8.

Funtowicz, S. O. and J. R. Ravetz. 1992. The good, the true and the postmodern. *Futures* 24:963–976.

——. 1994. Uncertainty, complexity and post-normal science. *Environmental Toxicology and Chemistry* 13:1881–1885.

Gates, J. 1998. *The Ownership Solution*. Reading, MA: Addison-Wesley.

Geertz, C. 1963. *Agricultural Involution: The Process of Agricultural Change in Indonesia*. Berkeley: University of California Press.

Gibbon, E. 1776–88. *The Decline and Fall of the Roman Empire*. New York: Modern Library.

Gibson, J. J. 1979. *The Ecological Approach to Visual Perception*. Boston: Houghton Mifflin.

Goldstone, J. A. 1991. *Revolution and Rebellion in the Early Modern World*. Berkeley: University of California Press.

Goodall, J. 1988. *In the Shadow of Man*. Boston: Houghton Mifflin.

Gordon, R. L. 1981. *An Economic Analysis of World Energy Problems*. Cambridge: Massachusetts Institute of Technology Press.

Green, S. F., G. P. C. King, V. Nitsiakos, and S. E. van der Leeuw. 1998. Landscape perceptions in Epirus in the late 20th century. Pp. 330–359 in S. E. van der Leeuw (ed.), *The Archaeomedes Project: Understanding the Natural and Anthropogenic Causes of Land Degradation and Desertification in the Mediterranean Basin*. Luxembourg: Office for Official Publications of the European Communities.

Greig-Smith, P. 1971. Application of numerical methods to tropical forests. Pp. 149–166 in G. Patil, E. Pielou, and W. Waters (eds.), *Statistical Ecology*, Volume 3. College Park: Pennsylvania State University Press.

Grene, M. 1974. *The Knower and the Known*. Berkeley: University of California Press.

Gunderson, L. H., C. S. Holling, and S. S. Light. 1995. *Barriers and Bridges to Renewal of Ecosystems and Institutions*. New York: Columbia University Press.

Haldon, J. F. 1990. *Byzantium in the Seventh Century: The Transformation of a Culture*. Cambridge, England: Cambridge University Press.

——. 1999. *Warfare, State, and Society in the Byzantine World, 565–1204*. London: UCL Press.

Hall, C. A. S., C. J. Cleveland, and R. Kaufmann. 1992. *Energy and Resource Quality: The Ecology of the Economic Process*. Niwot: University Press of Colorado.

Hammond, M. 1946. Economic stagnation in the early Roman Empire. *Journal of Economic History, Supplement* 6:63–90.

Harl, K. W. 1996. *Coinage in the Roman Economy, 300 B.C. to A.D. 700.* Baltimore, MD: Johns Hopkins University Press.

Harper, J. L. 1967. A Darwinian approach to plant ecology. *Journal of Ecology* 55:247–270.

———. 1977. *Population Biology of Plants.* New York: Academic Press.

Harrington, J. P. 1942. *Culture Element Distributions: XIX, Central California Coast.* University of California Anthropological Records VII(1). Berkeley: University of California Press.

Hart, H. 1945. Logistic social trends. *American Journal of Sociology* 50:337–352.

Haviland, W. A. 1967. Stature at Tikal, Guatemala: Implications for classic Maya demography and social organization. *American Antiquity* 32:316–325.

———. 1970. Tikal, Guatemala, and Mesoamerican urbanism. *World Archaeology* 2:186–197.

Hawkens, P. 1983. *The Next Economy.* New York: Ballantine.

Heerwagen, J. H. and G. H. Orians. 1993. Humans, habitats, and aesthetics. Pp. 138–172 in S. R. Kellert and E. O. Wilson (eds.), *The Biophilia Hypothesis.* Washington, DC: Island Press.

Hegmon, M. 1996. Variability in food production, strategies of storage and sharing, and the pithouse-to-pueblo transition in the northern Southwest. Pp. 223–250 in J. A. Tainter and B. B. Tainter (eds.), *Evolving Complexity and Environmental Risk in the Prehistoric Southwest.* Santa Fe Institute, Studies in the Sciences of Complexity, Proceedings Volume 24. Reading, MA: Addison-Wesley.

Heintzelman, D. S. 1984. *Guide to Watching North American Owls.* Piscataway, NJ: Winchester Press.

Hitchens, C. 1988. *Prepared for the Worst.* New York: Hill & Wang.

Hodges, R. and D. Whitehouse. 1983. *Mohammed, Charlemagne and the Origins of Europe.* Ithaca, NY: Cornell University Press.

Hodgett, G. A. J. 1972. *A Social and Economic History of Medieval Europe.* London: Methuen.

Hoekstra, T. W. and M. Shachak (eds.). 1999. *Arid Lands Management: Towards Ecological Sustainability.* Urbana: University of Illinois Press.

Holling, C. S. 1986. The resilience of terrestrial ecosystems: Local surprise and global change. Pp. 292–317 in W. C. Clark and R. E. Munn (eds.), *Sustainable Development of the Biosphere.* Cambridge, England: Cambridge University Press.

———. 1992. Cross-scale morphology, geometry and dynamics of ecosystems. *Ecological Monographs* 62:447–502.

References

Hong, S., J.-P. Candelone, C. C. Patterson, and C. F. Bouton. 1994. Greenland ice evidence of hemispheric lead pollution two millennia ago by Greek and Roman civilizations. *Science* 265:1841–1843.

Horst, S. 2001. Slipping through our fingers. *Albuquerque Tribune* July 28:C1–C2.

Hughes, J. D. 1975. *Ecology in Ancient Civilizations*. Albuquerque: University of New Mexico Press.

Humbolt, A. von and A. Bonpland. 1809. *Essai sur la Geographie des Plantes*. Paris; reprinted New York: Arno Press, 1977.

Hutchinson, G. E. 1959. Homage to Santa Rosalia: Or, why are there so many kinds of animals? *American Naturalist* 93:145–159.

International Joint Commission, Great Lakes Regional Office. 1978. *Revised Great Lakes Water Quality Agreement*. Windsor, Ontario: International Joint Commission. Great Lakes Regional Office.

Isaac, J. P. 1971. *Factors in the Ruin of Antiquity: A Criticism of Ancient Civilization*. Toronto: Bryant Press.

Ives, A. R. 1998. Population ecology. Pp. 235–313 in S. I. Dodson, T. F. H. Allen, S. R. Carpenter, A. R. Ives, R. L. Jeanne, J. F. Kitchell, N. E. Langston, and M. J. Turner (eds.), *Readings in Ecology*. New York: Oxford University Press.

James, W. H. 1980. Secular trend in reported sperm counts. *Andrologia* 12:381–388.

Janzen, D. H. 1999. Gardenification of tropical conserved wildlands: Multi-tasking, multicropping, and multiusers. *Proceedings of the National Academy of Sciences* 96(11):5987–5994.

Jones, A. H. M. 1964. *The Later Roman Empire, 284–602: A Social, Economic and Administrative Survey*. Norman: University of Oklahoma Press.

———. 1974. *The Roman Economy: Studies in Ancient Economic and Administrative History*. Oxford, England: Basil Blackwell.

Kaplan, R. D. 1996. *The Ends of the Earth: A Journey at the Dawn of the 21st Century*. New York: Random House.

Kay, J. J., J. A. Regier, M. Boyle, and G. Francis. 1999. An ecosystem approach for sustainability addressing the challenge of complexity. *Futures* 31:721–742.

Kennedy, P. 1987. *The Rise and Fall of the Great Powers: Economic Change and Military Conflict from 1500 to 2000*. New York: Random House.

Kerr, R. A. 1998. Acid rain control: Success on the cheap. *Science* 282:1024–1027.

King, C. E. 1982. Issues from the Rome mint during the sole reign of Gallienus. Pp. 467–485 in *Actes du 9ème Congrès International de Numismatique*. Louvain-la-Neuve: Association Internationale des Numismates Professionels.

King, C. 1971. Chumash inter-village economic exchange. *Indian Historian* 4(1):30–43.

436

King, L. 1969. The Medea Creek cemetery (LAn-243): An investigation of social organization from mortuary practices. *Archaeological Survey Annual Report* 1968–1969:23–68. Los Angeles: University of California Press.

Kitchell, J. F. and L. B. Crowder. 1986. Predator–prey interactions in Lake Michigan: Model predictions and recent dynamics. *Environmental Biology of Fishes* 16:205–211.

Knox, E. B. and J. D. Palmer. 1998. Chloroplast DNA evidence on the origin and radiation of the giant lobelias in eastern Africa. *Systematic Botany* 23:109–149.

Koestler, A. 1967. *The Ghost in the Machine*. Chicago: Gateway.

Kohler, T. A. and C. Van West. 1996. The calculus of self-interest in the development of cooperation: Sociopolitical development and risk among the northern Anasazi. Pp. 169–196 in J. A. Tainter and B. B. Tainter (eds.), *Evolving Complexity and Environmental Risk in the Prehistoric Southwest*. Santa Fe Institute, Studies in the Sciences of Complexity, Proceedings Volume 24. Reading, MA: Addison-Wesley.

Kroeber, A. L. 1925. *Handbook of the Indians of California*. Bureau of American Ethnology, Bulletin 78. Washington, DC: Bureau of American Ethnology, Smithsonian Institution.

Landberg, L. C. W. 1965. *The Chumash Indians of Southern California*. Southwest Museum Papers 19. Los Angeles: Southwest Museum.

Lee, R. B. 1968. What hunters do for a living, or, how to make out on scarce resources. Pp. 30–48 in R. B. Lee and I. DeVore (eds.), *Man the Hunter*. Chicago: Aldine.

———. 1969. Eating Christmas in the Kalahari. *Natural History* 78(10):14, 16, 18, 21–22, 60–63.

Le Gentilhomme, P. 1962. Variations du titre de l'antoninianus au IIIe siècle. *Revue Numismatique* 6(4):141–166.

Leopold, A. 1949. *A Sand County Almanac*. New York: Oxford University Press.

Levins, R. 1970. Extinction. Pp. 77–107 in M. Gerstenhaber (ed.), *Some Mathematical Questions in Biology, Second Symposium on Mathematical Biology*. Providence, RI: American Mathematics Society.

Lévy, J.-P. 1967. *The Economic Life of the Ancient World* (trans. by J. G. Biram). Chicago: University of Chicago Press.

Lewis, H. T. 1973. *Patterns of Indian Burning in California: Ecology and Ethnohistory*. Ballena Press Anthropological Papers 1. Ramona, CA: Ballena Press.

Likens, G. E., F. H. Bormann, N. M. Johnson, D. W. Fisher, and R. S. Pierce. 1970. Effects of forest cutting and herbicide treatment on nutrient budgets in the Hubbard Brook watershed-ecosystem. *Ecological Monographs* 40:23–47.

References

Lindeman, R. L. 1942. The trophic-dynamic aspect of ecology. *Ecology* 23:399–418.

Longden, R. P. 1936. Nerva and Trajan. Pp. 188–222 in S. A. Cook, F. E. Adcock, and M. P. Charlesworth (eds.), *The Cambridge Ancient History*, Volume II, *The Imperial Peace*, A.D. 70–192. Cambridge, England: Cambridge University Press.

Loucks, O. L. 1970. Evolution of diversity, efficiency and community stability. *American Zoologist* 10:17–25.

Lovelock, J. E. 1979. *Gaia: A New Look at Life on Earth*. Oxford, England: Oxford University Press.

Luttwak, E. N. 1976. *The Grand Strategy of the Roman Empire from the First Century A.D. to the Third*. Baltimore, MD: Johns Hopkins University Press.

Luvall, J. and H. R. Holbo. 1989. Measurements of short term thermal responses of coniferous forest canopies using thermal scanner data. *Remote Sensing of Environment* 27:1–10.

Mac, M. J. 1988. Toxic substances and the survival of Lake Michigan salmonids: Field and laboratory approaches. Pp. 389–401 in M. S. Evans (ed.), *Toxic Contaminants and Ecosystem Health: A Great Lakes Focus*. Toronto: Wiley.

Machlup, F. 1962. *The Production and Distribution of Knowledge in the United States*. Princeton, NJ: Princeton University Press.

MacMullen, R. 1976. *Roman Government's Response to Crisis*, A.D. 235–337. New Haven, CT: Yale University Press.

March, J. G. and H. A. Simon. 1958. *Organizations*. New York: Wiley.

Marcus, J. 1976. *Emblem and State in the Classic Maya Lowlands: An Epigraphic Approach to Territorial Organization*. Washington, DC: Dumbarton Oaks Research Library and Collection.

———. 1993. Ancient Maya political organization. Pp. 111–183 in J. A. Sabloff and J. S. Henderson (eds.), *Lowland Maya Civilization in the Eighth Century A.D.* Washington, DC: Dumbarton Oaks Research Library and Collection.

Matheny, R. T. 1976. Maya lowland hydraulic systems. *Science* 193:639–646.

———. 1978. Northern Maya lowland water-control systems. Pp. 185–210 in P. D. Harrison and B. L. Turner II (eds.), *Pre-Hispanic Maya Agriculture*. Albuquerque: University of New Mexico Press.

Mattingly, H. 1960. *Roman Coins* (second edition). Chicago: Quadrangle.

Mazzarino, S. 1966. *The End of the Ancient World* (trans. by G. Holmes). London: Faber & Faber.

May, R. M. 1974. *Stability and Complexity in Model Ecosystems* (second edition). Princeton, NJ: Princeton University Press.

McCain, G. and E. M. Segal. 1973. *The Game of Science* (second edition). Monterey, CA: Brooks/Cole.

McCauley, J. F., G. G. Schaber, C. S. Breed, M. J. Grolier, C. V. Haynes, B. Issawi, C. Elachi, and R. Blom. 1982. Subsurface valleys and geoarcheology of the eastern Sahara revealed by shuttle radar. *Science* 218:1004–1020.

McCune, B. and T. F. H. Allen. 1985. Forest dynamics in the Bitterroot Canyons, Montana. *Canadian Journal of Botany* 63:377–383.

McEachern, A. K., M. L. Bowles, and N. B. Pavlovic. 1994. A metapopulation approach to Pitcher's thistle (*Cirsium pitcheri*) recovery in southern Lake Michigan dunes. Pages 194–218 in M. L. Bowles and C. J. Whelan (eds.), *Restoration of Endangered Species*. Cambridge, England: Cambridge University Press.

McGuire, R. H. 1983. Breaking down cultural complexity: Inequality and heterogeneity. Pp. 91–142 in M. B. Schiffer (ed.), *Advances in Archaeological Method and Theory*, Volume 6. New York: Academic Press.

McIntosh, R. J., J. A. Tainter, and S. K. McIntosh (eds.). 2000. *The Way the Wind Blows: Climate, History, and Human Action*. New York: Columbia University Press.

McNeill, W. H. 1976. *Plagues and Peoples*. Garden City, NY: Anchor/Doubleday.

Meyer, J. L. 1987. The monetary reforms of Aurelian and Diocletian. *Roman Coins and Culture* 3(2):20–42.

Miller J. G. 1978. *Living Systems*. New York: McGraw-Hill.

Milne, B. T., M. G. Turner, J. A. Wiens, and A. R. Johnson. 1992. Interactions between the fractal geometry of landscapes and allometric herbivory. *Theoretical Population Biology* 41:337–353.

Minister of Supply and Services Canada. 1991. *Toxic Chemicals in the Great Lakes and Associated Effects*, Volume 2, *Effects*. Toronto: Environment Canada.

Murphy, M. and J. Vogel. 1985. Looking out of the isolator: David's perceptions of the world. *Pediatrics* 6:118–121.

Neilson, R. P. 1986. High-resolution climatic analysis and Southwest biogeography. *Science* 232:27–34.

——. 1987a. Biotic regionalization and climatic controls in western North America. *Vegetatio* 70:135–147.

——. 1987b. On the interface between current ecological studies and the paleobotany of pinyon–juniper woodlands. Pp. 93–98 in R. L. Everett (comp.), Proceedings, Pinyon–Juniper Conference. Ogden, UT: USDA Forest Service, Intermountain Forest and Range Experiment Station, General Technical Report INT-215.

Neilson, R. P. and L. H. Wullstein. 1983. Biogeography of two southwest

References

American oaks in relation to atmospheric dynamics. *Journal of Biogeography* 10:275–297.

———. 1986. Microhabitat affinities of Gambel oak seedlings. *Great Basin Naturalist* 46:294–298.

O'Connell, J. J. F. and K. Hawkes. 1981. Alyawara plant use and optimal foraging theory. Pp. 99–125 in B. Winterhalder and E. A. Smith (eds.), *Hunter-Gatherer Foraging Strategies: Ethnographic and Archeological Analyses.* Chicago: University of Chicago Press.

Odum, E. P. 1969. The strategy of ecosystem development. *Science* 164:262–270.

Odum, H. T. and E. P. Odum. 1976. *Energy Basis for Man and Nature.* New York: McGraw-Hill.

O'Neill, R. V. 1971. Systems approaches to the study of forest floor arthropods. Pp. 441–478 in B. C. Patten (ed.), *Systems Analysis and Simulation in Ecology,* Volume 1. New York: Academic Press.

O'Neill, R. V., D. DeAngelis, J. Waide, and T. F. H. Allen. 1986. *A Hierarchical Concept of Ecosystems.* Monographs in Population Biology 23. Princeton, NJ: Princeton University Press.

Orians, G. H. 1998. Human behavioral ecology: 140 years without Darwin is too long. *Bulletin of the Ecological Society of America* 79(1):15–28.

Orians, G. H. and J. H. Heerwagen. 1992. Evolved responses to landscapes. Pp. 555–579 in J. H. Cosmides and J. Tooby (eds.), *The Adapted Mind.* New York: Oxford University Press.

Ostrogorsky, G. 1969. *History of the Byzantine State* (revised). New Brunswick, NJ: Rutgers University Press.

Parker, G. 1988. *The Military Revolution: Military Innovation and the Rise of the West, 1500–1800.* Cambridge, England: Cambridge University Press.

Parsons, W., T. W. Swetnam, and N. L. Christensen. 1999. Invited feature: Uses and limitations of historical variability concepts in managing ecosystems. *Ecological Applications* 9:1177–1178.

Parton, W. J., D. S. Schimel, C. V. Cole, and D. S. Ojima. 1987. Analysis of factors controlling soil organic matter levels in Great Plains grasslands. *Soil Science Society of America Journal* 51:1173–1179.

Pattee, H. 1972. The evolution of self-simplifying systems. Pp. 31–41 in E. Lazlo (ed.), *The Relevance of General Systems Theory.* New York: Braziller.

———. 1978. The complementarity principle in biological and social structures. *Journal of Social and Biological Structures* 1:191–200.

Patten, B. C. (ed.). 1971. *Systems Analysis and Simulation in Ecology,* Volume 1. New York: Academic Press.

———. 1972. *Systems Analysis and Simulation in Ecology*, Volume 2. New York: Academic Press.

———. 1975. *Systems Analysis and Simulation in Ecology*, Volume 3. New York: Academic Press.

———. 1976. *Systems Analysis and Simulation in Ecology*, Volume 4. New York: Academic Press.

Pearce, D. W., G. D. Atkinson, and W. R. Dubourg. 1994. The economics of sustainable development. *Annual Review of Energy and Environment* 19:457–474.

Peri, D. W. and S. M. Patterson. 1993. "The basket is in the roots: That's where it begins." Pp. 173–193 in T. C. Blackburn and K. Anderson (eds.), *Before the Wilderness: Environmental Management by Native Californians*. Ballena Press Anthropological Papers 40. Menlo Park, CA: Ballena Press.

Peterson, G., C. R. Allen, and C. S. Holling. 1998. Ecological resilience, biodiversity and scale. *Ecosystems* 1:6–18.

Petit-Maire, N., J. C. Celles, D. Commelin, G. Delibrias, and M. Raimbault. 1983. The Sahara in northern Mali: Man and his environment between 10,000 and 3500 years bp. *The African Archaeological Review* 1:105–125.

Petroski, H. 1993. Predicting disaster. *American Scientist* 81:110–113.

Pfister, C. 1913. Gaul under the Merovingian Franks. Pp. 132-158 in H. M. Gwatkin and J. P. Whitney (eds.), *The Cambridge Medieval History*, Volume 2, *The Rise of the Saracens and the Foundation of the Western Empire*. Cambridge, England: Cambridge University Press.

Piaget, J. 1963. *The Origins of Intelligence in Children*. New York: Norton.

Polybius. 1979. *The Rise of the Roman Empire* (*The Histories*, trans. by I. Scott-Kilvert). Harmondsworth, England: Penguin.

Posposil, L. 1963. Kapauku Papuan economy. *Yale University Publications in Anthropology* 67. New Haven, CT: Department of Anthropology, Yale University.

Pound, R. and F. E. Clements. 1898. II. A method of determining the abundance of secondary species. *Minnesota Botanical Studies*, Series 2, Part 1:19–24.

———. 1900. *The Phytogeography of Nebraska* (second edition). Lincoln, NE: The Seminar.

Preston, F. W. 1962. The canonical distribution of commonness and rarity, Part 1. *Ecology* 43:185–215, 431–432.

Price, D. de Solla. 1963. *Little Science, Big Science*. New York: Columbia University Press.

Puleston, D. E. 1977. The art and archaeology of hydraulic agriculture in the

Maya lowlands. Pp. 449–467 in N. Hammond (ed.), *Social Process in Maya Prehistory: Studies in Honour of Sir Eric Thompson*. London: Academic Press.

Puth, L. M. 1997. Finding a powerful definition of corridors: Lessons from the spread of an exotic crayfish in northern Wisconsin. Unpublished M.S. thesis, University of Wisconsin, Madison.

Pyne, S. J. 1998. Forged in fire: History, land, and anthropogenic fire. Pp. 64–103 in W. Balée (ed.), *Advances in Historical Ecology*. New York: Columbia University Press.

Raish, C. 1992. *Domestic Animals and Stability in Pre-State Farming Societies*. British Archaeological Reports International Series 579. Oxford: British Archaeological Reports.

Randsborg, K. 1991. *The First Millennium A.D. in Europe and the Mediterranean: An Archaeological Essay*. Cambridge, England: Cambridge University Press.

Rasler, K. and W. R. Thompson. 1989. *War and State Making: The Shaping of the Global Powers*. Boston: Unwin Hyman.

Ravetz, J. R. 1999. What is post-normal science? *Futures* 31:647–653.

Renfrew, C. 1979. Systems collapse as social transformation: Catastrophe and anastrophe in early state societies. Pp. 481–506 in C. Renfrew and K. L. Cooke (eds.), *Transformations: Mathematical Approaches to Culture Change*. New York: Academic Press.

Rescher, N. 1978. *Scientific Progress: A Philosophical Essay on the Economics of Research in Natural Science*. Pittsburgh: University of Pittsburgh Press.

———. 1980. *Unpopular Essays on Technological Progress*. Pittsburgh: University of Pittsburgh Press.

Rosen, R. 1981. The challenges of system theory. Presidential Address 26th Annual Conference SGSR, Toronto, Canada. *General Systems Bulletin* 11(2):2–5.

———. 1989. Similitude, similarity, and scaling. *Landscape Ecology* 3:207–216.

———. 2000. *Essays on Life Itself*. New York: Columbia University Press.

Rostovtzeff, M. 1926. *The Social and Economic History of the Roman Empire*. Oxford, England: Oxford University Press.

Rostow, W. W. 1980. *Why the Poor Get Richer and the Rich Slow Down*. Austin: University of Texas Press.

Roughgarden, J. E. and S. Pacala. 1989. Taxon cycling among *Anolis* lizard populations: Review of the evidence. In D. Ote and J. Endler (eds.), *Speciation and Its Consequences*. Sunderland, MA: Sinauer.

Russell, J. C. 1958. Late ancient and medieval population. *Transactions of the American Philosophical Society* 48(3).

Sahlins, M. 1972. *Stone Age Economics*. Chicago: Aldine.

Sanders, W. T. 1981. Classic Maya settlement patterns and ethnographic analogy. Pp. 351–369 in W. Ashmore (ed.), *Lowland Maya Settlement Patterns.* Albuquerque: University of New Mexico Press.

Sato, R. and G. S. Suzawa. 1983. *Research and Productivity: Endogenous Technical Change.* Boston: Auburn House.

Saul, F. P. 1972. *The Human Skeletal Remains of Altar de Sacrificios: An Osteobiographic Analysis.* Papers of the Peabody Museum of Archaeology and Ethnology, Harvard University 63(2). Cambridge, MA: Peabody Museum of Archaeology and Ethnology, Harvard University.

——. 1973. Disease in the Maya area: The Pre-Columbian evidence. Pp. 301–324 in T. P. Culbert (ed.), *The Classic Maya Collapse.* Albuquerque: University of New Mexico Press.

Schmidt-Nielsen, K. 1984. *Scaling: Why Is Animal Size So Important?* Cambridge, England: Cambridge University Press.

Schmookler, J. 1966. *Invention and Economic Growth.* Cambridge, MA: Harvard University Press.

Schneider, E. and J. J. Kay. 1994. Life as a manifestation of the second law of thermodynamics. *Mathematical and Computer Modelling* 19:25–48.

Schneider, M. L. and C. F. Moore. 2000. Effect of prenatal stress on development: A nonhuman primate model. Pp. 201–243 in C. Nelson (ed.), *Minnesota Symposium on Child Psychology.* Mahwah, NJ: Erlbaum.

Schneider, M. L., E. C. Roughton, A. J. Koehler, and G. R. Lubach. 1999. Growth and development following prenatal stress in primates: An examination of ontogenetic vulnerability. *Child Development* 70:263–274.

Schumpeter, J. A. 1950. *Capitalism, Socialism and Democracy.* New York: Harper.

Sellers, P. J., Y. Mintz, Y. C. Sud, and A. Dalcher. 1986. A simple biosphere model (SiB) for use within general circulation models. *Journal of the Atmospheric Sciences* 43:505–531.

Shugart, H. H., D. C. West, and W. R. Emanuel. 1981. Patterns and dynamics of forests: An application of simulation models. Pp. 74–94 in D. C. West, H. H. Shugart, and D. Botkin (eds.), *Forest Succession: Concepts and Applications.* New York: Springer Verlag.

Shulka, J., C. Nobre, and P. Sellers. 1989. Amazon deforestation and climate change. *Science* 247:1322–1325.

Siemens, A. H. 1982. Prehispanic agricultural use of the wetlands of northern Belize. Pp. 205–225 in K. V. Flannery (ed.), *Maya Subsistence: Studies in Memory of Dennis E. Puleston.* New York: Academic Press.

Simberloff, D. 1983. Competition theory, hypothesis testing, and other community ecology buzzwords. *American Naturalist* 122:626–635.

References

——. 1988. The contribution of population and community biology to conservation science. *Annual Review of Ecology and Systematics* 19:473–511.

Simberloff, D., J. A. Farr, J. Cox, and D. Mehlman. 1992. Movement corridors: Conservation bargains or poor investments? *Conservation Biology* 6:493–504. (Reprinted in 1995 as pp. 59–70 in D. Ehrenfeld (ed.), *The Landscape Perspective*. London: Blackwell.)

Simon, H. A. 1962. The architecture of complexity. *Proceedings of the American Philosophical Society* 106:467–482.

——. 1997. *Administrative Behavior: A Study of Decision-Making Processes in Administrative Organizations* (fourth edition). New York: Free Press.

Snyder, N. F. R., S. R. Derrickson, S. R. Beissinger, J. W. Wiley, T. B. Smith, W. D. Toone, and B. Miller. 1994. Limitations of captive breeding in endangered species. *Conservation Biology* 10:338–346.

Snyder, N. F. R. and H. Snyder. 2000. *The California Condor: A Saga of Natural History and Conservation*. San Diego: Academic Press.

Sollins, P., W. F. Harris, and N. T. Edwards. 1976. Simulating the physiology of a temperate deciduous forest. Pp. 173–219 in B. C. Patten (ed.), *Systems Analysis and Simulation in Ecology*, Volume 4. New York: Academic Press.

Sparks, B. W. and R. G. West. 1972. *The Ice Age in Britain*. London: Methuen.

Stevens, M. R. 1996. A rhapsody in problem solving. Unpublished M.S. thesis, University of Wisconsin, Madison.

Stevenson, G. H. 1934a. The army and navy. Pp. 218–238 in S. A. Cook, F. E. Adcock, and M. P. Charlesworth (eds.), *The Cambridge Ancient History*, Volume 10, *The Augustan Empire, 44 B.C.–70 A.D.* Cambridge, England: Cambridge University Press.

——. 1934b. The imperial administration. Pp. 182–217 in S. A. Cook, F. E. Adcock, and M. P. Charlesworth (eds.), *The Cambridge Ancient History*, Volume 10, *The Augustan Empire, 44 B.C.–70 A.D.* Cambridge, England: Cambridge University Press.

Steward, J. H. 1955. *Theory of Culture Change*. Urbana: University of Illinois Press.

Sundberg, U., J. Lindegren, H. T. Odum, and S. Doherty. 1994. Forest EMERGY basis for Swedish power in the 17th century. *Scandinavian Journal of Forest Research*, Supplement 1.

Taber, R. D. 1953. Studies of black-tailed deer reproduction on three chaparral cover types. *California Fish and Game* 39:177–186.

Tainter, J. A. 1988. *The Collapse of Complex Societies*. Cambridge, England: Cambridge University Press.

——. 1992. Evolutionary consequences of war. Pp. 103–130 in G. Ausenda (ed.), *Effects of War on Society*. San Marino: Center for Interdisciplinary Research on Social Stress.

——. 1995. Sustainability of complex societies. *Futures* 27:397–407.
——. 1996a. Complexity, problem solving, and sustainable societies. Pp. 61–76 in R. Costanza, O. Segura, and J. Martinez-Alier (eds.), *Getting Down to Earth: Practical Applications of Ecological Economics*. Washington, DC: Island Press.
——. 1996b. Introduction: Prehistoric societies as evolving complex systems. Pp. 1–23 in J. A. Tainter and B. B. Tainter (eds.), *Evolving Complexity and Environmental Risk in the Prehistoric Southwest*. Santa Fe Institute, Studies in the Sciences of Complexity, Proceedings Volume 24. Reading, MA: Addison-Wesley.
——. 1997. Cultural conflict and sustainable development: Managing subsistence hunting in Alaska. Pp. 155–161 in *Sustainable Development of Boreal Forests: Proceedings of the 7th International Conference of the International Boreal Forest Research Association*. Moscow: All-Russian Research and Information Center for Forest Resources.
——. 1999. Post-collapse societies. Pp. 988–1039 in G. Barker (ed.), *Companion Encyclopedia of Archaeology*. London: Routledge.
——. 2000. Global change, history, and sustainability. Pp. 331–356 in R. J. McIntosh, J. A. Tainter, and S. K. McIntosh (eds.), *The Way the Wind Blows: Climate, History, and Human Action*. New York: Columbia University Press.
——. 2001. Sustainable rural communities: General principles and North American indicators. Pp. 347–361 in C. J. P. Colfer and Y. Byron (eds.), *People Managing Forests: The Links Between Human Well-Being and Sustainability*. Washington, DC and Bogor, Indonesia: Resources for the Future Press and Center for International Forestry Research.
Tainter, J. A. and G. J. Lucas. 1983. Epistemology of the significance concept. *American Antiquity* 48:707–719.
Tainter, J. A. and B. B. Tainter (eds.). 1996a. *Evolving Complexity and Environmental Risk in the Prehistoric Southwest*. Santa Fe Institute, Studies in the Sciences of Complexity, Proceedings Volume 24. Reading, MA: Addison-Wesley.
——. 1996b. Riverine settlement in the evolution of prehistoric land-use systems in the middle Rio Grande valley, New Mexico. Pp. 22–32 in *Desired Future Conditions for Southwestern Riparian Ecosystems: Bringing Interests and Concerns Together*. General Technical Report RM-GTR-272. Fort Collins, CO: USDA Forest Service, Rocky Mountain Forest and Range Experiment Station.
Tansley, A. G. 1926. Succession: Its concept and value. Pp. 677–686 in B. M. Duggar (ed.), *Proceedings of the International Congress of Plant Sciences*. Menasha, WI: Banta.
——. 1935. Use and abuse of vegetational concepts and terms. *Ecology* 16:284–307.

References

Temple, S. A. (ed.). 1978. *Endangered Birds: Management Techniques for Preserving Threatened Species*. Madison: University of Wisconsin Press.

Thom, R. 1976. *Structural Stability and Morphogenesis*. (Trans. D. H. Fowler). Reading, MA: W. A. Benjamin.

Thomas, D. 1954. *Quite Early One Morning*. New York: New Directions.

Tilman, D. 1988. *Plant Strategies and the Structure and Dynamics of Plant Communities*. Princeton, NJ: Princeton University Press.

Tilman, D. and D. Wedin. 1991. Plant traits and resource reduction for five grasses growing on a nitrogen gradient. *Ecology* 72:685–700.

Timbrook, J., J. R. Johnson, and D. D. Earle. 1982. Vegetation burning by the Chumash. *Journal of California and Great Basin Anthropology* 4:136–186.

Toumey, C. P. 1996. *Conjuring Science: Scientific Symbols and Cultural Meanings in American Life*. New Brunswick, NJ: Rutgers University Press.

Toynbee, A. J. 1962. *A Study of History* (12 volumes). Oxford, England: Oxford University Press.

Transeau, E. N. 1926. The accumulation of energy by plants. *Ohio Journal of Science* 26: 1–10.

Treadgold, W. 1988. *The Byzantine Revival, 780–842*. Stanford, CA: Stanford University Press.

——. 1995. *Byzantium and Its Army, 284–1081*. Stanford, CA: Stanford University Press.

——. 1997. *A History of the Byzantine State and Society*. Stanford, CA: Stanford University Press.

Tul'chinskii, L. I. 1967. Problems in the profitably of investments in public education. *Soviet Review* 8(1):46–54.

Turner, B. L., II. 1974. Prehistoric intensive agriculture in the Mayan lowlands. *Science* 185:118–124.

——. 1978. Ancient agricultural land use in the central Maya lowlands. Pp. 163–183 in P. D. Harrison and B. L. Turner II (eds.), *Pre-Hispanic Maya Agriculture*. Albuquerque: University of New Mexico Press.

——. 1979. Prehispanic terracing in the central Maya lowlands: Problems of agricultural intensification. Pp. 103–115 in N. Hammond and G. R. Willey (eds.), *Maya Archaeology and Ethnohistory*. Austin: University of Texas Press.

Tyler, P. 1975. The Persian wars of the 3rd century A.D. and Roman imperial monetary policy, A.D. 253–68. *Historia, Einzelschriften* 23.

Uexkull, J. von. 1957. A stroll through the worlds of animals and men: A picture book of invisible worlds. In C. H. Schiller (ed. and trans.), *Instinctive Behavior*. New York: International Universities Press.

Ulanowicz, R. 1997. *Ecology, the Ascendent Perspective*. New York: Columbia University Press.

U.S. Bureau of the Census. 1983. *Statistical Abstract of the United States: 1984* (104th edition). Washington, DC: U.S. Government Printing Office.

U.S. Fish and Wildlife Service. 1982. *Snail Darter Recovery Plan.* Ashville, NC: U.S. Fish and Wildlife Service.

Vallentyne, J. R. 1983. Implementing an ecosystem approach to management of the Great Lakes Basin. *Environmental Conservation* 10:273–274.

van der Leeuw, S. E. 1998a. Introduction. Pp. 2–22 in S. E. van der Leeuw (ed.), *The Archaeomedes Project: Understanding the Natural and Anthropogenic Causes of Land Degradation and Desertification in the Mediterranean Basin.* Luxembourg: Office for Official Publications of the European Communities.

——. 1998b. Multidisciplinarity, policy-relevant research and the non-linear paradigm. Pp. 25–41 in S. E. van der Leeuw (ed.), *The Archaeomedes Project: Understanding the Natural and Anthropogenic Causes of Land Degradation and Desertification in the Mediterranean Basin.* Luxembourg: Office for Official Publications of the European Communities.

Van Meter, D. 1991. *The Handbook of Roman Imperial Coins.* Nashua, NH: Laurion Numismatics.

Vinogradoff, P. 1913. Foundations of society (origins of feudalism). Pp. 630–654 in H. M. Gwatkin and J. P. Whitney (eds.), *The Cambridge Medieval History,* Volume 2, *The Rise of the Saracens and the Foundation of the Western Empire.* Cambridge, England: Cambridge University Press.

Virtual Elimination Taskforce. 1993. *A Strategy for Virtual Elimination of Persistent Toxic Chemicals. Appendices,* Volume 2. Windsor, Ontario: International Joint Commission.

Vogel, S. 1988. *Life's Devices: The Physical World of Animals and Plants.* Princeton, NJ: Princeton University Press.

Waines, D. 1977. The third century internal crisis of the Abbasids. *Journal of the Economic and Social History of the Orient* 20:282–306.

Walker, D. R. 1976. *The Metrology of the Roman Silver Coinage,* Part 1, *From Augustus to Domitian.* British Archaeological Reports, Supplementary Series 5. Oxford: British Archaeological Reports.

——. 1977. *The Metrology of the Roman Silver Coinage,* Part 2, *From Nerva to Commodus.* British Archaeological Reports, Supplementary Series 22. Oxford: British Archaeological Reports.

——. 1978. *The Metrology of the Roman Silver Coinage,* Part 3, *From Pertinax to Uranius Antoninus.* British Archaeological Reports, Supplementary Series 40. Oxford: British Archaeological Reports.

Walters, C. 1986. *Adaptive Management of Renewable Resources.* New York: Macmillan.

References

Watt, A. S. 1947. Pattern and process in the plant community. *Journal of Ecology* 35:1–22.

Weber, M. 1976. *The Agrarian Sociology of Ancient Civilizations* (trans. by R. I. Frank). London: NLB.

Webster, D. 1976. *Defensive Earthworks at Becan, Campeche, Mexico.* Middle American Research Institute, Publication 44. New Orleans: Middle American Research Institute, Tulane University.

———. 1977. Warfare and the evolution of Maya civilization. Pp. 335–372 in R. E. W. Adams (ed.), *The Origins of Maya Civilization.* Albuquerque: University of New Mexico Press.

Wedin, D. 1990. Nitrogen cycling and competition among grass species. Unpublished Ph.D. diss., University of Minnesota, Minneapolis.

Weinberg, G. 1975. *An Introduction to Systems Thinking.* New York: Wiley.

West, R. G. 1970. Pleistocene history of the British flora. Pp. 1–11 in D. Walker and R. G. West (eds.), *Studies in the Vegetational History of the British Isles.* Cambridge, England: Cambridge University Press.

White, C. S., S. R. Loftin, and R. Aguilar. 1997. Application of biosolids to degraded semiarid rangeland: Nine-year responses. *Journal of Environmental Quality* 26:1663–1671.

White, L. A. 1949. *The Science of Culture.* New York: Farrar, Straus & Giroux.

———. 1959. *The Evolution of Culture.* New York: McGraw-Hill.

Whittaker, R. H. 1956. Vegetation of the Great Smoky Mountains. *Ecological Monographs* 26:1–80.

Wicken, J. S. 1979. The generation of complexity in evolution: A thermodynamic and information-theoretical discussion. *Journal of Theoretical Biology* 77:349–365.

———. 1980. A thermodynamic theory of evolution. *Journal of Theoretical Biology* 87:9–23.

———. 1987. *Evolution, Thermodynamics, and Information.* New York: Oxford University Press.

Wickham, C. 1981. *Early Medieval Italy: Central Power and Local Society 400–1000.* London: Macmillan.

———. 1984. The other transition: From the ancient world to feudalism. *Past and Present* 103:3–36.

Wilken, G. C. 1971. Food-producing systems available to the ancient Maya. *American Antiquity* 36:432–448.

Wilkinson, R. G. 1973. *Poverty and Progress: An Ecological Model of Economic Development.* London: Methuen.

Willey, G. R. 1980. Towards an holistic view of ancient Maya civilisation. *Man* 15:249–266.

———. 1981. Maya Lowland settlement patterns: A summary view. Pp. 385–415 in W. Ashmore (ed.), *Lowland Maya Settlement Patterns*. Albuquerque: University of New Mexico Press.

Williams, S. 1985. *Diocletian and the Roman Recovery*. New York: Methuen.

Winterhalder, B. 1981. Foraging strategies in the boreal forest: An analysis of Cree hunting and gathering. Pp. 66–98 in B. Winterhalder and E. A. Smith (eds.), *Hunter-Gatherer Foraging Strategies: Ethnographic and Archeological Analyses*. Chicago: University of Chicago Press.

Winterhalder, B. and E. A. Smith (eds.). 1981. *Hunter-Gatherer Foraging Strategies: Ethnographic and Archeological Analyses*. Chicago: University of Chicago Press.

Wiseman, F. W. 1983. Subsistence and complex societies: The case of the Maya. Pp. 141–189 in M. B. Schiffer (ed.), *Advances in Archaeological Method and Theory*, Volume 6. New York: Academic Press.

Wolf, S. A. and T. F. H. Allen. 1995. Recasting alternative agricultural planning as a management model: The value of adept scaling. *Ecological Economics* 12:5–12.

Wolfle, D. 1960. How much research for a dollar? *Science* 132:517.

World Commission on Environment and Development. 1987. *Our Common Future*. Oxford, England: Oxford University Press.

Worthington, N. L. 1975. National health expenditures, 1929–74. *Social Security Bulletin* 38(2):3–20.

Wright, S. 1970. Random drift and the shifting balance theory of evolution. Pp. 1–31 in K. Kojima (ed.), *Mathematical Topics in Population Genetics*. Berlin: Springer-Verlag.

Yoffee, N. 1988. The collapse of ancient Mesopotamian states and civilization. Pp. 44–68 in N. Yoffee and G. L. Cowgill (eds.), *The Collapse of Ancient States and Civilizations*. Tucson: University of Arizona Press.

Zedler, J. B. (ed.) 2001. *Handbook for Restoring Tidal Wetlands*. Boca Raton, FL: CRC Press.

Zhu, Y., H. Chen, J. Fan, Y. Wang, Y. Li, J. Chen, J. X. Fan, S. Yang, L. Hu, H. Leung, T. Mew, P. Teng, A. Wang, and C. Mundt. 2000. Genetic diversity and disease control in rice. *Nature* 406:718–722.

Zimmerer, K. S. 1998. The ecogeography of Andean potatoes. *BioScience* 48:445–454.

Index

Index

Greece (*continued*)
   intensity of use, 224
   population decline, 221–223

H

Hierarchies
   and conflicting management actions, 370
   defined, 353
   flat, 354
   and instability, 306
   of metapopulation(s), 248–251
   of Native California, 207–208
   and vertical differentiation, 354
Holism, 48
   vs. reductionism, 43
   and scale, 49
Holon(s), 193
Homolog(s), 186, 187
Horizontal differentiation
   and complexity, 354–355
   tensions associated with, 358
Hunter-gatherers, 68, 82, 363
   landscape preferences, 204–205
   population densities, 206
Hydrogen, 365, 366, 367, 368
Hysteresis
   defined, 261

I

Iltis, Hugh, 3
Industrial Revolution, 3, 30
   and acid rain, 31
   and development of new technology, 357
   emerging from resource depletion, 80
   in England, 79–80, 364, 388
   not repeatable, 369
Industrialism, 410
Information
   and energy, 367
   era, 365

   exchange of in human societies, 156
   revolution, 424
   technology, 368
Intercropping
   leading to complex organized vegetation, 350
Intussusception
   growth by, 408, 409
Irrigation
   as an increased energy input, 405
   potential for salting, 406–408
Island biogeography, 37
   used as a basis for design of nature reserves, 38

K

Knowledge
   as the basis of supply-side sustainability, 391
   production of, 83–95
      subject to the law of diminishing returns, 85
Kyoto agreement, 12

L

Landscape(s)
   discussed, 197–235
   explained by narrative, 233
   and importance of scale, 199–200
   of the Maya, 211–217
   of the Middle Rio Grande Basin, 202–205
   of Native California, 205–211
   policy implications on, 227–230
   preferences of humans, 187–189
   preferences of hunter-gatherers, 204–205
   of Roman Greece, 217–225
   and sustainability, 198
Little Ice Age, 290
Livestock
   effect on development of sociopolitical complexity, 77–78